Introduction to Cell Mechanics and Mechanobiology

Introduction to Cell Mechanics and Mechanobiology

Christopher R. Jacobs

Hayden Huang

Ronald Y. Kwon

Garland Science
Taylor & Francis Group
NEW YORK AND LONDON

Garland Science
Vice President: Denise Schanck
Editor: Summers Scholl
Senior Editorial Assistant: Allie Bochicchio
Production Editor and Layout: Natasha Wolfe
Illustrator: Laurel Muller, Cohographics
Cover design: Andrew Magee
Copyeditor: Christopher Purdon
Proofreader: Mary Curioli
Indexer: Indexing Specialists (UK) Ltd

ISBN 978-0-8153-4425-4

Library of Congress Cataloging-in-Publication Data

Jacobs, C. R. (Christopher R.)
　Introduction to cell mechanics and mechanobiology/Christopher R. Jacobs, Hayden Huang, Ronald Y. Kwon.
　　p. cm.

　Includes bibliographical references.
　Summary: "Introduction to Cell Mechanics and Mechanobiology teaches advanced undergraduate students a quantitative understanding of the way cells detect, modify, and respond to the physical properties within the cell environment. Coverage includes the mechanics of single molecule polymers, polymer networks, two-dimensional membranes, whole-cell mechanics, and mechanobiology, as well as primer chapters on solid, fluid, and statistical mechanics"–Provided by publisher.

　ISBN 978-0-8153-4425-4 (pbk.)
I. Huang, Hayden.　II. Kwon, Ronald Y.　III. Title.
　[DNLM: 1. Cell Physiological Phenomena. 2.　Biomechanics. QU 375]
　572–dc23
　　　　　　　　　2012018504

Published by Garland Science, Taylor & Francis Group, LLC, an informa business,
711 Third Avenue, New York, NY, 10017, USA, and 2 Park Square, Milton Park, Abingdon, OX14 4RN, UK.

Printed in the United States of America

15　14　13　12　11　10　9　8　7　6　5　4　3　2　1

Visit our website at http://www.garlandscience.com

Preface

In recent years, mechanical signals have become widely recognized as being critical to the proper functioning of numerous biological processes. This has led to the emergence of a new field called cellular mechanobiology, which merges cell biology with various disciplines of mechanics (including solid, fluid, statistical, computational, and experimental mechanics). Cellular mechanobiology seeks to uncover the principles by which the sensation or generation of mechanical force alters cell function. *Introduction to Cell Mechanics and Mechanobiology* presents students from a wide variety of backgrounds with the physical and mechanical principles underpinning cell and tissue behavior.

This textbook arose from a cell mechanics course at Stanford University first offered by two of us in 2005. Over several iterations, we taught from a set of course notes and chapter excerpts—having found no textbook to cover the necessary breadth of topics. Our colleagues had similar experiences teaching with the same adhoc approach, which convinced us of the need for a comprehensive instructional tool in this area. Another reason we felt compelled to write this text is that cell mechanics provides an excellent substrate to introduce many types of mechanics (solid, fluid, statistical, experimental, and even computational). These topics are traditionally covered in separate courses with applications largely focused on engineering structures. As authors, we have varied backgrounds, but share a common fondness for the insights mechanical engineering brings to cell biology.

Introduction to Cell Mechanics and Mechanobiology is intended for advanced undergraduates and early graduate students in biological engineering and biomedical engineering, including those not necessarily in a biomechanics track. We do not assume an extensive knowledge in any area of biology or mechanics. We do assume that students have a mathematics background common to all areas of engineering and quantitative science, meaning exposure to calculus, ordinary differential equations, and linear algebra.

The field of cell mechanics encompasses advanced concepts, such as large deformation mechanics and nonlinear mechanics. We do not expect our audience to have a strong background in the advanced mathematics of continuum mechanics. Our intent is to avoid graduate-level mathematics wherever possible. In our approach, the treatment of tensor mathematics—central to large deformation mechanics (common in cell mechanics)—poses unique difficulties. To show simplified mathematical derivations measuring mechanical parameters in the context of living cells, we present tensors "by analogy" as matrices, rather than introducing them in a fully rigorous fashion. For example, we skip index notation entirely. Admittedly, this approach may be less satisfying to mechanicians, which we also consider ourselves. However, we hope that the advantages of this approach will outweigh our oversimplifications.

The book is grouped into two parts: (I) Principles and (II) Practices. We have written the chapters to allow instructors flexibility in presentation, depending on the level of students and the length of the course. After introducing cell mechanics as a framework in Chapter 1, we provide a review of cell biology in Chapter 2. The next four chapters establish the necessary concepts in mechanics with enough depth that the student attains a basic competency and appreciation for each topic. Chapter 3 covers solid mechanics—including rigid and deformable bodies as well as a short overview of large-deformation mechanics. Fluid mechanics (Chapter 4) is important for cell mechanics not only in cytoplasmic flow, but also as a physical signal that regulates cell mechanobiological behavior. Chapter 5 dives into statistical mechanics, with descriptions of energy, entropy, and random walks, common themes for understanding the aggregate behavior of systems composed of many objects. In Chapter 6, we describe experimental methods, an area that is always changing, but is essential in demonstrating how theory may be reconciled with actual experiments. These fundamentals in Part I are followed by cell mechanics proper in Part II. Chapters 7–9 begin with a discussion of an aspect of cell biology followed by analysis of the mechanics. We undertake polymer mechanics in Chapter 7 from a continuum and a statistical viewpoint and examine situations in which both need to be considered simultaneously. These tools are applied to individual cytoskeletal polymers as well as to other polymers such as DNA. Polymer networks are presented in Chapter 8, with a focus on the role of the cytoskeleton in regulating physical properties, such as red blood cell shape and limitations on cell protrusion lengths. Chapter 9 examines the bilayer membrane, from both the perspective of matter floating around within it (diffusion) as well as a mechanical perspective of bending and stretching. The last two chapters address mechanobiology. Chapter 10 is focused on cellular force generation and the related processes of adhesion and migration. Chapter 11 discusses the process of mechanosensing or mechanotransduction and intracellular signaling. These last chapters do not have as much rigorous mechanical engineering mathematics, but are an integral part of cell biomechanics.

Given the varied backgrounds of our students and the interdisciplinary nature of the subject, we have attempted to provide some guidance on the treatment of variables and units. At the start of the book, we present a master list of all the variables used in the text that specifies exactly what each variable is used for in a particular chapter. We have retained the "contextual" usage in each chapter, accepted within each field, to prepare students for reading the literature. Three types of boxes supplement the main text: "Advanced Material" challenges readers to think critically and problem-solve; interesting and noteworthy asides are denoted as "Nota Bene"; "Examples" provide in-depth solved calculations and explanations. Each chapter concludes with a set of Key Concepts, Problems that can be used as homework sets, and Annotated References that guide students for further study.

Online Resources

Accessible from www.garlandscience.com/cell-mechanics, Student and Instructor Resource websites provide learning and teaching tools created for *Introduction to Cell Mechanics and Mechanobiology*. The Student Resources site is open to everyone, and users have the option to register in order to use book-marking and note-taking tools. The Instructor's Resource site requires registration; access is available to instructors who have assigned the book to their course. To access the Instructor's Resource site, please contact your local sales representative or email science@garland.com. Below is an overview of the resources available for this book. Resources may be browsed by individual chapters and there is a search engine. You can also access the resources available for other Garland Science titles.

For students:

- Computer simulation modules in two formats: ready-to-run simulations that simulate the mechanical behavior of cells and tutorial MATLAB modules on simulation of cell behavior with the finite element method.
- Color versions of several figures are available, indicated by the figure legend in the text.

- A handful of animations and videos dynamically illustrate important concepts from the book.
- Solutions to selected end-of-chapter problems are available to students.

For instructors:

- In addition to color versions of several figures, all of the images from the book are available in two convenient formats: Microsoft PowerPoint® and JPEG. They have been optimized for display on a computer. Figures are searchable by figure number, figure name, or by keywords used in the figure legend from the book.
- The animations and videos that are available to students are also available on the Instructor's Resource website in two formats. The WMV-formatted movies are created for instructors who wish to use the movies in PowerPoint presentations on computers running Windows®; the QuickTime®-formatted movies are for use in PowerPoint for Apple computers or Keynote® presentations. The movies can easily be downloaded to your personal computer using the "download" button on the movie preview page.
- Solutions to selected end-of-chapter problems are available to qualified adopters.

The origin of the book is rooted in teaching from sections of outstanding books by David H. Boal, Jonathon Howard, and Howard C. Berg. We thank Roger Kamm, Vijay Pande, and Andrew Spakowitz, who taught some of us at various times and have unselfishly shared course materials and handouts and, in the case of Dr. Kamm, unpublished drafts of his own textbook. With their permission, we have incorporated their approach to some topics in Chapters 4, 5, 7, 8, and 9 and adapted several problems into sections of our book. We are grateful for their amazing willingness to share their intellectual product in the name of improving the educational experience of students around the world. We also thank reviewers Roland R. Kaunas and Peter J. Butler, who shared notes from their own courses in cell mechanics. We are profoundly appreciative of the tireless work of those who have preceded us, without whom we never could have completed this task. We thank the additional reviewers of the book, Dan Fletcher, Christian Franck, Wonmuk Hwang, Paul Janmey, Yuan Lin, Lidan You, and Diane Wagner, for their valuable insight and critiques of our drafts. We are also grateful to Summers Scholl and the editorial and production teams at Garland who took a chance on three textbook neophytes and guided us unerringly through uncharted waters. Finally we are each deeply indebted to our families, including Roberta, Jolene, VH, YYH, LHH, Joyce, Melody, Tae, and Cynthia. Without your support, patience, and understanding—as this project took us away from you on so many nights and weekends—we never could have contemplated this undertaking, much less completed it.

Christopher R. Jacobs
Hayden Huang
Ronald R. Kwon

Detailed Contents

PART I: PRINCIPLES

CHAPTER 1

Cell Mechanics as a Framework

Biological cells are the smallest and most basic units of life. The field of cell biology, which seeks to elucidate cell function through better understanding of physiological processes, cellular structure, and the interaction of cells with the extracellular environment, has become the primary basic science for better understanding of human disease in biomedical research. Until recently, the study of basic problems in cell biology has been performed almost exclusively within the context of biochemistry and through the use of molecular and genetic approaches. Pathological processes may be considered disruptions in biochemical signaling events. The regulation of cell function by extracellular signals may be understood from the point of view of binding of a molecule to a receptor on the cell surface. Basic cellular processes such as cell division are considered in terms of the biochemical events driving them. This emphasis on biochemistry and structural biology in cell biology research is reflected in typical curricula and core texts traditionally used for cell biology courses.

Recently, there has been a shift in paradigm in the understanding of cell function and disease primarily within the analytical context of biochemistry. In particular, it has become well established that critical insights into diverse cellular processes and pathologies can be gained by understanding the role of mechanical force. A rapidly growing body of science indicates that mechanical phenomena are critical to the proper functioning of several basic cell processes and that mechanical loads can serve as extracellular signals that regulate cell function. Further, disruptions in mechanical sensing and/or function have been implicated in several diseases considered major health risks, such as osteoporosis, atherosclerosis, and cancer. This has led to the emergence of a new discipline that merges mechanics and cell biology: cellular mechanobiology. This term refers to any aspect of cell biology in which mechanical force is generated, imparted, or sensed, leading to alterations in cellular function. The study of cellular mechanobiology bridges cell biology and biochemistry with various disciplines of mechanics, including solid, fluid, statistical, experimental, and computational mechanics.

The primary goal of this introductory chapter is to motivate the study of cell mechanics and cellular mechanobiology by: (1) demonstrating its role in basic cellular and pathological processes; and (2) showing how cell mechanics provides an ideal framework for introducing a broad mechanics curriculum in an integrated manner. We first present cell mechanics in the context of human disease by providing a survey of physiological and pathological processes that are mediated by cell mechanics and can be better understood through mechanical analyses. Next, we propose cell mechanics as an ideal substrate for introducing principles of solid, fluid, statistical, experimental, and even computational mechanics, and put forth the argument that cell mechanics may be the grand challenge of applied mechanics for the twenty-first century. Finally, we present a simple model problem: micropipette aspiration, in which a cell is partly "sucked" into a narrow tube by a vacuum. This example will help you develop a feeling for how cell mechanics is studied and demonstrate how a relatively simple approach can give important insight into cell mechanical behavior (and how this behavior can dictate cellular function).

1.1 CELL MECHANICS AND HUMAN DISEASE

Most of our understanding of biomedicine, both in terms of health and disease, is biological or biochemical in nature. There are some exceptions, of course, such as the component of mechanics at the tissue or whole organism level when we think about fracture of a bone, soft tissue trauma, or surgical repair. Further, when you think of your senses that involve mechanics, such as hearing and touch, the fact that mechanically specialized cells are involved is unsurprising. By contrast, we do not typically think about mechanics of cells in relation to cancer, malaria, or viral infections—but they are related. What may be even more of a surprise is that many of the causes of human suffering involve cell mechanics to some degree or other.

For instance, the health of several tissues, particularly tissues of the skeleton (bone and cartilage) and of the cardiovascular system (the heart and arteries), is heavily dependent on mechanical loading, which in turn comes from physical activity and the environment (gravity). To be clear, we are not simply saying that these physiological systems have a mechanical function (bones support the body and the heart pumps blood)—which they do. We are also saying that these systems actively change and respond to changes in mechanical forces at the cellular level—bones will reinforce certain regions and actively degrade others. In this chapter we hope to convince the reader that mechanics is in fact involved in virtually every aspect of life, although its influence may be subtle or indirect.

Understanding human health and disease often requires an understanding of biomechanics and mechanobiology at the cellular level, for example:

- When bone cells do not experience proper mechanical stimulation, bone formation ceases and bone resorption is initiated. So, in prolonged space travel, where gravity is virtually nonexistent, astronauts face major bone loss, even with rigorous exercise regimens.
- In coronary artery disease, changes in the temporal and spatial patterns of fluid shear stress on endothelial cells are linked to the formation of atherosclerotic plaques.
- The pathogenesis of osteoarthritis occurs due to changes in physical loading that lead to altered mechanical signals experienced by chondrocytes.
- Lung alveolar epithelial cells and airway smooth muscle cells are regulated by cyclic mechanical stretch during breathing, and hypersensitization due to airborne pathogens that can lead to sustained hypercontractility, which in turn can cause asthmatic attacks.
- Infection can be initiated from mechanical disruption of the cell membrane by viruses delivering foreign genetic material. This is a serious problem—if we could deliver genes as easily as viruses, we could potentially cure many genetic diseases by having cells express the corrected version (of the mutated gene). But the cell membrane is actually an excellent mechanical barrier.
- Metastatic cancer cells must be able to migrate through tissue and attach at distant sites to spread. Why certain cancers appear to metastasize preferentially to particular locations is still a mystery.
- Mechanical stimuli regulate fibroblast behavior during wound healing. Further, there is a difference between "normal" wound healing, where the wound is grown over, and the development of scar tissue.
- Physical forces are also known to be a critical factor in the regulation of the tissue-specific differentiation of adult and embryonic stem cells. For example, it is thought that the beating of some mammalian embryonic hearts is more for shaping the

heart muscle rather than for functional pumping, given that the heart does not need to pump blood in any serious manner *in utero.*

- Post-birth, brain development and angiogenesis all centrally involve cells' ability to interact with their dynamic mechanical environment.

- Cardiovascular diseases such as hypertension and heart failure often result from long-term mechanical influences. Indeed, cardiac hypertrophy is one of the most common responses to changes in forces. The distinction between healthy hypertrophy (resulting from exercise) versus pathological hypertrophy (resulting from poor health) is still not well understood.

- The fundamental cellular processes of membrane trafficking, endocytosis and exocytosis (the ways in which a cell engulfs or expels substances, respectively), microtubule assembly and disassembly, actin polymerization and depolymerization, dynamics of cell–matrix and cell–cell adhesions, chromosome segregation, kinetochore dynamics (such as DNA motion during cell division), cytoplasmic protein and vesicle sorting and transport, cell motility, apoptosis ("programmed cell death"), invasion (motion of a cell to where it is not usually located), and proliferation and differentiation (specialization of a cell to a phenotype with a particular function) are all regulated, at least in part, by mechanical forces.

In the sections that follow, we examine a few of these examples in more detail.

Specialized cells in the ear allow you to hear

At its most basic level, hearing is a process of transduction (transduction being the conversion of a signal from one type to another). A physical signal in the form of sound (pressure) waves is converted into electrical impulses along a nerve. Mechanotransduction (transduction in which the incoming signal is mechanically based) occurs in the ear via a specialized cell called the inner ear hair cell. This cell has small hairs called *cilia* (singular: cilium) extending from the apical (top) surface of the cell into the lumen of the cochlea. Sound in the form of pressure waves caused by vibrations of the inner ear bones travels through the fluid in the cochlea.

Investigators have recently deduced the remarkable mechanism of transduction in the hair cell. Filaments (fibers) of the cytoskeletal protein actin were identified linking the tip of one cilium to the side of an adjacent cilium (Figure 1.1). The actin filaments are anchored to proteins that span the cell membrane and form small holes or pores known as channels. These channels are normally closed, but, when open, permit the passage of small ions (in the case of hair cells, calcium

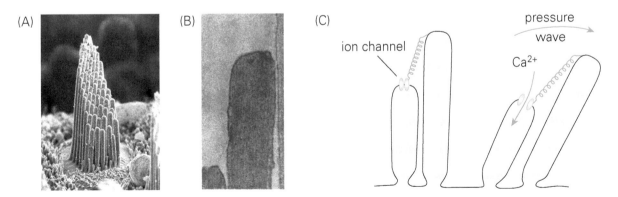

Figure 1.1 Hearing occurs via mechanotransduction by the inner ear hair cell. (A) The bundle of cilia extending from the apical surface of the cell is deflected by pressure waves in the cochlear lumen. (B) Tiny actin bundles called tip links are stretched as the cilia deflect due to the pressure wave. (C) The tip links are attached to calcium channels or pores that open and allow calcium into the cell where it eventually leads to a nerve impulse. (A, Courtesy of Dr. David Furness; B, from, Jacobs RA, Hudspeth AJ (1990) *Symp. Quant. Biol.* 55, 547–561. With permission from Cold Spring Harbor Press.)

ions) along their concentration gradient. In the resting state, the cell keeps its internal calcium level extremely low (<1 mM) relative to the calcium concentration outside the cell. When sound is transmitted to the inner ear, the vibrations cause the cilia to deflect, which in turn stretches the actin filaments. This stretching creates tension that is transmitted to the channels, causing them to open. So, when the channel is opened, calcium flows down its concentration gradient, and the intracellular calcium concentration increases. The kinetics of signaling proteins inside the cell are altered by this change in concentration, and a cascade of biochemical events is initiated that eventually leads to a depolarization of the cell and a nerve impulse.

As you might imagine, mechanics is very important in this process. The cilia need to have the right mechanical characteristics to stand upright, but remain flexible enough that they can be deflected by sound waves. The actin tip links need to be strong enough to open the channel and to have the appropriate polymer mechanics behavior so that they are stretched by cilium deflection, but are not affected by thermal noise (recall that these are very small objects, so the soup of molecules floating around will periodically collide with them, and can generate some forces that need to be ignored). In this text, our goal is to build a foundation and present a framework so that you can consider these questions effectively.

Hemodynamic forces regulate endothelial cells

Blood vessels are not passive piping for the blood. They are very responsive and are constantly changing their radius (via vascular tone or under the influence of vasodilators and vasoconstrictors) and leakiness. The cells lining these vessels are called *endothelial* cells (or collectively, the *endothelium*). Endothelial cells are very responsive to mechanical forces generated by the circulatory system, including the shear from flow, stretch from the distension of the (larger) vessels, and transmural pressure differences (pressure differences between the inside of the vessel and outside). The response of the endothelial cells is varied—they can change shape to align their long axis in the direction of flow, alter their internal structure (the cytoskeleton and adhesive plaques), and release a variety of signaling molecules. These actions help maintain blood flow and homeostasis (maintenance of physiological conditions at some baseline), and there is strong evidence that pathophysiological changes (such as atherosclerosis) occur in regions where mechanical signaling is disrupted.

To keep bone healthy, bone cells need mechanical stimulation

Physical loading is critical for skeletal health. Indeed, one of the most important factors in keeping bone healthy is for it to receive normal mechanical stimulation. When bone is not loaded it is said to be in a state of partial disuse, perhaps owing to a sedentary lifestyle, or complete disuse, which might occur because of bed rest or during long-duration spaceflight. In these extreme latter cases bone loss has been documented to occur at rates as high as 1–2% of total bone mass per month. Bone loss puts people at increased risk of fracture, even when trauma is absent or relatively low. These *fragility* or *osteoporotic* fractures can be devastating both to individuals and as a public health issue, costing billions of dollars annually. In fact, one-half of all women and one-quarter of all men older than 50 today will experience an osteoporotic fracture in their lifetime. Hip fractures are the most devastating result of low bone mass, and for most patients the first step in a downward spiral of lost ambulation, lost independence, institutionalization, and secondary medical morbidity and mortality. Shockingly, within 1 year of a hip fracture, 50% of patients will be unable to walk unaided, 25% will be institutionalized, and 20% will have died.

The good news is that physical loading on your bone from staying active will protect you from losing bone, although some activities appear to be better than others. Ballet is better than swimming, presumably because of the impact loading involved. In fact, it has been shown that high-level athletes can actually build bone specifically in regions of the skeleton that experience higher loading during their sport. Despite its critical importance for human health and its status as a compelling scientific question, the mechanism that allows bone cells (*osteocytes* and *osteoblasts* primarily) to sense and respond to loading by coordinating the cellular response remains basically unknown. It has been suggested that the sensing mechanism might involve the cytoskeleton, focal adhesions, adherens junctions, membrane channels, and even the biophysical behavior of the membrane itself. Indeed there is evidence for each of these and many others, so it seems likely that several cellular sensors exist, perhaps forming a redundant system.

The cells that line your lungs sense stretch

During respiration, the lung is exposed to constant oscillatory stresses arising from expansion and contraction of the basement membrane. These mechanical signals are postulated to play an important role in maintaining normal lung function and morphology. Stretch regulates pulmonary epithelial cell growth and cytoskeletal remodeling, as well as secretion of signaling molecules and phospholipids. These mechanical loads may be increased, for example, when a patient is subjected to mechanical ventilation. The physiological consequences of altered cellular function in response to such perturbations in mechanical loading are not yet fully understood.

Pathogens can alter cell mechanical properties

Malaria provides an interesting example of subtle mechanical alterations at the cellular level. Malaria is a mosquito-transmitted parasite that infects *red blood cells* (RBCs). Because the parasite resides in the RBCs during a large part of its lifetime, it is generally protected from the immune system. Because infected RBCs can be destroyed by the spleen, the parasite causes the infected RBC to increase its stickiness by inducing the expression of adhesive surface proteins on the RBC membrane. This allows the RBC to stick to the vessel walls and avoid being filtered in the spleen. Because there are many variations of this class of malaria surface proteins, the immune system is slow to adapt and remove these infected RBCs. As you can imagine, there has to be some deftness in the change in adhesion so that the cells will tend to stick a bit more, but not so much more that all the blood clumps together. Indeed, one effect of having stickier RBCs is that occasionally there will be an accumulation of RBCs in smaller blood vessels, resulting in a hemorrhage.

Other pathogens can use cell mechanical structures to their advantage

Bacteria of the genus *Listeria* act similarly by hiding within cells to evade the immune system. To invade other cells, the bacteria take over part of the cell's *actin* machinery (part of the cell *cytoskeleton*). Actin is polymerized to form fibers within the cell to provide structure and anchorage. The bacteria "sit" on the tip of the growing actin polymer and wait for a polymer to grow long enough for the bacteria to be pushed out of the cell and into an adjacent cell. Once the host is infected, the bacteria can spread throughout the host's body without ever exposing themselves to the immune system. For this mechanism to work, the bacteria must be able to achieve sufficient force to break through two cell membranes. The bacteria have to be able to "know" where to sit on the actin filament to be propelled—this is a source of active investigation for use in generating molecular machines.

Cancer cells need to crawl to be metastatic

Cancer metastasis is the process whereby an individual cancer cell(s) detaches from the main tumor, enters the bloodstream, reattaches at some new location, exits the blood vessel, and starts growing in its new location. Metastasis causes most cancer deaths, but many aspects of this process have yet to be fully understood. Cell migration is a critical component that is mediated by mechanical processes such as adhesion and intracellular force generation. Changes to aspects of these processes (such as the migratory speed of the cells) are generally tied to the long-term prognosis of the cancer, but the ways in which this occurs are not well understood. Adhesion may not only be important for allowing cancer cells to migrate but also for them to home in on a particular location. Not all tumor cells metastasize the same way; certain tumors will preferentially metastasize to specific regions or tissues. Whether this is due to selective adhesion at the preferred sites or diminished survival at other sites is not clear.

Solid tumors, as a whole, also exhibit altered physiology. Not only do the cells within tumors exhibit increased ("out-of-control") division rates, but tumors can redirect blood flow to allow themselves to grow faster. Further, many primary solid tumors tend to be stiffer than the surrounding tissues, even though they generally originate from the same tissue mass. Whether this increased stiffness alters cancer cell function through a mechanosensing function is not known, but it does have a practical use—many superficial (i.e., close to the skin) tumors can be detected by performing a self-examination, by feeling for a "lump" or "bump" that is somewhat harder compared to the surrounding tissue.

Viruses transfer their cargo into cells they infect

When a cell is invaded by a virus, the viral cargo of genetic material must be introduced into the cell. There are two mechanisms by which this can occur: endocytosis or membrane fusion. In the case of the former, the binding of proteins (called *ligands*) on the surface of the virus to proteins (called *receptors*) on the surface of the cell initiates a process called receptor-mediated endocytosis. In this process, the virus is enveloped by the cell, allowing it to deliver its genetic cargo and replicate. This process relies on a coordinated sequence of mechanical events, including adhesion, membrane pinching, and generation of cytoskeletal force that may provide potential targets for therapeutics aimed at inhibiting viral invasion. In addition, given the highly efficient means by which viruses invade cells, there is interest in understanding these mechanical processes for purposes of biomimicry, such as virus-based methodologies for nanoparticle delivery into cells.

1.2 THE CELL IS AN APPLIED MECHANICS GRAND CHALLENGE

In the twentieth century, the state of the art for applied mechanics was structural analysis on the large scale. Amazing achievements were realized in construction, such as high-rise buildings and beautiful bridges. Architecture was allowed to move beyond bulky stone and brick to elegant steel and glass. Transportation was revolutionized with cars and trains and modern aircraft representing outstanding examples of highly efficient structures that could only be created once their mechanical behavior had been analyzed in detail—engines, streamlining, brakes, lift, power, heat, etc., all having to be characterized and then applied together. Mechanics also played a major role in allowing people to reach the moon and explore the planets. Mechanics was and is key to military advances (missile technology, armor, advanced planes and drones, robotics, etc.). However, the theories and analysis required to design and build these impressive structures are, to a great extent, mature. For instance, much of car body design is based on computational analysis and not on the development of new laws or principles. Although

Nota Bene

Membrane fusion is a chemical process by which a pore is introduced into the cell membrane at the point where the membranes of the virus and the target cell are fused together. The mechanism underlying this is not understood well and may not be as mechanically dependent as endocytosis. But a certain degree of adhesion and membrane bending will invariably have to occur.

applied mechanics experienced great growth in the past, more recently it has undergone some degree of contraction.

Comprehensive mechanical analysis of a cell is extremely complex. There are one-dimensional linear elements in the cytoskeleton and two-dimensional curved shells in the cell membrane. There are also three-dimensional solids and enormous potential for pressure effects and fluid–solid interaction. Indeed, the overall cell is part–solid, part–liquid, something we call *viscoelastic*. Not only that, but the properties of the cellular "substance" change depending on the frequency with which forces are applied. On top of this, cellular structures are so small that thermal and entropic effects can play an important role in their mechanics, often requiring the analytical framework of statistical mechanics to understand their behavior. Thus, in terms of difficult challenges in applied mechanics with potential for critical new advances and fundamental insight, it is hard to imagine a more compelling problem than the cell. The overall picture we wish to present to you is that much can be explained using basic mechanics, but there is still much to be done using only slightly more advanced mechanical analysis.

Computer simulation of cell mechanics requires state-of-the-art approaches

Just as cell mechanics is a compelling challenge in applied mechanics, it is also a difficult, but rewarding, challenge in computational mechanics. For example, multi-scale modeling involves coupling a large-scale simulation with another simulation representing microscopic behavior. For cells one might simulate the behavior of individual actin and tubulin polymers and couple that to models of cytoskeletal networks or even the whole cell. The full mechanical behavior of the cell is a synthesis of solid, fluid, and statistical mechanics such that there is an opportunity for multi-physics formulations. There is also potential for fluid-structure problems, contact, even nonlinear material models. There is hardly an area within advanced computational mechanics that does not have application within cell mechanics.

1.3 MODEL PROBLEM: MICROPIPETTE ASPIRATION

We conclude this chapter with a simple model problem to give you a first glimpse into approaches for investigating cell mechanics and what sort of understanding we can gain from these analyses. Micropipette aspiration was one of the first methods used to examine cellular behavior and is responsible for some remarkably important and surprising insights into cellular behavior. Micropipette aspiration involves relatively simple instrumentation, and the experimental analysis can be very straightforward. The introduction of micropipette aspiration here is meant to foreshadow the level of abstraction and rigor that will follow throughout the text.

What is a typical experimental setup for micropipette aspiration?

Some of the earliest mechanical measurements of cell membranes were made using micropipette aspiration experiments. These measurements were based partly on the concept that cells were pouches with fluid interiors (use of RBCs eliminated the problem of the nucleus, because RBCs have none). The versatility of micropipette aspiration and ease of interpretation of experimental results continue to make this an important experimental approach for studying the mechanics of cells (and not just RBCs). As we will see, these experiments not only allow one to make measurements of cell membrane mechanical properties, but they also provide insight into the mechanical behavior of whole cells.

Figure 1.2 Red blood cell being drawn into a micropipette. (Courtesy of Richard Waugh, University of Rochester.)

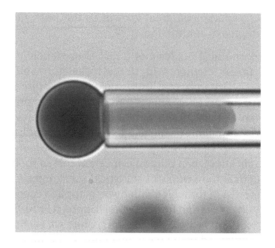

A micropipette is a rigid tube (usually glass) that tapers to a diameter of several micrometers at the tip (near the tip, the diameter is constant). It is hollow all the way through its length and a suction (negative) pressure is applied to the interior (the *lumen*). If the end is brought in proximity to a cell while suction is applied, a seal will form, and the cell will be drawn into the micropipette, forming a protrusion (Figures 1.2 and 1.3). The negative pressure can be applied in a variety of ways. One way is to apply the suction by mouth—this method actually provides a lot of control, and researchers commonly do this when forming the seal.

Another common way is to connect the micropipette to tubing that runs to a water-filled reservoir with controllable height. In this case, decreasing the height of the fluid surface in the reservoir relative to the height of the fluid surface in the dish in which the cells are cultured creates a suction pressure within the micropipette. In theory, the minimum suction pressure that can be applied is determined by the minimum change in height of the fluid reservoir that can be achieved (typically on the order of ~0.01 Pa). In practice, the resolution is worse (usually on the order of ~1 Pa), owing to drift caused by water evaporating from the reservoir. Typically, the maximum pressure that can be applied is on the order of atmospheric pressure, resulting in a wide range of forces, from ~10 pN to ~100 nN.

Once the cell is drawn into the micropipette, the morphology of the cell relative to the pipette can be divided into three regimes, as in Figure 1.4. The first regime is when the length of the protrusion of the cell into the pipette L_{pro} is less than the radius of the pipette R_{pip}, or $L_{pro}/R_{pip} < 1$. The second regime is when the protrusion length is equal to the pipette radius, or $L_{pro}/R_{pip} = 1$, and the protrusion is hemispherical. The third regime is when $L_{pro}/R_{pip} > 1$, and the protrusion is cylindrical with a hemispherical cap. The radius of the hemispherical cap is R_{pip}, since the radius of the protrusion cannot change once the hemispherical cap is formed.

Figure 1.3 A micropipette aspiration experiment. The micropipette tip is placed in the proximity to the cell, and suction pressure is applied. A seal forms between the cell and the micropipette, forming a cell protrusion into the micropipette.

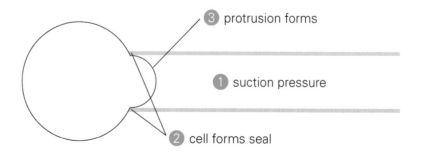

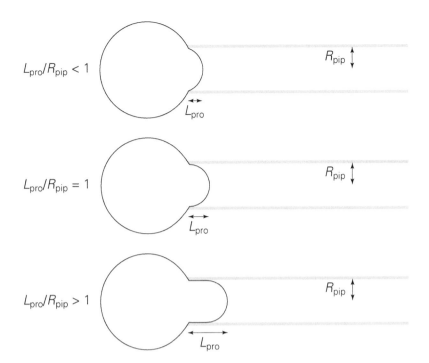

Figure 1.4 Three regimes of cell aspiration into a micropipette.
(Upper) The length of the protrusion of the cell into the pipette is less than the radius of the pipette, or $L_{pro}/R_{pip} < 1$. (Middle) $L_{pro}/R_{pip} = 1$. In this case, the protrusion is hemispherical. (Lower) $L_{pro}/R_{pip} > 1$. The protrusion is cylindrical with a hemispherical cap of radius R_{pip}.

At this point one should be able to see that geometrically the radius of the protrusion of the cell in the first regime is larger than R_{pip}.

The liquid-drop model is a simple model that can explain some aspiration results

When researchers performed early micropipette aspiration experiments on cells such as *neutrophils* (a type of white blood cell), they noticed that after the micropipette pressure exceeded a certain threshold, the cells would continuously deform into the micropipette (in other words, the cells would rapidly "rush into" the pipette). This observation led to the development of the *liquid-drop model*. In this model, the cell interior is assumed to be a homogeneous Newtonian viscous fluid, and the surrounding membrane is assumed to be a thin layer under a constant surface tension, and without any bending resistance. Further, it is assumed that there is no friction between the cell and the interior walls of the pipette.

Surface tension has units of force per unit length, and can be thought of as the tensile (stretching) force per unit area (*stress*) within the membrane, integrated through the depth of the membrane. For example, if a membrane of thickness d is subject to a tensile stress σ that is constant through its depth, then we can model the membrane as having a surface tension of $n = \sigma d$. If the membrane is very thin compared to the radius of the cell, we can ignore the membrane thickness for analysis and rely exclusively on the surface tension n (Figure 1.5).

Why is this model called the liquid-drop model? A drop of cohesive liquid (such as water) suspended in another less cohesive fluid (such as air) has a thin layer of water molecules at the surface of the drop, and, because of the imbalance of intermolecular forces at the surface, packs together and causes surface tension. In brief, every molecule exerts, on average, an attractive force on every other molecule. So, a molecule deep within the drop is pulled in every direction about equally and has no net average force on it. However, a molecule at the surface is pulled with a net force into the "bulk" of the drop, resulting in a spherical shape for the drop and some resistance to deformation at the surface, which we call

Figure 1.5 A membrane of thickness d subject to a tensile stress σ that is constant through its depth can be modeled as infinitely thin with a surface tension of $n = \sigma d$.

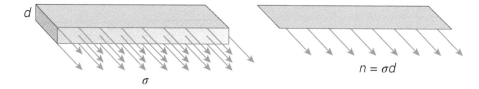

surface tension. This surface tension is what allows some types of insects to walk on the surface of water.

The Law of Laplace can be applied to a spherical cell

By modeling the cell as a liquid drop, we can analyze micropipette aspiration experiments using the *Law of Laplace*, which relates the difference in pressure between the inside and outside of a thin-walled pressure vessel with the surface tension within the vessel wall. It can be derived from a simple analysis using a *free body diagram*, a topic discussed in detail in Chapter 3. Consider a spherical thin-walled vessel with radius R, a pressure of P_i inside the vessel, and a pressure of P_o outside the vessel (**Figure 1.6**).

If we cut the sphere in half, then there are two equal and opposite resultant forces acting on the cut plane. The first is due to pressure, and it is calculated as $F_p = (P_i - P_o)\pi R^2$. The second resultant force is due to surface tension on the wall. If the surface tension is given by n, then the resultant force due to surface tension F_t is s $F_t = n2\pi R$ (the surface tension multiplied by the length of the edge exerting such tension, which would be the circumference of the circle). Setting $F_p = F_t$, we arrive at the Law of Laplace,

$$P_i - P_o = \frac{2n}{R}. \tag{1.1}$$

Note that we set the forces equal because we assume that the droplet is not accelerating. In this case, Newton's second law implies that the forces due to pressure and surface tension must sum to zero. Because they act in opposite directions, the forces are equal.

Micropipette aspiration experiments can be analyzed with the Law of Laplace

We can use the Law of Laplace to analyze micropipette aspiration experiments by relating suction pressure to the morphology of the cell as it enters the pipette. Consider the configuration in **Figure 1.7**, where P_{atm} is the pressure of the environment, P_{cell} is the pressure within the cell, P_{pip} is the pressure within the pipette, R_{cell} is the radius of the cell outside the pipette, R_{pip} is the radius of the pipette,

Nota Bene

History of Young and Laplace. The Law of Laplace is also known as the Young–Laplace equation in honor of Thomas Young, who made initial qualitative observations of the curvature of liquid menisci in 1804, and Pierre-Simon Laplace, who later introduced the mathematical formalism. It is also sometimes called the Young–Laplace–Gauss equation.

Figure 1.6 A spherical vessel with an inner pressure exceeding the outer pressure balanced by a membrane tension n (left) and the resulting free-body diagram (right).

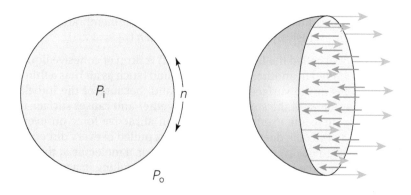

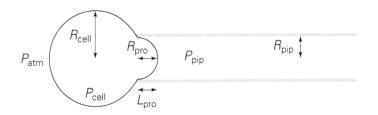

R_{pro} is the radius of the protrusion, L_{pro} is the length of the protrusion into the micropipette and R_{pip} is the radius of the micropipette.

For the portion of the cell that is not in the micropipette, from Equation 1.1 we know that

$$P_{cell} - P_{atm} = \frac{2n}{R_{cell}},$$ (1.2)

where n is the surface tension of the cell. For the protrusion, or the portion of the cell in the pipette,

$$P_{cell} - P_{pip} = \frac{2n}{R_{pro}}.$$ (1.3)

Combining equations 1.2 and 1.3, we obtain

$$P_{atm} - P_{pip} = \Delta P = 2n\left(\frac{1}{R_{pro}} - \frac{1}{R_{cell}}\right),$$ (1.4)

which relates the difference between the pressure in the surroundings and in the pipette with the radius of the cell inside and outside of the pipette for a cell with a given surface tension. Note that we assume the surface tension is constant throughout the cell, even in the "crease" region where the cell contacts the micropipette, and in the protrusion area and the cell membrane outside the micropipette.

How do we measure surface tension and areal expansion modulus?

One can now easily measure surface tension from Equation 1.4. Once the cell is drawn into the micropipette such that the protrusion is hemispherical ($L_{pro} = R_{pip}$), then the radius of the protrusion also equals the radius of the pipette, $R_{pro} = R_{pip}$. Here,

$$\Delta P = 2n\left(\frac{1}{R_{pip}} - \frac{1}{R_{cell}}\right).$$ (1.5)

The pressure ΔP is controlled by the user, and R_{pip} is also known. The cell radius R_{cell} can be measured optically under a microscope, allowing one to calculate the surface tension n. Evans and Yeung performed this experiment using different micropipette diameters and found a surface tension of (~35 pN/μm) in neutrophils. This tension was found to be independent of the pipette diameter, which supports the validity of the technique. We will see in the next section that surface tension does not stay exactly the same if the cell is further deformed.

For a liquid drop, the surface tension will remain constant as it is aspirated into a micropipette. In reality, cells do not behave like a perfect liquid drop. This is because the cell membrane area increases as it is aspirated, resulting in a slight increase in surface tension. The increase in tension per unit areal strain is given by what is called the *areal expansion modulus*. Needham and Hochmuth quantified the areal expansion modulus in neutrophils by aspirating them through a tapered pipette (Figure 1.8). Applying progressively higher pressures, caused the cell to advance further into the taper, increasing its surface area while maintaining constant volume (maintaining constant volume is a condition called *incompressibility*). The radii at either end of the cell, R_a and R_b, the total volume V, and the *apparent* surface area A were measured from the geometry. By apparent we mean that the surface area of the cell is approximated to be smooth, and we ignore small folds and undulations. The surface tension was calculated using the Law of Laplace as

$$\Delta P = 2n\left(\frac{1}{R_a} - \frac{1}{R_b}\right). \tag{1.6}$$

The "original" radius R_0 of the cell was calculated from the volume (assuming the volume remained constant) as $V = 4/3\pi R_0^3$, allowing the "original" or undeformed apparent surface area to be calculated as $A_0 = 4\pi R_0^2$. The areal strain $(A - A_0)/A_0$ and surface tension could therefore be found for the same cell as the pressure was increased, and the cell advanced through the taper. The surface tension was plotted as function of areal strain, and the data was fitted to a line. The areal expansion modulus was found from the slope of the line and was calculated to be 39 pN/μm. Extrapolating the fit line to zero areal strain resulted in a resting surface tension of 24 pN/μm in the undeformed state.

Figure 1.8 Cell being aspirated within a tapered pipette. The radius of the pipette opening is 4 μm. (A) A cell was aspirated into the tapered pipette and allowed to recover to its resting spherical shape. A positive pipette pressure was then applied, and the cell was driven down the pipette. Final resting configuration after (B) $\Delta P = 2.5$ Pa, (C) $\Delta P = 5.0$ Pa, and (D) $\Delta P = 7.5$ Pa. (Adapted from, Needham D & Hochmuth RM (1992) *Biophys J.* 61, 1664–1670.)

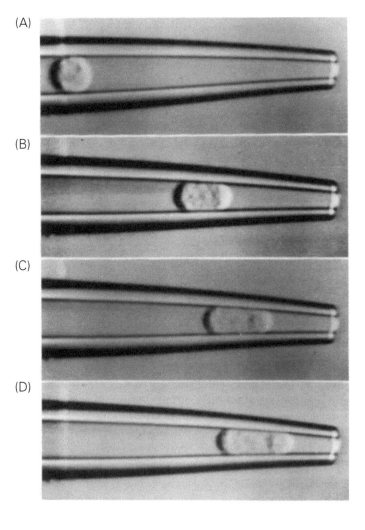

(A)

(B)

(C)

(D)

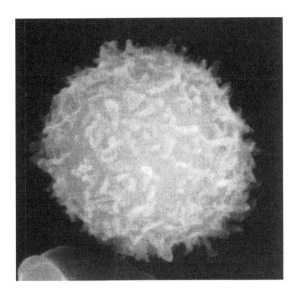

Figure 1.9 Electron micrograph of a neutrophil. Ruffles in the membrane can be clearly seen. (Adapted from, Needham D & Hochmuth RM (1992) *Biophys J.* 61, 1664–1670.)

Why is this tension in the undeformed state important in neutrophils? Remember that neutrophils circulate in the blood and therefore need to squeeze through small capillaries with diameters smaller than the cells themselves, similar to RBCs. As neutrophils squeeze through capillaries, the shape of the cells transform from a sphere into a "sausage" (i.e., a cylinder with hemispherical caps at both ends). Because the cells contain mostly fluid, they are usually incompressible and so must maintain constant volume during this shape change. The surface area of the "sausage" is greater than a sphere of the same volume, and the increase in surface area grows larger as the radius of the "sausage" decreases. However, we will see later that biomembranes are quite inextensible. So how do neutrophils undergo this increase in surface area when squeezing through small blood vessels?

As can be seen in Figure 1.9, neutrophils contain many microscopic folds in their membrane. What this means is that their "apparent" surface area is much less than the actual surface area of the membrane if one were to take into account all the folds and ruffles. The folds allow neutrophils to substantially increase their apparent surface area without actually increasing the surface area of the membrane, as long as the folds are not completely smoothed out. The tension within the cortex of these cells has a crucial role: it allows the cells to have folds and ruffles in the membrane while maintaining their spherical shape.

Why do cells "rush in"?

Remember that the liquid-drop model was developed in large part in response to the observation that some cells would "rush in" after applying any pressure greater than the critical pressure at which $L_{pro} = R_{pip}$. Why do liquid drops do this? Remember that Equation 1.4 is a relation that must be satisfied for equilibrium. Suppose we apply the critical pressure such that $L_{pro} = R_{pip}$, and then we increase ΔP. Let us examine what happens to the terms on the right-hand side of Equation 1.4. We already learned that n is constant for liquid drops, and approximately constant for neutrophils as they are aspirated. The radius of the protrusion, R_{pro}, will also remain constant, because the radius of the hemispherical cap will be equal to R_{pip} for any $L_{pro} > R_{pip}$. R_{cell} cannot increase, meaning that the volume of the cell remains essentially constant over the time of the experiment. We therefore have increased the left-hand side of Equation 1.4 with no way to increase the right-hand side. The result is that equilibrium cannot be satisfied. This produces an instability, and the result is the cell will rush into the pipette.

Figure 1.10 ΔP as a function of $L_{pro} = R_{pip}$ for a cell that behaves like an elastic solid (dotted line) and a cell that behaves like a liquid drop (solid line). When $L_{pro} = R_{pip} = 1$, an instability occurs for the cell that behaves like a liquid drop, and the cell rushes into the micropipette. By contrast, a cell that behaves like an elastic solid will not have this instability.

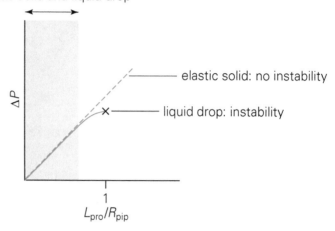

Cells can behave as elastic solids or liquid drops

Micropipette aspiration is an extremely versatile technique for measuring the properties of membranes. It can apply a wide range of forces, and the experiments are conducive to mechanical analysis. In addition to making mechanical measurements of membranes, micropipette aspiration experiments play perhaps an even more important role in understanding cell mechanics, in that they easily allow insight into the fundamental mechanical behavior of different cell types. For example, when researchers performed experiments on endothelial cells or chondrocytes, they found that they would not rush into the pipette, even after $L_{pro} = R_{pip}$. Why? Put simply, these cells do not behave like a liquid drop. Subsequent experiments and analyses showed that their mechanical behavior is much more like that of an elastic solid, so it will not have this instability. Therefore, by observing whether a particular cell does or does not rush into the pipette after $L_{pro} = R_{pip}$, one can easily distinguish whether its mechanical behavior is more like a liquid drop or an elastic solid. Note that this simple method for classification is only possible if the critical pressure is exceeded. If the experiment is terminated before $L_{pro} = R_{pip}$, then one cannot (as easily) distinguish between elastic solid and liquid-drop behavior, as can be seen in Figure 1.10.

Key Concepts

- The study of cellular mechanobiology bridges cell biology and biochemistry with various disciplines of mechanics, including solid, fluid, statistical, experimental, and computational mechanics.

- A wide variety of devastating human diseases such as osteoporosis, heart disease, and even cancer involve cell mechanics in a fundamental way.

- Cell mechanics is an excellent substrate for introducing students to a wide variety of cutting-edge approaches in mechanics.

- Understanding the mechanical behavior of cells presents a grand challenge in theoretical, computational, and experimental mechanics.

- Micropipette aspiration is an early and straightforward approach for investigating cell mechanics. By modeling the cell as a liquid drop, we can analyze micropipette aspiration experiments using the Law of Laplace.

- Liquid-drop cells exhibit an instability when the radius of the aspirated protrusion is equal to the radius of the pipette. At this point the cell cannot resist increased pressure and rushes into the pipette.

- By observing the movement of cells within the micropipette at the instability point, one can distinguish two types of cellular behavior cells: those whose behavior is dominated by membrane tension (similar to a liquid drop), and those that act as a continuum solid.

Problems

1. As we have seen, when a liquid drop–type cell such as a neutrophil is subjected to micropipette aspiration, it becomes unstable. This occurs when the radius of the protrusion is equal to the radius of the pipette. For a cell whose behavior is better approximated by a continuum such as a chondrocyte, do you think there is a maximum pressure that can be exerted on the cell? What configuration would the cell be in at this point?

2. In our analysis we ignored the friction between the inside of the pipette and the cell wall. Is this a reasonable assumption? Why? How might friction affect the results of an aspiration experiment if it were large?

3. The classic micropipette aspiration experiment can be modified to address the instability of a cell with liquid-drop behavior so that membrane behavior can be measured. This is done with a tapered pipette (**Figure 1.8**). With this approach the pipette radius changes along its length. Derive the relationship between membrane tension and the pressure in the micropipette assuming that the two sides of the cell have radii of R_a and R_b and pipette pressures are P_a and P_b.

4. There are other structures similar to cells that can be treated with the liquid-drop model. The bronchial passages of the lungs terminate in small spherical sacs known as alveoli. There are around 150 million alveoli in your lungs. During respiration the alveoli are filled and emptied through the action of the diaphragm and intercostal muscles in the chest wall. During inhalation the pressure outside the alveoli can drop by up to 200 Pa. The layer of cells that line the alveoli is constantly hydrated. So, we can model them as a bubble of air surrounded by water, assuming that the surface tension of water is 70 dyn/cm. With this information, what is the radius of the alveoli? In actuality, the radius of an alveoli is about 0.2 mm. They can be this small because the epithelial cells secrete a protein called *surfactant* that reduces the water surface tension. If fact, there is a developmental condition known as infant respiratory distress syndrome (IRDS) that occurs when insufficient surfactant protein is produced. What must the surface tension be to keep the alveoli inflated?

5. Consider two adjacent alveoli and the end of a bronchus such that air can pass easily between them as well as to the outside. Assume that they have the same radius and that they are in a state of equilibrium with respect to pressure and surface tension. What would happen if a small volume of air moved from one alveolus to the other? What must be true of surfactant as a result?

Annotated References

Evans E & Yeung A (1989) Apparent viscosity and cortical tension of blood granulocytes determined by micropipet aspiration. *Biophys. J.* 56, 151–160. *An early report on micropipette aspiration can be used to determine membrane tension.*

Hochmuth RM (2000) Micropipette aspiration of living cells. *J. Biomechanics* 33, 15–22. *An excellent review of the use of micropipette aspiration to characterize the fundamental behavior of cells and membranes.*

Huang H, Kamm RD & Lee RT (2004) Cell mechanics and mechanotransduction: pathways, probes, and physiology. *Am. J. Physiol. Cell Physiol.* 287, C1–11. *An overview of mechanosignaling with some discussion of techniques as well as an overview of cell-force interactions.*

Jacobs CR, Temiyasathit S & Castillo AB (2010) Osteocyte mechanobiology and pericellular mechanics. *Annu. Rev. Biomed. Eng.* 12, 369–400. *An extensive review of how mechanics and biology interact at the level of the cell to regulate the skeleton in health and disease. Includes extensive supplemental online material.*

Jacobs J (2008) The Burden of Musculoskeletal Disease in the United States. Rosemont, IL. *This monograph is an extensive collection documenting the prevalence of, as well as the social and economic costs associated with, musculoskeletal disease.*

Janmey PA & Miller RT (2011) Mechanisms of mechanical signaling in development and disease. *J. Cell Sci.* 124, 9–18. *This article discusses what is known about the ability of cells to sense local stiffness and includes a discussion of diseases and developmental processes.*

Krahl D, Michaelis U, Peiper HG et al. (1994) Stimulation of bone growth through sports. *Am. J. Sports Med.* 22, 751–157. *Early evidence that loading of bone through physical activities leads to bone formation.*

Malone AM, Anderson CT, Tummala P et al. (2007) Primary cilia mediate mechanosensing in bone cells by a calcium-independent mechanism. *Proc. Natl. Acad. Sci. USA* 104, 13325–13330. *Early evidence that primary cilia act as mechanosensors in bone cells.*

Needham D & Hochmuth RM (1992) A sensitive measure of surface stress in the resting neutrophil. *Biophys. J.* 61, 1664–1670. *An early demonstration of how micropipette aspiration can be used to determine membrane properties.*

CHAPTER 2

Fundamentals in Cell Biology

When thinking about cell mechanics, it is easy to get caught up in models and mathematics, and thinking about cells as somewhat exotic material that can be subjected to the same sort of testing as is performed on inert metals and plastics. However, an equally important part of cell mechanics is the study of how mechanics interacts with the biological behavior of cells. The former is termed "biomechanics"; the latter is increasingly referred to as "mechanobiology," as mentioned in Chapter 1. The former is meant to emphasize biomechanics as the subdiscipline of mechanics that considers the mechanical properties of biological structures and mechanobiology as the subdisipline of biology that is focused on how mechanics regulates biological processes, or how biological processes generate and regulate physical forces. Oftentimes this distinction is artificial and may not be useful, as some of the most fascinating problems cannot be understood without considering biology and mechanics with equal depth and rigor.

Just as our treatment of cell biomechanics is built on a firm understanding of the fundamental principles of mechanics, to appreciate mechanobiology fully we must take a step back and form an understanding of the fundamentals of biology. This must include not only the underpinnings of cell and molecular biology, but also some of the modern experimental techniques that allowed these insights to be made. Our goal is not only to provide you with enough biology background to understand cell mechanics, but to allow you to read and understand the basics of modern biology scientific publications.

Modern biology is undergoing a revolution of understanding that began in 1953 with Watson and Crick's determination of the double-helix structure of DNA. This has led to the sequencing of the human genome, modern genetics, and, in an amazingly short period of time in terms of the history of science, to dramatic advances in biology. This explosion of knowledge, known as the molecular revolution, has given rise to the field of molecular biology sometimes referred to as *modern biology*. J. Craig Venter, one of the pioneering founders of the high-throughput approaches to the study of genes and their regulation (who also participated via Celera in the Human Genome Project), declared this to be the opening of the "Century of Biology." Many have speculated that the discoveries currently being made will lead to the next wave of social change following the Industrial Revolution and the Information Revolution.

2.1 FUNDAMENTALS IN CELL AND MOLECULAR BIOLOGY

Formal writing in biology is heavy on terminology, in part due to the need to be precise in descriptions while being succinct. Let us say you develop a hairline fracture on one of your ribs. If the location of the fracture is closer to the sternum, we call it a *medial* fracture (closer to the "middle" of the body). If it is closer to your sides/arms/shoulders, we call it a *lateral* fracture. The terms medial and lateral

Table 2.1 The monomer subunits and resulting polymers that make up the biochemical constituents of the cell.

Monomer	Polymer	
nucleic acid	RNA, DNA	genes
amino acid	peptide, protein	gene products
fatty acid	lipid	not coded by genes
sugar	polysaccharide, carbohydrate	not coded by genes

(along with other terms) can be used to quickly describe locations without having to reference specific sites, much like north and south can be used if you are not familiar with local landmarks.

In a similar way, there are a few terms that recur in biomedical writing with which you should be familiar. We may not use them all in this book, but the diligent student will want to get to know these terms. Some common ones are:

- **Cell culture:** the process of extracting live cells from biological tissue, and the maintenance and growth of those cells, typically *in vitro*, to study the physiological behavior of the cells under controlled conditions.
- ***In vitro***: in a laboratory dish; not inside an organism; literally, "in glass".
- ***Ex vivo***: sometimes interchanged with *in vitro*, but occasionally referring to entire tissues that are cultured in a laboratory dish.
- ***In vivo***: inside a living organism, typically but not always at the natural location.
- ***In situ***: inside a living organism, in the natural location. Growing an ear in a mouse's back would be an *in vivo* but not an *in situ* experiment.
- **Amino acids:** the fundamental building blocks (monomers) of proteins.
- **DNA:** deoxyribonucleic acid, the "hard copy" of genes and their regulatory components.
- **RNA:** ribonucleic acid, the "working copy" of genes that assembles proteins.
- **Gene:** sequence of DNA that encodes a protein.
- **Promoter:** sequence of DNA that regulates when a particular gene is expressed. Typically, but not always, near the gene being regulated.
- **Probe:** A fragment of DNA or RNA that is complementary to a specific target sequence. The probe is able to bind to the target in a process called *hybridization*.

Many cellular components are polymers, assembled from subunits called monomers (Table 2.1). The major exception to this is the most abundant constituent of cells, water, which makes up roughly 70% of a cell's mass. Inorganic ions are another exception and are critical to cell metabolism and signaling ("organic" compounds have carbon, "inorganic" compounds do not). These components are vital for the non-inert response of cells to mechanical forces. In some cases, biological polymers form the physical structure of the cell, and the biomechanical characterization of the cell relies on the properties of these polymers. In other cases, the molecules are responsible for signaling. Such molecules can be important for signaling in response to mechanical stresses (mechanosignaling), which we discuss further in Chapter 11.

Proteins are polymers of amino acids

In general, proteins consist of many (> 50) amino acids. Short chains of amino acids or fragments of proteins are called peptides. Each amino acid shares a common backbone structure of an amino group (H_2N), a carbon atom, and a carboxyl group (COOH) (Figure 2.1). The central carbon, known as the α-*carbon*, attaches to a side chain. The diversity of charges, sizes, and interactions of these side chains

Nota Bene

Inorganic carbon compounds.
There are several carbon compounds that are inorganic. Diamond and graphite are obvious examples. However, there are no hard-and-fast rules to make the distinction.

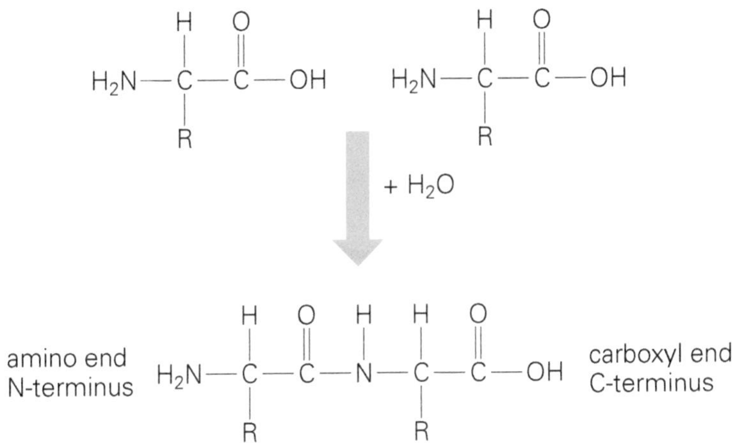

Figure 2.1 Chemical structure of a polypeptide. Note the repeating backbone structure and the residues that distinguish one amino acid from another.

distinguishes one amino acid from another. The properties of these side chains additionally determine the hydrophilic/hydrophobic nature of the local peptide sequence, which is useful for determining regions of proteins that span membranes and the orientation of the protein. Remarkably, there are only 20 common amino acids (although there are several more in nature, plus some synthetic ones that have been created in laboratories). The sequence, or order, in which these amino acids are assembled gives rise to the remarkable diversity of structure and function found in proteins. The amino acid sequence is formally referred to as the primary (1°) structure of the protein.

Amino acids are asymmetric because one side has an amino group and the other has a carboxyl group. This directionality is preserved in peptides and proteins, with one end terminated with the amino group, the N-terminal, and the other terminated with the carboxyl group, the C-terminal. When amino acids are assembled, one oxygen from the carboxyl group and two hydrogens (one from each group) form a molecule of water. This process of assembly is called *dehydration synthesis* or *condensation*. When these bonds are broken, a molecule of water is utilized in a reaction called *hydrolysis*.

Once a protein is assembled, it can fold based on the distribution of charges and steric interactions, the latter being space available for movement based on the size of the side groups. Some common local folding patterns include α-helices (single spring-like structures) and β-sheets (mostly flat loops) (Figure 2.2). These patterns are typically referred to as the secondary (2°) structure, and are mainly a result of hydrogen bonding. The global folding pattern of the entire protein is referred to as the tertiary (3°) structure. How proteins fold is important for their function, because the local shape/charge distribution determines which other molecules the proteins can interact with. The prediction of folding patterns based on amino acid sequence is an open problem currently, with computational models only able to handle short peptide fragments. Finally, the association of multiple proteins is termed the quaternary (4°) structure. Molecular dynamics computer simulations are capable of generating information regarding secondary and tertiary structures, but are known for being computationally expensive. Simulation of quaternary structures, which would be a boon for pharmaceutical design, but are also extremely computationally expensive.

The mechanics involved in folding or unfolding of proteins is a subject of intense investigation, because such unfolding may occur in your cells. Not only does the resistance to unfolding contribute to the "stiffness" of the molecule but there are signaling pathways that may be activated by protein unfolding, as discussed in later chapters. Knowing the charges and sterics of proteins and how a protein folds is not only important in biology, but also in mechanobiology.

Figure 2.2 Schematic of the two most common protein secondary folding structures, (A) an α-helix and (B) a β-sheet. (Adapted from, Alberts B, Johnson A, Lewis J et al. (2008) Molecular Biology of the Cell, 5th ed. Garland Science.)

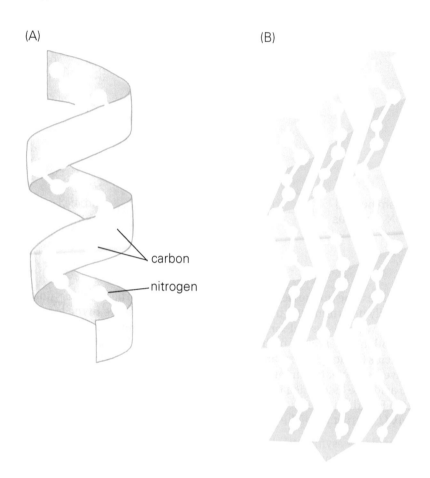

(A)

(B)

carbon

nitrogen

Example 2.1: What determines how hard it is to fold a protein?

Consider the effects of side chains on protein folding and unfolding. Start with a long protein that is entirely uncharged with polar side chains. It takes some force to coil it up, because of the bending of the bonds between adjacent amino acids.

If you somehow charge the protein with the same charge throughout the amino acids, is it easier or more difficult to fold?

Next, take the original uncharged, polar protein, but in its folded configuration.

If you introduce hydrophobic side chains such that large sections of these side chains are clustered together in the folded protein, and keep the protein in an aqueous environment, is the protein harder or easier to unfold?

In the first case, with all same charges throughout the protein, the protein becomes harder to fold compared with the uncharged state. This is because of like charges repelling, so that to fold the protein, not only do you have to work against the bonds between the amino acids bending, but also in bringing charges closer.

In the second case, the protein becomes harder to unfold compared with the original protein with polar side chains, because hydrophobic groups tend to cluster together in an aqueous environment. As a result, it will take more work to separate the groups from each other to extend the protein.

DNA and RNA are polymers of nucleic acids

Nucleotides are the monomers that make up the genetic polymers DNA (deoxyribonucleic acid) and RNA (ribonucleic acid). As one can deduce from the names, in RNA the sugar backbone is ribose, and in DNA it is deoxyribose. In both cases, there is a side group known as a *base* for its ability to bind hydrogen in aqueous conditions. For both RNA and DNA there are four types of bases: cytosine (C), guanine (G), adenine (A), and thymine (T) in DNA or uracil (U) in RNA. Bases are

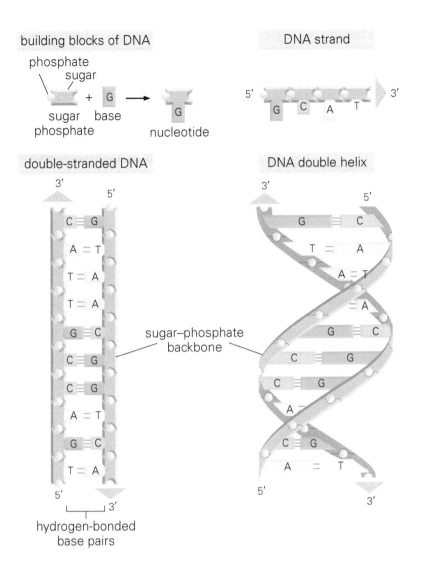

Figure 2.3 The various molecular units that combine to form the double-stranded DNA polymer are shown at different scales, from the components making up a nucleotide to the assembled double helix. (Adapted from, Alberts B, Johnson A, Lewis J et al. (2008) Molecular Biology of the Cell, 5th ed. Garland Science.)

able to form hydrogen bonds with each other, but only in a complementary fashion: C with G, and A with T (DNA) or U (RNA). Long nucleic acids can bind to each other in the famous double-helix arrangement, but only when they have complementary sequences (Figure 2.3). Although nucleotides can have many roles in the cell, they are primarily responsible for the storage of genetic information through the sequence in which the bases are assembled and replicated during cell division (Figure 2.4).

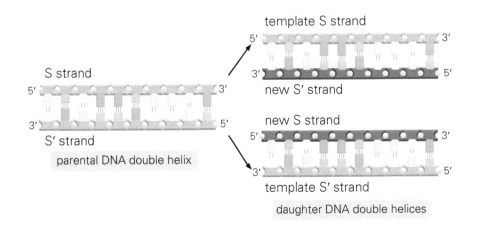

Figure 2.4 The ability to replicate is one key property of DNA, allowing it to pass on genetic sequence information. The parental S strand and its complement S' strand are separated and new (daughter) S and S' strands are synthesized. (Adapted from, Alberts B, Johnson A, Lewis J et al. (2008) Molecular Biology of the Cell, 5th ed. Garland Science.)

The forces required to unfold nucleic acid polymers are of interest for several reasons. DNA is usually tightly coiled around proteins called histones to conserve space. Further, because DNA is maintained in a double helix, there is some twisting to the molecule as it unwinds to allow access for DNA replication or transcription (more on transcription shortly). Finally, DNA activation sometimes depends on binding or association of one part of the gene to another, and the two parts that require interaction may be separated by a long distance. The distance between these two parts is guided, at least in part, by how flexible the DNA is. Understanding the mechanism by which DNA (and RNA) works requires some comprehension of the mechanical behavior underlying these molecules.

Advanced Material: DNA and RNA are directional

Each of the carbons in the sugar backbone are labeled numerically 1′ through 5′. Adjacent rings are bound to each other through a phosphate group. This phosphate covalently links the 5′ carbon of one monomer to the 3′ carbon of the next. Similar to proteins, this gives a directionality to nucleotides and their chains, nucleic acids. The end of the polymer where the phosphate binds the 5′ carbon is known as the 5′ end. Genetic sequence is usually given from the 5′ end to the 3′ end in a series of single-letter codes corresponding to the bases. Many processes involving nucleic acids are also directional, as seen with the polymerase chain reaction later in this chapter.

Polysaccharides are polymers of sugars

Another class of cellular polymers is the complex sugars. Familiar to us as an (sweet) energy source, sugars also have important structural roles—the most abundant organic molecule on the planet is cellulose from the plant cell wall. The "crunchiness" of vegetables and the strength of wood are derived from cellulose. Also, many proteins are decorated with sugars in their functional form. A single sugar molecule is known as a monosaccharide. More complex oligosaccharides can be formed as disaccharides, trisaccharides, etc. Large sugar polymers are known as polysaccharides. All sugars and the complex polymers made from them are known collectively as carbohydrates. Similar to proteins, bonds between monosaccharides are formed through condensation and broken through hydrolysis.

Fatty acids store energy but also form structures

The final major class of molecules introduced here is the fatty acids. A fatty acid is made up of a long hydrocarbon-rich chain terminated with an acidic carboxyl group. Again, we primarily think of fats or lipids as a more concentrated source of energy than sugars. When their hydrocarbon tails are broken down, they produce many times more energy than carbohydrates on a per-weight basis. Fatty acids also perform critical biochemical and structural functions in the cell. Although they do not form linear polymers such as nucleotides, amino acids, or monosaccharides, they are able to aggregate to form globular and sheet structures, such as the cell membrane. These structures not only help to segregate the interior of the cell from the exterior, but also help to maintain the cell's shape. Because fats and water do not mix well, deforming a cell requires work to overcome the tendency of fats to aggregate. Fatty acids are partly responsible for the mechanical resistance of a cell to external forces.

Correspondence between DNA-to-RNA-to-protein is the central dogma of modern cell biology

Our interest in the biological aspects of cells extends beyond passive structural components. As we mentioned in the introduction to this chapter, we also wish to consider the molecular response of cells to mechanical forces. But what responses can there be, other than molecules being deformed? Different signaling pathways are engaged by different cells in response to different types of mechanical stimuli. A cell will change what proteins it makes and what part of the cell they are transported to in response to mechanical stimuli. But to understand this, one must first understand how a cell synthesizes proteins and how that process is regulated. We now briefly review the central dogma of molecular biology, which is the link between the genetics of a cell to eventual protein creation.

Once the structure of DNA was discovered in the 1950s, it quickly became clear that the sequence of nucleotides in DNA holds information that is passed from cell to cell. There are two key properties of DNA that allow this to happen. The first is replication, in which the entire sequence of DNA is copied resulting in two separate strands with identical sequences. The original DNA double-strand is pulled apart, and the complementary base sequences are assembled using the separated original strands as templates. This results in two identical copies of the DNA polymer.

The second critical property of DNA is that its sequence holds the instructions for assembly of amino acids in a particular sequence to form specific proteins. The region of DNA that contains the sequence information for a particular order is defined as a *gene* (Figure 2.5). Fundamentally there are two steps in the creation of a protein from a gene. The first is to create a polymer of complementary RNA from the gene in a process known as *transcription*. This RNA leaves the nucleus carrying its sequence or "message". This type of RNA is known as *messenger* RNA or simply mRNA. The mRNA then docks with a *ribosome* that is responsible for assembling the peptide sequence specified by the mRNA in a process called *translation* (Figure 2.6). The base pairs of the mRNA strand are grouped into triplets, each of which is known as a *codon* that corresponds to a particular amino acid in the forming peptide (Table 2.2). A useful memory tool to distinguish transcription from translation is that transcription uses the same "alphabet"—the sequence of bases in the DNA and RNA nucleic acids—but translation involves converting from sequences of bases to sequences of amino acids, or a different alphabet. This flow of information from DNA to RNA to protein was termed *the central dogma of molecular biology* by Francis Crick, co-discoverer of the structure of DNA, in 1958.

One might wonder why a cell requires RNA. Why not just simply build a protein straight from the DNA template? It turns out that only about 5–10% of the

Nota Bene

Combinatorics of nucleic acids and amino acids. Since there are four base pairs, and 20 amino acids, a triplet codon is the minimum necessary to have at least one unique sequence per amino acid. However, a triplet codon has 64 possible combinations, whereas only 20 are needed for each amino acid. Most amino acids, therefore, are represented by several codon triplets. Additionally, there are three triplets for a "stop" codon that does not code for a particular amino acid. This codon designates the end of the protein. The "start" codon that designates the start of the protein corresponds to the amino acid methionine (Met). Note that Met can appear in the middle of a peptide or protein, so not all instances of Met (ATG) represent the start of a new gene.

The redundancy is not evenly spread among the amino acids (see Figure 2.8). There is a loose pattern in which an amino acid appears to be coded by sequences that differ only in the last base pair. Current thinking is that the redundancy takes advantage of common errors to build in some sort of protection against mutations so that key amino acids are not as affected by such errors.

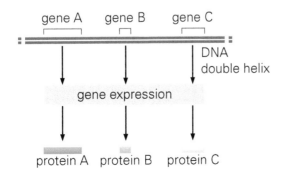

Figure 2.5 Genes are units of sequence information that encode for specific proteins. This correspondence between genes and proteins is known as the central dogma of molecular biology. (Adapted from, Alberts B, Johnson A, Lewis J et al. (2008) Molecular Biology of the Cell, 5th ed. Garland Science.)

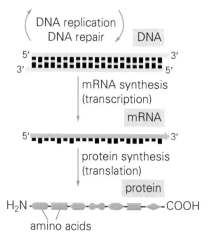

Figure 2.6 Transcription is the production of mRNA from DNA; translation is the production of a protein from mRNA. (Adapted from, Alberts B, Johnson A, Lewis J et al. (2008) Molecular Biology of the Cell, 5th ed. Garland Science.)

Table 2.2 The 20 common amino acids and their codon sequences. Three codons code for the "stop" sequence. Note that there is significant redundancy.

Ala-Alanine	A	GCA GCC GCG GCU
Arg-Arginine	R	AGA AGG CGA CGC CGG CGU
Asp-Aspartic acid	D	GAC GAU
Asn-Asparagine	N	AAC AAU
Cys-Cysteine	C	UGC UGU
Glu-Glutamic acid	E	GAA GAG
Gln-Glutamine	Q	CAA CAG
Gly-Glycine	G	GGA GGC GGG GGU
His-Histidine	H	CAC CAU
Ile-Isoleucine	I	AUA AUC AUU
Leu-Leucine	L	UUA UUG CUA CUC CUG CUU
Lys-Lysine	K	AAA AAG
Met-Methionine	M	AUG
Phe-Phenylalanine	F	UUC UUU
Pro-Proline	P	CCA CCC CCG CCU
Ser-Serine	S	AGC AGU UCA UCC UCG UCU
Thr-Threonine	T	ACA ACC ACG ACU
Trp-Tryptophan	W	UGG
Tyr-Tyrosine	Y	UAC UAU
Val-Valine	V	GUA GUC GUG GUU
Stop		UAA UAG UGA

entire genomic DNA strand encodes genes. These encoding regions are called exons ("e" for encoding, Figure 2.7). Noncoding regions are called introns; their function is not precisely known, but some intron regions are known to be important for alternative splicing. Other intron regions are thought to be important for DNA folding. mRNA is a compact subset of the genome that contains the gene of interest, but also can be manipulated without disturbing the original template stored in the DNA. Further, multiple mRNAs can be generated from a single DNA template, allowing for a relative amplification of protein expression at the mRNA level than if DNA were used for protein synthesis directly.

Figure 2.7 Exons (blue), the units of sequence that encode amino acid sequences, are typically a small fraction of the entire gene. (Adapted from, Alberts B, Johnson A, Lewis J et al. (2008) Molecular Biology of the Cell, 5th ed. Garland Science.)

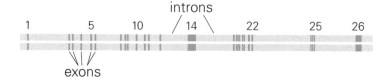

Example 2.2: Transcription and translation

The general idea underlying protein synthesis, following the central dogma of cell biology, starts from a hard-coded gene and ends in a protein (Figure 2.8). Suppose you have a DNA sequence known to contain a gene, without introns, as follows:

5′-ATACCCTATGAACAGATGAACC-3′

What is the sequence of the complementary strand and peptide generated from this sequence?

The complementary strand would be:

3′-TATGGGATACTTGTCTACTTGG-5′.

Note that the 3′ and 5′ are reversed. This is important, because if the directionality were not reversed, then the strand would not hybridize to the original sequence, and would not be a complement.

We start from the 5′ end in the original sequence and look for a start codon, and then separate into triplets:

5′-ATACCCT ATG AAC AGA TGA ACC-3′

So we get Met-Asn-Arg. That is it, since the TGA is a stop codon. Of course, most proteins are far longer than this three-amino-acid example peptide.

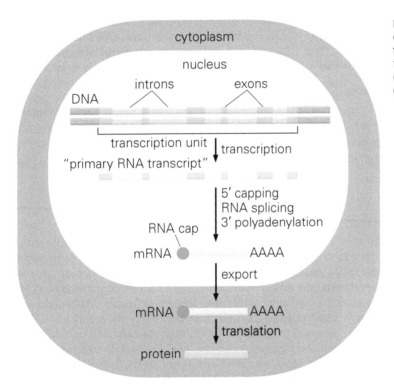

Figure 2.8 Summary of the process of protein synthesis from DNA to fully functional protein. (Adapted from, Alberts B, Johnson A, Lewis J et al. (2008) Molecular Biology of the Cell, 5th ed. Garland Science.)

Phenotype is the manifestation of genotype

The collection of genes represented in a cell is known as its *genome*. With few exceptions, all of the cells in a particular organism contain the same DNA sequence and therefore the same genes. Almost all of the cells in a person's body contain the same DNA sequence and are said to be of the same *genotype* (notable exceptions include the red blood cells, which lack nuclei and genomic DNA, and reproductive cells). Of course, your genetic makeup likely differs from others as you probably have different genotypes. What, then accounts for the difference between the behavior of one of your liver cells from one of your bone cells if they are genotypically identical? The difference in cellular behavior is a function of which genes are being transcribed or *expressed* at any particular time. The word *phenotype* was coined in 1911 by Wilhelm Johannsen to describe the behavior or character of a cell or organism in contrast with its genotype, which is inherited. During development or certain repair processes, cells change their phenotype and become more specialized in a process known as *differentiation*.

Advanced Material: Phenotype can be elusive

By contrast with genotype, a cell's phenotype can be much more difficult to define and is often only the result of long-term consensus. This is confounded by the fact that the behavior of a cell changes considerably over time as a result of changes in the extracellular environment. In some ways, phenotypic change and differentiation is meant to imply more enduring changes to a cell's behavior than simple responses to external stimuli. When these stimuli persist for a sufficient time, they can lead to changes in phenotype. For example, in bone, the cell responsible for forming new bone is called the osteoblast. An osteoblast may or may not form bone in response to biochemical and physical signals, though it remains an osteoblast. Occasionally, an osteoblast becomes embedded in the same bone that it is forming, and becomes more quiescent than an osteoblast. This cell is said to have differentiated into a different cell type called an osteocyte.

It is difficult to determine when a cell has changed its behavior sufficiently to have differentiated into another cell type. These changes must, in some sense, be irreversible and permanent. Yet, in some situations cells can lose the phenotypic markers of differentiation in a process known as *dedifferentiation*. Organisms that can regenerate lost limbs are thought to be able to induce a dedifferentiation process. In some conditions osteocytes can be induced to make bone. Does this mean that they have dedifferentiated into osteoblasts again? Cells often lose differentiation markers of their original cell type. Have they dedifferentiated? These questions, although they appear simple on the surface, can be scientifically ambiguous and are often based on functional definitions. These issues can become quite contentious and vary substantially from field to field.

Nota Bene

RNA undergoes maturation modifications. The newly formed mRNA transcript often undergoes *post-transcriptional modification* before translation into protein. For instance, the 3′ end of the mRNA is usually supplemented with a polyadenosine tail (poly(A) tail), which allows protein complexes to bind to the mRNA transcript. These protein complexes can splice out the introns of the gene to leave behind only the exons for translation. It is also possible to splice out some exons that the introns flank, thereby creating several protein variants from the same DNA source gene. This modification is referred to as alternative splicing.

Progenitor cells have the potential to differentiate into multiple different cell types, thereby acquiring different phenotypes. A cell that has the potential to differentiate into multiple cell types is known as *multipotent* or *pluripotent*. Pluripotent cells can differentiate into cells or all three germ layers (endoderm, mesoderm, and ectoderm), while multipotent cells can differentiate into multiple, but more limited, lineages. Cells that can differentiate into any cell type in an organism are referred to as *totipotent*. Progenitor cells that can differentiate into more specialized cell types, but can also undergo cell division without differentiating, are known as *stem cells*. Both embryonic stem cells (currently only obtainable from a blastocyst) and adult stem cells (obtainable from a variety of tissues, including bone marrow, umbilical cord blood, fat, and muscle) are being studied for their potential use in regenerative medicine and tissue engineering. One key challenge in working with stem cells is that once the desired cell is generated, it must be conditioned to work with existing tissues. For example, if a partial tissue replacement is required, the cells must integrate with the host cells and exhibit the correct strength and durability to continue functioning. Many of these characteristics depend on cell mechanical properties, as well as mechanobiological adaptation, and understanding and manipulating these characteristics remains a fundamental challenge in tissue engineering.

Transcriptional regulation is one way that phenotype differs from genotype

One of the ways in which the cell regulates its behavior is by altering which genes are transcribed and how often. RNA polymerase is the enzyme responsible for transcribing DNA into RNA. Its activity is regulated by controlling its proclivity to bind to a particular region of DNA, known as a *promoter*, usually proximal to the coding sequence (the region that encodes the amino acid sequence) for a particular gene, and begin transcription. The cell is not constrained to up- or down-regulate (increase or decrease) transcription of all genes equally. To regulate function and behavior, the cell needs to control transcription of specific genes without affecting transcription of other genes. DNA-binding proteins known as *transcription factors* recognize specific DNA sequences and regulate the binding of RNA polymerase. Given the complexity of cellular function, there is a dizzying array of transcription factors. Indeed, most genes require a complex of transcription factors to form a protein cluster for transcripts to be produced. Finally, transcription factors are themselves coded from genes, so their availability is also subject to transcriptional regulation.

Cell organelles perform a variety of functions

The interior of a cell is not a simple amorphous soup of materials. It is organized into a wide variety of small subcellular units with particular functions, some of which we have already discussed. They are called *organelles* by way of analogy to the organs of a larger organism, a term coined in 1884 by Karl Möbius (not of the Möbius strip). In our discussion of protein synthesis, we have already introduced a few critical organelles—the nucleus, the ribosome, the Golgi apparatus, and the endoplasmic reticulum. Table 2.3 lists major organelles and their functions for eukaryotic cells.

The "central dogma" is important for cell mechanics and mechanobiology for several reasons. First, the proteins that exist to support cells structurally do not usually pre-exist within the cell. They have to be made by the cell, typically in an ongoing process involving rather intricate pathways. The cell produces proteins from amino acids using instructions from genes. So, unlike the framework of a building, the skeleton of the cell is constantly shifting and rebuilding, and anything that disrupts this process can result in significant changes to cell properties.

Second, for mechanobiology in particular, cells respond to several signals, including physical forces, by altering their expression profile. The phenotype of the cell depends in part on mechanical influences acting upon the cell. How the phenotype adjusts is a matter of much interest—cells will change morphology, increase levels of some proteins, decrease others, and alter their overall behavior (motion, activity, proliferation, etc.). To study these responses at the molecular level, we need to know what to examine. Now that we know the dogma we can

Nota Bene

Often proteins need some final tweaks. To become fully functional, many proteins need to be further modified after translation. This is known as *post-translational modification*. Modifications may include formation of disulfide bonds within a protein (linking together two parts of the protein normally far apart) and attachment of non-amino-acid functional groups such as carbohydrates, lipids, phosphates, etc. During post-translational modification, the protein is often folded to acquire its secondary and tertiary structure, typically in the Golgi apparatus and endoplasmic reticulum (ER). X-ray crystallography is used to determine the structure of a protein experimentally.

Table 2.3 Some of the eukaryotic cellular organelles and their functions. These functions are described to provide an idea of their general function, but in some cases they may be involved in other tasks.

Organelle	Primary function(s)
nucleus	DNA repository, site of transcription
ribosome	responsible for mRNA-to-protein translation
mitochondrion	responsible for energy production by conversion of glucose metabolites into ATP; has its own maternal DNA
endoplasmic reticulum	site of translation and some post-translational modification and folding of proteins
golgi apparatus	post-translational modification and sorting of proteins; important for secreting proteins
lysosome	degradation of proteins and carbohydrates; it is important that the degradative enzymes responsible remain sequestered from the cytoplasm
vesicle	intracellular transport and trafficking

Nota Bene

Bacteria do not have nuclei. A bacterium is a prokaryotic cell—a cell without a nucleus. Eukaryotic cells have a nucleus plus many of the other organelles. Prokaryotic cells are nearly always unicellular and do not form complex organisms. Originally they were thought to have no intracellular organelles, but is now known that they do have some (such as ribosomes). Their DNA is often organized into a single closed loop known as a *plasmid*.

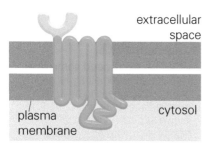

Figure 2.9 A receptor typically contains several membrane-spanning domains, an extracellular component that binds to the target or ligand, and an intracellular component that initiates intracellular signaling or other forms 25 of cellular regulation. (Adapted from, Alberts B, Johnson A, Lewis J et al. (2008) Molecular Biology of the Cell, 5th ed. Garland Science.)

consider several strategies. Examination of mRNA levels would be an indication of the activation of certain genes. One might complement mRNA profiles with protein measurements—and typically protein measurements are made after mRNA is known to change. This timing is selected because proteins come from mRNA, so one would normally expect the changes in protein concentration to lag behind the changes in mRNA, which we know only because of the central dogma.

Finally, there is the challenge of determining the exact mechanism by which cells sense and respond to mechanical forces. Although it is established that, stretching a cell will lead to up-regulation of mRNA and protein from certain genes, less is known about how the up-regulation occurs. There are many hypotheses, including proteins and nucleic acids being directly unfolded by the forces, the cell membrane itself being flexed, and others. The central dogma provides a much better idea of what to look for if we know roughly the pathway such up-regulation takes.

2.2 RECEPTORS ARE CELLS' PRIMARY CHEMICAL SENSORS

Receptors are proteins that bind to specific target molecules or *ligands* and initiate a cellular response, typically in the form of a biochemical intracellular signaling cascade. Often receptors are suspended in the cell membrane, but they can also be found in the cytoplasm or the nuclear membrane. Membrane receptors can be identified by one or more hydrophobic membrane-spanning domains (**Figure 2.9**). The specific structure of a receptor can be determined by X-ray crystallography or by computer simulation predictions based on the protein's sequence. Of course there is an enormous diversity of receptors and corresponding ligands. A cell can change its sensitivity to a given ligand by up-regulating or down-regulating the number of receptors on its surface. Often this regulation occurs in response to activation of the receptor itself, thereby creating a form of feedback mechanism. When a cell senses a small number of a particular ligand, it commonly makes more receptors for that ligand. This is a positive feedback loop. It allows the cell to be highly sensitive to a huge number of extracellular signals and to do so very efficiently.

By being adaptive to its environment, the cell need not maintain large numbers of receptors for every conceivable signal, but at the same time, it does not sacrifice sensitivity. The converse is also true. When a particular ligand is abundant, the cell may decrease its sensitivity to that signal selectively. Receptors may become internalized upon activation and no longer be available for ligand binding. Receptors can also become desensitized or less likely to bind their ligand if they are repeatedly activated, providing a mechanism of cellular *accommodation*. For a receptor to function properly it must undergo a change upon binding its ligand that allows this information to be propagated. This occurs through the process of biochemical signaling, which is a broader subject not limited to receptors.

Cells communicate by biochemical signals

Cellular signaling concerns the flow of information across time and space. It is fundamental to how cells regulate their behavior and interact with their environment. Temporal flow simply refers to chains of events in signaling cascades or pathways, with prior events being upstream and subsequent events being downstream. Spatial information flow can take the form of molecules diffusing, being transported, crossing membranes, or being sequestered. Information flow in the world of the cell occurs by molecular information carriers moving from place to place. We refer to these as biochemical *signals* to reflect their role as carriers of information. It is also possible for information to be transmitted by physical means, such as electric fields and physical forces. These physical signals can play a fundamental role in many aspects of cell biology and have only recently begun to be investigated. Physical signaling and the emerging field of cellular mechano-biology will be discussed in some detail in Chapter 11. We begin here by first considering traditional biochemical signaling.

Signaling between cells can occur through many different mechanisms

Extracellular biochemical signals are one of the most important ways that cells sense and respond to changes in their environment. They are critical to coordinating the behavior of individual cells so they can assemble into tissues, organs, and ultimately entire organisms. *Hormones* are substances that can have a dramatic effect on cell behavior, often in very low concentrations. They are distinguished from other biochemical signals in that they must be secreted by a specific organ and circulate throughout the organism. Examples include adrenaline, parathyroid hormone, insulin, melatonin, and sex hormones such as estrogen and testosterone. *Cytokines* are distinguished from hormones in that they do not originate from a particular organ. Rather, cytokines mediate cell-to-cell communication and tend to affect a nearby collection of cells, although some can have systematic effects. Cytokines are not only important in the functioning of the adult organism, they are central mediators of cell-to-cell regulation of development. Extracellular signals are further classified in terms of the spatial extent of their effects. Substances that have effects throughout an organism are known as *endocrine* signals, substances that are responsible for cells regulating the behavior of nearby cells are known as *paracrine signals*, and cells regulating themselves are mediated by *autocrine signals* (Figure 2.10). Of course, it is nearly impossible for a cell to selectively signal to itself, so nearly all autocrine signals are also paracrine signals.

Signal transduction is the process of one type of chemical (or physical) signal being changed or *transduced* into another. The most common type of transduction is when an extracellular biochemical signal arrives at the cell membrane and is transduced into an intracellular signal (Figure 2.11). The vast majority of extracellular signals are transduced by membrane receptors. The intracellular domain of a receptor typically undergoes a conformational change as a result of ligand binding. This might enhance enzymatic activity within the structure of the receptor or expose a previously hidden binding site with which another protein can interact. Signaling molecules will eventually become activated and leave the membrane to propagate the signal intracellularly and initiate a complex set of biochemical reactions sometimes termed a *signaling cascade*.

Nota Bene

Neurotransmitters are a special class of extracellular cell-to-cell signal that medicate communication between neurons and adjacent cells. They typically act over the exceedingly short distance of a neural synapse. *Steroids* are another particular type of extracellular signal that owing to their hydrophobic nature, cross the cell membrane to bind a cytoplasmic or nuclear receptor.

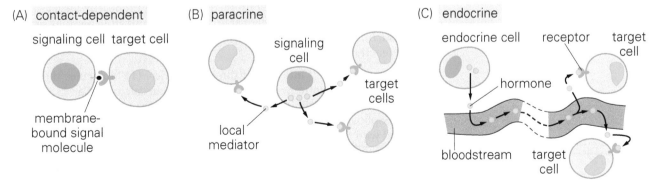

Figure 2.10 Signaling between cells can occur through many different mechanisms. (A) Cell–cell contact may or may not be required. (B) Autocrine and paracrine signaling occurs through the release of cell-secreted soluble chemicals. Paracrine signaling is depicted here, in which the target and signaling cells are different. (C) Endocrine signaling involves a soluble chemical produced in the cells of specialized organs and released into circulation. (Adapted from, Alberts B, Johnson A, Lewis J et al. (2008) Molecular Biology of the Cell, 5th ed. Garland Science.)

Figure 2.11 When an extracellular signal (or ligand) is bound by a receptor, the intracellular domain of the receptor changes conformation in such a way as to initiate an intracellular signaling cascade. This can eventually produce changes in cell metabolism, gene expression, or cell morphology or motility. (Adapted from, Alberts B, Johnson A, Lewis J et al. (2008) Molecular Biology of the Cell, 5th ed. Garland Science.)

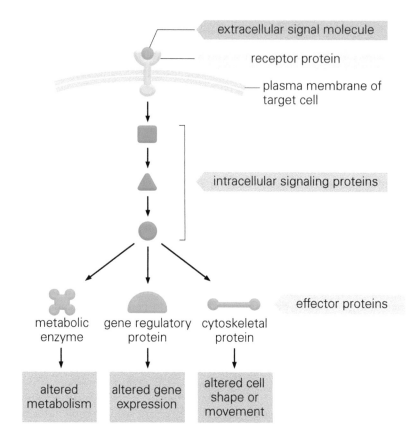

Intracellular signaling occurs via small molecules known as second messengers

Second messengers are rapidly generated, quickly diffuse though the cell to activate their targets, and then return to a low baseline concentration. The potency of a second messenger to deliver information is directly related to its ability to become rapidly elevated when activated and to remain at a low level otherwise. By analogy, this is a second messenger's signal-to-noise ratio. The level to which it can be raised is its "signal" and the background concentration its "noise". In addition, a molecule that is held at a very low concentration will diffuse more rapidly when elevated than one that is not. Second messenger systems are therefore always associated with active cellular processes that quickly return them to low concentrations after activation by removing them from the intracellular space, sequestering them, or degrading them. The target of a second messenger can be activation of an ion channel, activating another signaling protein such as protein kinases (discussed below), or affecting the kinetics of another reaction.

Second messengers fall into one of three categories. (1) Molecules that do not freely cross lipid membranes, such as intracellular calcium (Ca^{2+}), and cyclic nucleotides, such as cyclic adenosine monophosphate (cAMP) and cyclic guanosine monophosphate (cGMP). (2) Hydrophobic molecules that are typically associated with the cell membrane and form as the metabolic products of phospholipids; owing to their insolubility they tend to have proteins in the membrane as their targets. Diacylglycerol (DAG) and inositol triphosphaste (IP_3) are the classic examples. (3) Some dissolved gases such as nitric oxide (NO) and carbon monoxide (CO) can be potent second messengers, owing to their ability to quickly diffuse though the cytoplasm and across lipid membranes.

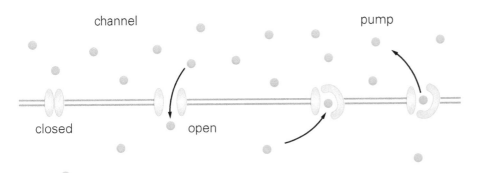

channel   pump

closed   open

Figure 2.12 Channels and pumps are examples of proteins or protein complexes that regulate the movement of second messengers across the plasma membrane and intracellular membranes that distinguish various intracellular organelles. Channels regulate transport along a concentration gradient and pumps against it. In the case of pumps, conversion of adenosine triphosphate (ATP) to adenosine diphosphate (ADP) is required to provide a source of energy.

Intracellular calcium signaling is one of the most studied and understood of the second messenger systems. Calcium is maintained at a very low level in the cytoplasm. This is done either by pumping it out of the cell or by sequestering it in the endoplasmic reticulum (sarcoplasmic reticulum in muscle cells), and to a lesser extent mitochondria. Transient increases in intracellular calcium can occur through several molecular mechanisms. Calcium channels in the cell membrane can allow calcium to flow from outside the cell into the cytoplasm in response to an extracellular signal. There are also channels to allow the sequestered intracellular stored calcium to be released (Figure 2.12). Calcium signaling can be readily studied, by using depletion (removing calcium) or via the use of fluorescent reporters (Figure 2.13).

Advanced Material: Calcium channels

Calcium channels fall into two categories, the IP$_3$ receptor and the ryanodine receptor. IP$_3$, as we have discussed, is a hydrophobic second messenger released into the cytoplasm from the membrane. The ryanodine receptor is sensitive to calcium concentration itself. This forms a highly sensitive positive feedback loop such that a small elevation in intracellular calcium induces a much larger and more rapid release from intracellular stores. This increase in calcium concentration is transient and sometimes called a wave or oscillation. It is typically finished in a matter of seconds, after which the calcium is again pumped out and sequestered in preparation for the next transient. Calcium is known to regulate the activity of several downstream proteins. One of the most well-understood of these proteins is calmodulin. Calmodulin, in turn, regulates the activities of other enzymatic proteins known as, not surprisingly, calmodulin-dependent protein kinases or CaM kinases.

(A)

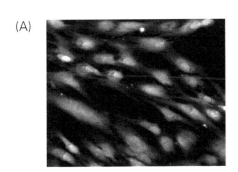

(B)

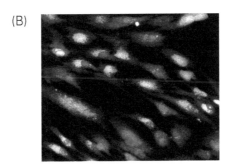

Figure 2.13 Cells filled with the dye Fura-2, whose fluorescent emission changes as a function of calcium concentration. The cells in (A) are in a baseline state and the cells in (B) have been stimulated with a calcium agonist. The ease with which calcium signaling can be studied optically has contributed to its being one of the better-understood second messenger systems. See www.garlandscience.com for a color version of this figure.

Large molecule signaling cascades have the potential for more specificity

Signaling cascades involving large molecules and proteins tend to be more slower acting than second messenger systems, but have the potential for more specific activity. Some complex but necessary nomenclature is required to understand the dynamics of these signaling cascades. The activity of many signaling and effector proteins is altered when they become *phosphorylated*, or experience the addition of a phosphate (PO_4) group. Typically, the phosphate is donated by a high-energy donor such as ATP. This is a reversible reaction that induces a conformation change, allowing a protein to switch from an inactive to an active state. Likewise, some proteins switch from an active to an inactive state when they are phosphorylated. Enzymes that phosphorylate other proteins are known as *kinases* and those that dephosphorylate other proteins are called *phosphatases* (Figure 2.14). In general phosphorylation only occurs on three specific amino acid residues within a protein: tyrosine, serine, or threonine. Whereas a kinase phosphorylates another protein, a "tyrosine kinase" is an enzyme that is restricted to adding a phosphate only to the tyrosine residue of another protein. The diversity and complexity of protein kinase signaling is immense. More than 500 different kinases have been identified in humans. The target of a kinase might itself be a kinase that becomes activated upon phosphorylation. These are known as kinase kinases. There are also kinase kinase kinases, but at this point it becomes so cumbersome that they begin to be referred to as 3-kinases, 4-kinases, etc.

As protein–protein signaling cascades progress, relatively small events, such as binding of a few receptors, can lead to larger and larger biochemical responses. This can occur through positive feedback loops at the receptor level or from catalytic reactions within the cascade itself, in which a single protein can catalyze a downstream reaction repeatedly, amplifying the signal. This intracellular amplification can be quite profound and sometimes results in cells being exquisitely sensitive to small extracellular signals. Protein–protein signaling pathways are critical because they allow for the intricate signaling cascades that the cell needs to orchestrate its varied and complex functions (Figure 2.15).

Figure 2.14 Sequential protein phosphorylations form signaling cascades or networks that are generally slower than second messenger signaling, but are potentially more specific. The ultimate target may be either regulation of protein activity or altered gene expression. Phosphate can be provided by guanosine triphosphate (GTP) or ATP. MAP kinase is mitogen activated protein kinase. (Adapted from, Alberts B, Johnson A, Lewis J et al. (2008) Molecular Biology of the Cell, 5th ed. Garland Science.)

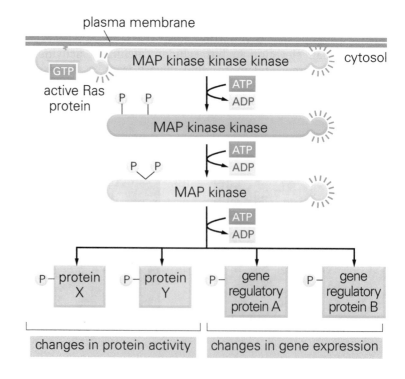

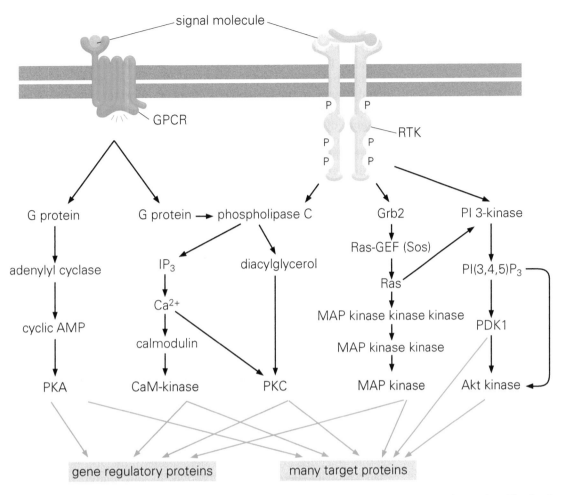

Figure 2.15 Intracellular signaling networks can be exceedingly complex. They can involve negative and positive feedback, cross-talk and even redundancies. Although the details of the pathways in this schematic are beyond the scope of this course, it illustrates most of the major signal systems at work in typical cellular regulation and allows cellular responses that are both sensitive and specific (Adapted from, Alberts B, Johnson A, Lewis J et al. (2008) Molecular Biology of the Cell, 5th ed. Garland Science.)

Advanced Material: Proteins may need to form complexes for signaling

Adaptor proteins do not have any enzymatic activity themselves, but are required to bind to other proteins to allow them to become active. For instance, they can bind to particular recognition sites on a protein and induce a conformational change that exposes a particular residue for phosphorylation. This may be necessary for the formation of complexes of multiple proteins that only function when bound to each other. Alternatively, protein complexes may be subsequently transported to other regions of the cell or their formation may be a signal for degradation of the protein complex. Adaptor proteins add another layer of regulation allowing additional specificity in cell signaling.

Receptors use several mechanisms to initiate signaling

As we have seen, receptors, as mediators of transduction, form an interface between the extracellular signals a cell receives and the complex signaling pathways the cell activates in coordinating its response. It might not be surprising that there are many molecular similarities between signaling pathways and the intracellular domains of receptors that initiate signaling pathways. Receptors fall into several classes depending on how they acheive this. One such class is the receptor tyrosine kinases. These are transmembrane proteins with an extracellular domain separated from an intracellular domain by one or more membrane-spanning domains. The extracellular domain is responsible for recognizing and specifically binding to the ligand target of the receptor. Ligand binding leads to a conformational change in the intracellular domain. Receptor tyrosine kinases, like many receptors, are actually dimers of two proteins. These dimers are only stable in the presence of the ligand. When a stable dimer forms, tyrosines in the cytoplasmic domains of the receptor are autophosphorylated and are thereby activated to initiate the intracellular signaling cascade. At this point, the specific signaling pathway that is activated will depend on the particular receptor.

Another class of membrane receptors comprises the G-protein-coupled receptors (GCPRs). They contain seven membrane-spanning domains and are each linked to a G-protein, as the name suggests. Before ligand binding, the G-protein coupled to the receptor is in its deactivated guanosine diphosphate (GDP)-bound form. Upon ligand binding, the receptor acts as a guanine nucleotide exchange factor (GEF) and allows a guanosine triphosphate (GTP) to substitute for the GDP, thereby activating the G-protein. The activated G-protein dissociates from the receptor and is free to diffuse and activate several downstream targets. A hydrophobic domain in the G-protein tends to keep it membrane-bound. Not surprisingly then, the targets of the liberated G-protein are other membrane-bound proteins. GCPRs also activate second messenger signaling systems. These include production of IP$_3$ and DAG through the action of the G-protein on phosphodiesterases and phospholipases and production of cAMP and cGMP by activation of membrane-bound adenylyl cyclases. GPCRs can even potentiate intracellular calcium increases by opening membrane calcium channels.

Integrins are a particular type of receptor that also act as mechanical linkers of the cell to its extracellular environment (although they are not the only receptors that do this, integrins are among the best-studied). Integrin ligands are extracellular matrix proteins such as collagen, fibronectin, and laminin. In addition to providing a mechanical connection between the cell and its environment, when integrins bind they also activate intracellular signaling through integrin-associated kinases. This signaling results in increased integrin production, recruitment and clustering of integrins, and the eventual assembly of compact attachments known as *focal adhesions*. In many cell types integrin is also an important cell survival signal as well as a regulator of proliferation and differentiation. As we shall see, it also appears to be important in helping the cell sense mechanical stress upon itself. This is known as *outside-in* integrin signaling, in contrast with *inside-out* signaling. In the case of inside-out signaling, cytoplasmic signals regulate the ability of an integrin to bind its ligand by exposing or hiding the ligand-binding domain.

2.3 EXPERIMENTAL BIOLOGY

Progress in cell biology has been facilitated by the development of increasingly sophisticated experimental techniques. The inner workings of the cell may be understood through many different approaches, each with its complement of possibly unfamiliar terms. Here we review the basic terminology so that you can read the cell mechanics research literature with confidence.

In experimental biology, microscopy is used because cells and subcellular components are typically too small to be seen unassisted. Many cell mechanical studies also require microscopy, which may involve the terms below:

- **Lens:** an optical component that is used to change the convergence of a light beam. A convex or converging lens focuses a beam of light, whereas a concave or diverging lens spreads out the beam of light.
- **Resolution:** the distance between two point sources of light whereby the two sources can be distinguished (optical microscope resolutions are typically at 0.1–1 μm).

- **Wavelength:** the peak-to-peak distance between adjacent crests in the electromagnetic wave representing light (optical microscope wavelengths typically range from 400 to 900 nm).
- **Index of refraction:** the change in the speed of light as the light beam encounters different media (for example air to water) (typical indices in microscopy range from 1 to 1.5).
- **Noise:** unwanted signal, regardless of whether the signal constitutes actual data.

An incredible amount of information can be gained just by looking at cells. The compound microscope was originally developed by Galileo in 1609 and consisted of one convex–concave lens. Using these early simple instruments, Robert Hooke (the same person who described the deformation of elastic bodies in Hooke's law), along with Antoine van Leeuwenhoek, described microscopic living structures in cork bark and coined the word *cell*. Since the structures that make up a cell are almost all transparent, simple light microscopy is limited as an investigative tool in cell biology, yet a great number of cell properties relevant to cell mechanics can be obtained through observation. Cell morphology—whether the cell is round, spread, whether it has many protrusions like a starfish—is one example of a major factor in cell mechanics. Cells that are spread out will tend to exert higher forces on the surfaces to which they are attached. How the cell is shaped can reveal information about the cell's tendency to move or respond to forces— some cells will actually align in the direction of, or perpendicular to, applied forces.

Other factors of interest include both active processes, such as cell migration, and structural components, such as the density of the cell's support skeleton (the cytoskeleton). Some of these can be visualized with minor adjustments to optical tools; others require more sophisticated techniques.

Optical techniques can display cells clearly

Many microscopy techniques are used to increase contrast. Cells are generally difficult to image with just plain illumination, owing to their thinness. There are several different ways to generate contrast (Figure 2.16), each with different characteristics and uses. Phase–contrast microscopy (for which Frits Zernike, a physicist, was awarded the Nobel Prize in 1953) solves the problem of cells being too thin to directly observe via white light illumination (also called brightfield) by adding contrast to the optical image based on small phase shifts introduced by the specimen. Light from the source is passed through a ring-shaped mask and a lens or *condenser* before passing though the specimen, magnifying objective, and eyepiece. In this way, there are two sets of focal planes. The one associated with the objective lens creates an image of the specimen on the observer's retina as it would in a standard microscope. In addition, the condenser creates an image of the mask inside the microscope itself, typically within the multiple lenses of the

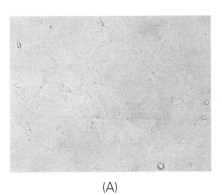
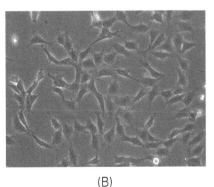
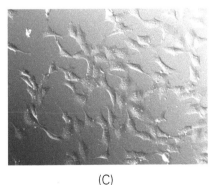

(A)   (B)   (C)

Figure 2.16 Differet types of contrast illumination available for imaging cells. (A) Brightfield, (B) phase contrast, (C) differential interference contrast (Courtesy of An Nguyen).

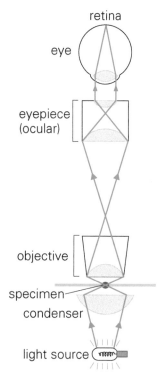

Figure 2.17 Optical schematic of a typical light microscope. In a phase-contrast approach masks are inserted in the condenser and objective. (Adapted from, Alberts B, Johnson A, Lewis J et al. (2008) Molecular Biology of the Cell, 5th ed. Garland Science.)

objective (Figure 2.17). The objective of a phase-contrast microscope has a second mask inserted in the objective that is the inverse of the mask in the condenser. When the microscope is properly aligned, this image comes into focus exactly on the mask in the objective. Generally, the objective mask is a ring, and the condenser mask is an annular slit. In the absence of a specimen, all light in a properly aligned phase-contrast microscope is blocked by one of the two masks. However, when an object is introduced it causes the light to bend slightly through diffraction and refraction, allowing those photons to pass through the objective. The resulting image has a greatly increased contrast when compared to brightfield microscopy. Objects in a phase-contrast image appear bright or dark depending not on their optical density, but rather on the phase shifts they introduce. Typically a structure will appear dark and be surrounded by a bright halo of light. Phase-contrast is so useful in visualizing cells that it is virtually ubiquitous in cell culture laboratories. Differential interference microscopy (DIC) further enhances contrast using polarized light to produce brighter diffraction halo around each object.

Unfortunately, basic optical techniques are subject to numerous limitations, including the inability to distinguish easily between different components within the cell. Some structures, such as microtubules, are not visible at all.

Fluorescence visualizes cells with lower background

Some molecules, when illuminated by specific wavelengths of light, exhibit a particular type of excitation whereby electrons absorb the incident photon and jump to a higher energy level. Some energy is lost to heat as a result of the jump "overshooting" the stable higher energy level (like throwing a ball higher than a roof in order to land it on the roof). After a short time, the electron will decay back to its original orbital and baseline energy level. This decay is accompanied by the release of a new photon, at a longer wavelength (lower energy). Such molecules are called fluorescent and can enhance the use of microscopy to examine cellular structures and processes.

Fluorescence microscopy is based on the excitation of molecules within the cell with fluorescent properties. Some endogenous molecules can fluoresce (called *autofluorescence*), but those are few and limited in scope. In general, flourescent molecules are introduced into the cell externally. These molecules are called *fluorophores* or *fluorescent dyes*. As was described in the previous paragraph, fluorescence occurs when a molecule absorbs a photon at one frequency (called the excitation) and emits a photon at a lower frequency (called the emission). This can be used to obtain very high-contrast images by illuminating the specimen at one wavelength and observing it at a different wavelength.

A typical epifluorescence microscope actually uses the same objective lens for illumination and observation (the prefix epi- means "beside", referring to the illumination and observation taking place on the same side of the specimen). A high-intensity excitation source (a mercury lamp, LEDs, or lasers) is filtered to produce a narrow band of wavelengths centered at or near the maximum absorbance wavelength of the dye being used. The light is then reflected by a dichroic mirror onto the specimen. The emitted fluorescent light, now at a longer wavelength, travels back though the objective, this time passing through the dichroic mirror.

Advanced Material: Dichroic mirrors change reflectance with wavelength

A dichroic mirror is the heart of an epifluorescence microscope. It is reflective below a critical wavelength and transparent to light above that wavelength. It allows the illumination light and the emitted light from the specimen to travel through the same objective, but for the emitted light to be separated and observed.

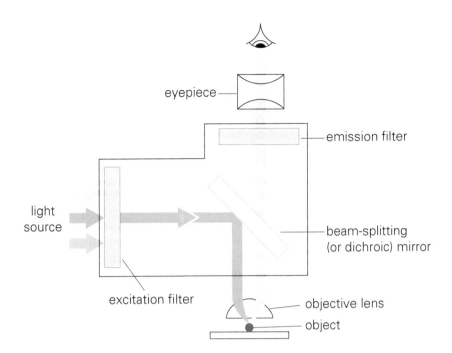

Figure 2.18 A light-path schematic of a typical epifluorescence microscope. The darker blue light path is the excitation and would be at a shorter wavelength. Fluorophores in the specimen absorb photons at the excitation wavelength and emit them at the emission wavelength. (Adapted from, Alberts B, Johnson A, Lewis J et al. (2008) Molecular Biology of the Cell, 5th ed. Garland Science.)

An emission filter helps improve the image by removing light at wavelengths different from a band centered around the dye's emission wavelength being used (Figure 2.18). In this way, high-quality fluorescence images can have a virtually black background with a very high signal-to-noise ratio.

Fluorophores can highlight structures

Fluorescence microscopy can be used to examine specific structures within cells. Certain compounds can be attached to molecules that only bind to particular proteins within the cell to determine their distribution. A broad way of doing this is using a technique called immunofluorescence. In immunofluorescence, an antibody is generated that is specific for the protein of interest. This antibody (called the primary antibody) can be linked to a fluorescent molecule. The cell is incubated with this antibody and then washed to remove unbound antibodies. Upon examination using fluorescence microscopy, only the proteins that were tagged with the fluorescent antibody would appear, providing information about the location of the protein, and to some extent its quantity. Alternatively, a nonfluorescent primary antibody can be followed with a fluorescent secondary antibody, which binds the primary antibody. This can also

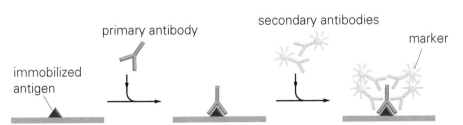

Figure 2.19 The use of a secondary antibody means that fluorophores or other means of visualization only need to be linked to one antibody. Additionally, there is potential for signal amplification, since multiple secondaries can bind to a single primary. (Adapted from, Alberts B, Johnson A, Lewis J et al. (2008) Molecular Biology of the Cell, 5th ed. Garland Science.)

Nota Bene

Cytoskeletal and nuclear fluorescence imaging. A common cell structure of interest is the cytoskeleton. Actin filaments can be visualized using antibodies against F-actin via immunofluorescence, or with various compounds such as phalloidin, (which is typically attached to, or *conjugated* to, a fluorescent molecule). Phalloidin is generally specific for actin (phalloidin staining is very fast compared with immunofluorescence, hence it is widely used despite requiring the cells to be fixed as with immunofluorescence). Unfortunately, the microtubules and intermediate filaments do not currently have widely available fluorescence agents and must be visualized using immunofluorescence or transfection. The nucleus can be visualized using several dyes; DAPI (4',6-diamidino-2-phenylindole) and propidium iodide are good for imaging fixed cells, Hoechst is a cell-permeant fluorescent dye. Nuclear labels are generally toxic to cells; these labels should not be used for long-term (over days) cell tracking. Other structures (mitochondria) can be highlighted using various compounds; an exhaustive listing would be impractical here.

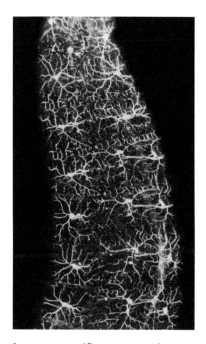

Figure 2.20 Epifluorescence image obtained by inserting the *GFP* gene driven by a promoter that is only turned on in a specific set of neurons in a fly. Thus, the cells that appear bright are expressing the promoter of interest. In this way, GFP can be used to interrogate function in living cells and tissues. See www.garlandscience.com for a color version of this figure. (Adapted from, Alberts B, Johnson A, Lewis J et al. (2008) Molecular Biology of the Cell, 5th ed. Garland Science.)

amplify the signal, as many secondary antibodies can attach to each primary antibody (Figure 2.19).

Fluorophores can probe function

Fluorescence microscopy is not limited to telling us what a cell looks like; it can also provide information about a cell's biological activity. This is because the fluorophors introduced into the cell can be designed to interact with the cell in specific ways. Green fluorescent protein (GFP), was originally isolated from a species of jellyfish (*Aequorea victoria*). Many fluorescent compounds are synthetically manufactured, which makes it difficult to introduce them into living cells. When the gene sequence for GFP was identified, it became possible to introduce this gene into nonfluorescent cells and make them fluoresce by hijacking the cell's own internal machinery. This in turn allowed living cells to be easily observed with fluorescence microscopy (Figure 2.20). Even more powerful artificial gene constructs can be created that allow fluorescent versions of normal proteins to be created by the cell. These can be used to determine their cellular location and trafficking. Constructs can also be created with a target gene's promoter placed upstream of GFP to determine gene transcription in real time. Variations of the original GFP have been created by mutating the protein to change its fluorescent properties. Pioneered by Roger Tsien, cyan (CFP), yellow (YFP), and red (dsRED) are commonly available and their numbers continue to grow (more recently, with the "fruit"-based labels, such as mCherry, mBanana, etc.).

In addition to providing enhanced contrast, fluorescent dyes may change properties in response to their environment. Aequorin's (also originally isolated from jellyfish) fluorescence increases as a function of calcium concentration. Calcium, as we have seen, is an important cellular second messenger and knowing its concentration in living cells in real time can give important insight into cell metabolism and signaling. A similar calcium reporter dye, Fura-2, also developed by Roger Tsien, relies on a wavelength shift in the absorbance of the molecule when

Example 2.3: Uses of fluorescence

List some possible reasons why one would use fluorescence microscopy to study cell- or subcellular-level biology. Then describe some disadvantages of using fluorescence.

Fluorescence is desirable because labeling the structure of interest with a fluorescent tag allows you to then image the structure with minimal interference from other parts of the cell. Unlike phase-contrast imaging, which illuminates multiple components, only the fluorescent tag is visible under fluorescence microscopy. Fluorescence imaging can be calibrated to yield quantitative results using specialty dyes and techniques.

Fluorescence labeling can be time-consuming, expensive, and difficult to achieve. Currently, only a limited number of fluorescent labels may be used with living cells.

it is bound to calcium. Fluorescent reporters for other ions, pH, and even voltage have been developed.

Atomic force microscopy can elucidate the mechanical behavior of cells

Originally a derivative of scanning tunneling microscopy (STM), atomic force microscopy (AFM) replaces the tunneling current with physical contact between the probe tip and the specimen. An image is formed either by dragging the tip across the sample or repeatedly pushing on the sample and withdrawing it. AFM may also expose cells to well-defined mechanical perturbations in studies of their mechanical behavior and response to mechanical stimulation. We discuss AFM in more detail in Chapter 6, since it is one elegant way to measure the stiffness of cells.

Gel electrophoresis can separate molecules

Suppose you stretch a group of cells and want to see if production of a particular protein that affects cell migration is altered in response. One way to do this is to stretch the cells, lyse them, and then measure the amount of RNA or protein of interest. Another way is to manufacture a short gene that links the promoter of the protein to a fluorescent molecule, and insert the gene (chemically, physically, or virally) into the cell. When the promoter is activated, the fluorescent molecule is generated and you can obtain a visual readout. Any one of these approaches, however, requires you to be able to isolate, purify, or quantify the molecules of interest.

Gel electrophoresis is used routinely to separate heterogeneous mixtures of DNA, RNA, and proteins from one another according to their molecular weight. A sample is forced through a polymer hydrogel and molecules that move quickly through the gel (typically smaller molecules) are able to travel further in a given amount of time than molecules that move more slowly (typically larger molecules). The molecules are driven by an externally applied electric field that produces a force on them proportional to their charge (Figure 2.21). Nucleotides (RNA and DNA) inherently contain a negative charge on their sugar backbone (from phosphates) and therefore travel toward the positive electrode. By contrast, proteins may contain a positive, negative, or no net charge. As a result, proteins are usually placed in a powerful detergent such as sodium dodecyl phosphate (SDS) that covers the proteins with negative charge and allows them to migrate through the gel. The SDS treatment also causes the proteins to unfold or *denature* when bonds that contribute to their tertiary and secondary structures are lost because of shielding of the protein's internal charge structure by the coating detergent.

Denaturing of proteins before gel electrophoresis has another important benefit. The speed at which a protein migrates through a gel is a function of how well it can slip through the polymer network of the gel. The drag that is exerted on the protein as it migrates is, in turn, related to the cross-sectional area that the protein presents to the network. Although cross-sectional area generally increases larger proteins, there is not a strict relationship between size and friction

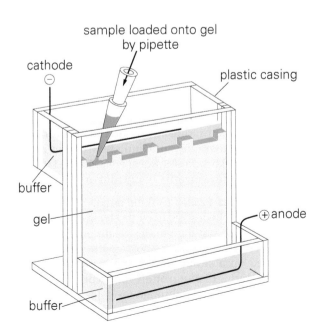

Figure 2.21 A common sodium dodecyl sulphate–polyacrylamide gel electrophoresis (SDS-PAGE) vertical format system. (Modified from, Alberts B, Johnson A, Lewis J et al. (2008) Molecular Biology of the Cell, 5th ed. Garland Science.)

Nota Bene

Antibodies can be used for protein detection and recognition. Antibodies bind to specific protein fragments known as *antigens* that contain a sequence of interest (the epitope). A polyclonal antibody is isolated from the blood of an animal that has been injected with a target protein. Polyclonal antibodies will bind to several epitopes on the target protein. A monoclonal antibody is produced in cell culture by immortalizing a single immune cell from the exposed animal. Monoclonal antibodies recognize a single specific epitope. Polyclonal antibodies can amplify better (more antibodies stuck to the same target) but variability each time you make more. Monoclonals are more restricted in targeting, but the repeatability is much higher.

during migration since large proteins can fold into a tightly packed small structure. Conversely, smaller proteins that are not well packed may actually migrate more slowly. In denaturing conditions, proteins acquire an unfolded linear structure and migrate inversely with their molecular weight. This means gels for proteins in denaturing conditions (such as SDS) separate samples according to their molecular weight. It should be noted that similar restrictions apply for nucleotide gels—if a DNA is in a circular plasmid configuration, it will appear smaller than a linear strand of the same number of base pairs (actually, you will tend to get multiple bands). Therefore, DNA is generally linearized before, it is run through a gel.

The vast majority of gels use one of two polymers to form the hydrogel network. High-density gels of polymerized *acrylamide* are used to separate proteins and smaller nucleic acid polymers. There are many easy-to-use *PAGE* (polyacrylamide gel electrophoresis) systems and they generally hold the gels between two glass or plastic plates that force the electric current to pass directly through the gel. Typically, these systems hold the gel vertically with the sample moving from the top positive electrode toward the bottom negative electrode. Higher–molecular-weight nucleotides (over a few hundred base pairs) are separated in lower density *agarose* gels. Agarose systems keep the gel horizontal because polymerized agarose is not strong enough to support its own weight. To keep the gel hydrated, the gel is submerged and a larger current passes through both the gel and the bathing fluid. Although automated systems that perform gel electrophoresis in tiny capillary tubes can greatly increase throughput, running agarose and SDS–PAGE gels remains a standard procedure in most molecular biology labs.

Visualizing gel-separated products employs a variety of methods

Most gel-running systems allow for several samples to undergo electrophoresis simultaneously in adjacent *lanes* of the gel. One lane is typically used to a molecular-weight standard that is a mixture of proteins or nucleic acids that have very distinct, known molecular weights. This standard separates into clearly identifiable bands that are then used to convert the distance migrated into the molecular weights of the samples. After a sample has been separated, a particular point in the gel now corresponds to a particular molecular weight. In this way gel electrophoresis can be used to isolate a fraction of a sample of a given molecular weight by physically cutting out a particular band and depolymerizing the surrounding hydrogel. Gels of this variety are known as *preparatory* gels. The purpose of running the gel can be to determine the molecular weight of an unknown sample. In these so-called *analytical* gels, the next step in the process is to visualize the sample. In some cases this can be accomplished by saturating the gel with a stain that provides color or fluorescence to all of the proteins or nucleotides in the lane, but more specific target detection is often needed. This can be accomplished with reagents that bind specifically to the target of interest (typically *antibodies* for proteins and *probes* for nucleotides). These can then be visualized by radioactively labeling the probe or antibody and detecting them via autoradiography, attaching fluorescent reporter molecules, or making them emit light via chemiluminescence. In the latter case, the light is detected by exposure to highly sensitive X-ray film or by digital cameras.

The binding of antibodies or probes to their targets can involve reactions that require hours to occur. The hydrogels used in electrophoresis are generally not structurally sound enough to stand up to this treatment. Therefore a second *blotting* step is required, in which the sample moves perpendicular to the plane of the gel and is immobilized on a strong membrane of nylon or nitrocellulose. The migration of the sample during blotting can also be accomplished by electrophoresis or by driving a large volume of water through the gel (Figure 2.22). The blotting of DNA is referred to as Southern blotting after its inventor, Edwin Southern. In a pun

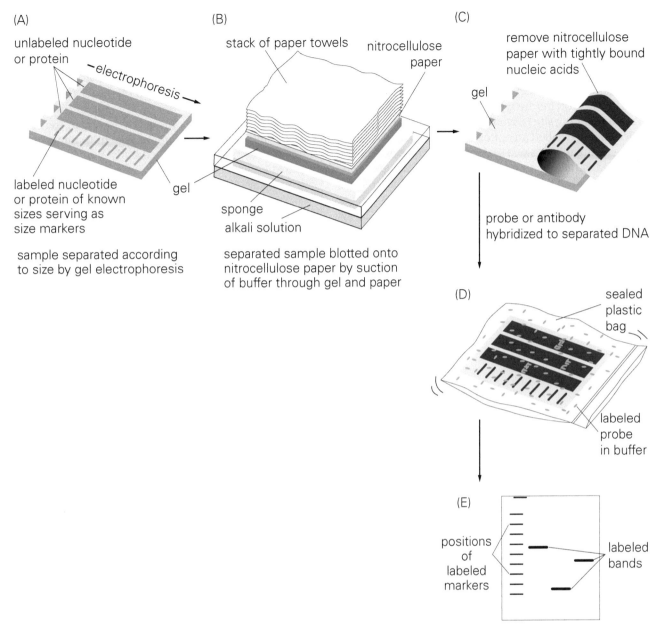

(A)

unlabeled nucleotide or protein

electrophoresis

labeled nucleotide or protein of known sizes serving as size markers

gel

sample separated according to size by gel electrophoresis

(B)

stack of paper towels

nitrocellulose paper

gel

sponge

alkali solution

separated sample blotted onto nitrocellulose paper by suction of buffer through gel and paper

(C)

remove nitrocellulose paper with tightly bound nucleic acids

gel

probe or antibody hybridized to separated DNA

(D)

sealed plastic bag

labeled probe in buffer

(E)

positions of labeled markers

labeled bands

Figure 2.22 Example of the steps involved to detect DNA by its radioactivity. To visualize proteins in a gel, you must first transfer or blot them onto a membrane. Although this can be done electrophoretically, it can also be accomplished simply by soaking water with paper towels. The membrane can then be incubated with tagged antibodies or nucleotide probes, and visualized with color or fluorescent dyes so that the presence of a desired protein or nucleic acid fragment can be detected. (Adapted from, Alberts B, Johnson A, Lewis J et al. (2008) Molecular Biology of the Cell, 5th ed. Garland Science.)

on the accepted name for the Southern blot, when the same approach was used with RNA in 1977 at Stanford, it was named the *northern* blot. The good-natured name-play continued with the development of the western blot for proteins. The tradition continues with, for instance, blotting of DNA-binding proteins being called a southwestern blot. Although there is no eastern blot, a method of blotting lipids developed in Japan was termed far-eastern blotting.

PCR amplifies specific DNA regions exponentially

The development of the polymerase chain reaction (PCR) by Kary Mullis in 1983 has reshaped modern biology. The dramatic increase in detection power that PCR

Advanced Material: Antibody uses in biology

Why are antibodies useful in experimental biology? Antibodies are proteins that have a "flexible" binding site. These proteins are typically drawn in the shape of a "Y". The binding site is usually targeted to a desired peptide sequence called an epitope, which is usually part of a foreign organism (say, a surface protein on some virus). The idea is that when a foreign organism invades your body, your immune system generates antibodies that are capable of binding key surface proteins on the foreign organism, thereby rendering the proteins inactive. Since there are multitudes of foreign organisms, the immune system must be capable of generating antibodies that stick to a wide variety of epitopes while not recognizing native proteins (so that the antibodies stick to unwanted stuff but not to parts of your own body). In some cases, this is not desirable; if you receive an organ transplant you do not want your immune system to attack it, but since it is technically a foreign body, you will generate antibodies against it, so immunosuppresants will generally be necessary). In other cases, there are mistakes—in autoimmune syndromes, your immune system generates antibodies against your own tissues, which can cause all sorts of complications (lupus is a classic example of an autoimmune dysfunction).

provides has led to it be a virtually ubiquitous tool in the modern biology lab. In recognition of the incredible importance of PCR, Mullis received one of very few Nobel prizes to be awarded for a technique rather than a scientific discovery. The key to PCR is to use a natural cellular enzyme, DNA polymerase, to replicate DNA *in vitro* rather than in the cell nucleus. As each new DNA strand is synthesized, it becomes in turn a template for subsequent synthesis in a chain reaction that produces an exponential increase in DNA and gives PCR its remarkable power to detect even a single molecule of DNA in the initial sample. At its core, PCR is made possible by two key components: the requirement of primers for DNA polymerase binding and the identification of thermally stable DNA polymerases.

PCR would be much less useful if it amplified all DNA equally, but in fact it can selectively amplify a target sequence of DNA (corresponding to a particular gene, say) and leave non-target DNA alone. This is done by exploiting a critical property of DNA polymerase. Specifically, DNA polymerase only adds nucleotides where a portion of single-stranded DNA already has some double-stranded DNA in place. During replication double-stranded DNA is separated into two single-strands. DNA polymerase converts each single-strand back to a double-strand by adding complementary bases to the 3′ end of the newly synthesized strand in a process known as *extension*. The fragment of complementary DNA that binds to the DNA being replicated and allows DNA polymerase to get started is known as a *primer*. During cell division these primers are made by specialized enzymes and allow the entire DNA strand to be replicated. By contrast, in PCR, primers are designed to flank a specific region of DNA and amplify only that DNA fragment. Since DNA polymerase only proceeds from the 3′ end of the primer and adds new base pairs in the 5′ direction, the flanking primers are directional. That is, a *forward primer* is designed to be complementary to a sequence of DNA that is unique and located upstream (in the 5′ direction) of the target. It will produce a long strand of complementary DNA starting at the forward primer and extending along the DNA template. A *reverse primer* is designed to bind uniquely to the complementary DNA at a point downstream of the target. The end result is that with each cycle of extension the number of fragments produced that are flanked by the primers doubles. In short order, the exponentially growing fragments vastly outnumber any other DNA sequence in the reaction (Figure 2.23).

In the 1970s scientists were already using DNA polymerase to replicate DNA. However, since DNA polymerase can only add bases to single-stranded DNA, they had to first separate the two strands by heating to the *melting* or *denaturing* temperature (typically 95°C) and allowing it to cool to around 65°C, the temperature

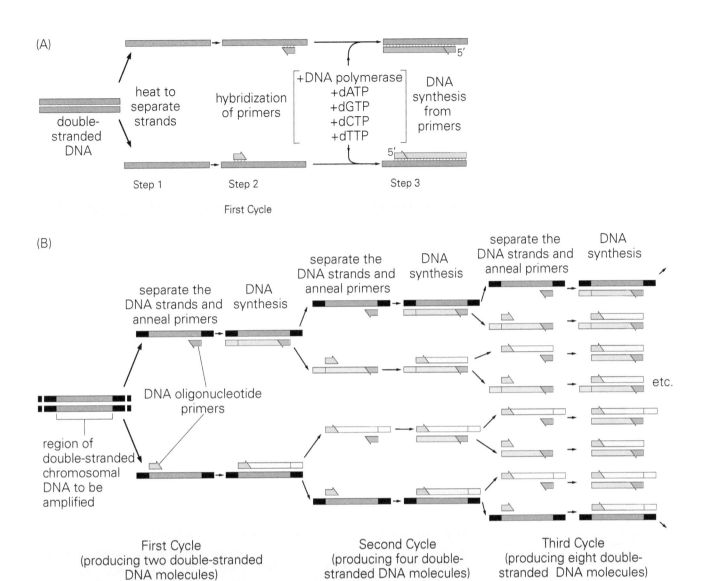

Figure 2.23 Schematic of DNA amplification in a typical polymerase chain reaction (PCR). The fragment flanked by the forward and reverse primers grows exponentially with the number of amplification cycles, while the other fragments grow linearly. Therefore the concentration of the fragment flanked by the primers quickly overwhelms all of the other species. (Adapted from, Alberts B, Johnson A, Lewis J et al. (2008) Molecular Biology of the Cell, 5th ed. Garland Science.)

at which extension can occur (the *annealing* temperature). The problem was that each time the sample was raised to the melting temperature, all of the DNA polymerase was denatured and needed to be re-added manually. This was very cumbersome and greatly limited yield. The answer came from the realization that thermophilic bacteria thrive in high-temperature environments. Therefore, they must have a form of DNA polymerase that can withstand high temperatures. One of the most common forms of DNA polymerase used today is Taq polymerase which was originally isolated from *Thermus aquatiqus*, a bacterium that naturally occurs in hot springs and hydrothermal vents. With this component in place, it was possible for modern PCR to be developed. In principle, it is simply a matter of mixing a sample, DNA polymerase, forward and reverse primers, and an abundance of DNA monomer in a tube and alternating quickly between high and low temperatures. This is typically done in a solid-state device developed just for this purpose known as a *thermal cycler*. The reaction will continue and the target is exponentially amplified until one of the reagents is exhausted.

2.4 EXPERIMENTAL DESIGN IN BIOLOGY

One striking difference between mechanics and biology is the vast diversity of biological systems. In the case of the cell, seemingly unending systems of regulation and control can be particularly intimidating to an engineer. The number of genes in a human (~40,000) means that there are about the same number of proteins that can potentially interact within a cell (more, when you consider alternative splicing and post-translational modifications). Although not all proteins are expressed in a particular cell at one time, the number of interactions is clearly too large for pen-and-paper analysis. Further, unlike most macroscopic mechanical systems, few components can be directly probed or visualized at the molecular level. As a result, most studies in cell biomechanics and mechanotransduction involve probing a limited aspect of overall cell behavior. Because of this, ad hoc snippets of behavior and knowledge predominate—a scientific paper entitled "X controls Y in a Z-dependent manner" suggests that the regulation of Y by X can be influenced by blocking or overexpressing Z. This situation is further exacerbated by the fact that currently so much of the cell's function and machinery remains virtually unknown. Of course, this is what makes the study of the cell and its diverse workings so exciting to many of us. Nonetheless, there are often fundamental differences in how experiments are conducted in biology versus mechanics, and fundamental differences in how the results can be interpreted. In cell mechanics it is not unusual for two different experiments to provide disparate answers to the same question. Consider the estimated value for cell stiffness across multiple techniques ranges from 100 Pa to 1 MPa. The fact that different methods were used on different cell types under presumably different cell culture conditions makes interpretation of these results all the more difficult.

In experimental mechanics the goal is generally to recreate the real-world situation with as much detail and fidelity as possible. If a material is to be characterized, one attempts to recreate its physical, chemical, and mechanical environment to the greatest extent possible. The environment of the cell is elaborate, making this task challenging indeed. When engineers first work with cells they can be tempted into visualizing working within a three-dimensional, multi-phase, physically and chemically representative model of a given cellular environment, with an abundance of variables to control. One should realize that many of the critical insights in cell biology have come from taking a more reductionist experimental approach, and engineers, although appropriately skeptical of how these findings might translate to more complex situations (such as an intact organism), should become familiar with them.

Reductionist experiments are powerful but limited

Reductionism refers to the process of understanding the very basic building blocks of a system, and then predicting macroscopic phenomena as a conglomerate of their behavior. Reductionist experimental design involves stripping the system of as much complexity as possible, while preserving the key behaviors under study. If one wants to study DNA replication, it is often sufficient to reproduce the biochemical reactions in a test-tube without having to reproduce the environment of an intact cell. Indeed, current cell culture techniques are based on simplifying the cell's external environment. Tissue matrix and substrates are replaced by plastic or glass, sometimes coated with adhesion molecules. The cell culture medium is often designed to encourage cell proliferation and activity, sometimes at the expense of realism and consistency. One frequently used supplement to encourage proliferation is bovine serum, which is used not only for culturing cow cells but cells from mice, rats, humans, etc.

These simple systems can yield important insights. Much of modern medicine and biology is based on discoveries made in the cell culture lab. But by its very nature, a simple system designed to produce straightforward results in an economical manner is scientifically limited. These systems do not address the question of whether the finding will translate when complexity is reintroduced. Verifying that a discovery made in a simple system holds true in a cell or organism is just as important as making the initial insight. Indeed, it has been reported that treatments and drugs that work in laboratory mice do not usually translate to humans—suggesting that creating a representative biological environment is no guarantee that the results will be any more useful than in simplified systems. We therefore consider simplified systems to be *models* of something much more complicated. One might say that the model system is useful in showing that something can occur, but it must be followed by a determination of whether it indeed occurs in reality. There is always the danger that a cell can be made to do something in the laboratory that it never does as part of a functional organism. There is also the alternative, that something can be done in an organism that cannot be reproduced in culture (for example, a hepatitis virus may freely infect and replicate in an organism, but this is actually difficult to achieve in cell culture). This latter issue is rarely considered, because random experimentation on whole organisms can be difficult, expensive, and can carry a higher ethical burden. Designing appropriate model systems and knowing their limitations is critical to successful science and is often as much the product of intuition and personal experience as rational logic and fact.

Another advantage to studying simple systems is that they more easily allow investigators to understand *mechanisms*. This concept is often proffered as a contrast to so-called *descriptive* experiments. This is a subtle argument because all science ultimately involves describing something or other. However, there is a bias, sometimes unspoken, that it is better to understand how something happens rather than merely to describe the fact that something happens. Showing that a particular mechanical stimulation may encourage stem cells to differentiate into a particular cell type might be descriptive, whereas determining the receptor that mediates this would provide a groundwork of the mechanism. One advantage to understanding the mechanism behind a particular behavior or response is that this knowledge takes us one step closer to being able to manipulate the response to a desired therapeutic end.

Investigations of proposed or hypothetical mechanisms generally involve an intervention or series of interventions applied to a system that models the desired behavior. Suppose we can show that, a given mechanical stimulation leads to cellular differentiation. We hypothesize that a particular receptor is responsible. Repeating the experiment in cells that have had this receptor removed and observing a failure to differentiate with the same stimulation would help support this hypothesis (hypotheses can be supported, but not proven). Another approach might be to use a biochemical blocker to suppress a particular pathway or target molecule. Currently, blockers exist with varying degrees of specificity for kinases and phosphatases, intracellular and extracellular receptors, ion channels, and some other assorted signaling molecules. "Blocking antibodies" bind to specific target proteins, but also inhibit their function—usually by interfering with the binding site for the ligand. A recently developed intervention is small inhibitory RNA (siRNA). It was discovered that as part of the cell's natural defense against certain viral attacks, the cell responds to the presence of short fragments of double-stranded RNA by producing RNA that is complementary to the introduced sequence. These fragments bind to the target mRNA and mark it for degradation. Production of a specific protein can be interrupted (or *knocked down*) at the post-transcriptional level by using knowledge of the gene sequence.

If a response to a stimulus is lost when a particular protein or signaling event is blocked, it is strong evidence that the blocked activity is involved in the pathway between the stimulus and the response. Often the activity is termed *necessary* for the response. An additional level of support can be shown if the activity can be supplied through exogenous means and the response observed even in the absence of the original stimulation. In this case, the activity is said to be *sufficient* for the response. When a step in a pathway is both necessary and sufficient, it is strong evidence indeed. While this is the current strategy behind biological and biomedical engineering approaches to mechanotransduction, it should be acknowledged that this strategy also has weaknesses. For instance, when a new blocker is discovered, it is impractical to test the blocker against every possible combination of molecules. As a result, only a few controls are used to assess the efficacy and specificity of the blocker. Blocking is also imperfect—siRNAs do not typically completely knock down the end protein expression, rather they suppress expression to varying degrees, depending on the cell type, design of the fragment, and method of introducing the siRNA to the cell.

Because of the nature of signaling, it is not entirely accurate to discuss elements in isolation. If stem cell differentiation by mechanical stimuli can be blocked by removing compound A, and restored by overexpressing compound A in the absence of stimuli, we might conclude that compound A, not mechanical stimuli, is the necessary (and sufficient) ingredient for differentiation (and if we show that mechanical stimuli up-regulate compound A, we might have the beginning of a signaling pathway). Such a linear approach is useful in the short term, but may be an oversimplification. For instance, there is no distinction between a major player (engine's spark plug) and a necessary component that is not a major player (ignition key). For engineering purposes, the former is far more useful than the latter, but often there is no way to tell which of these two is being investigated without far more comprehensive experiments. It is possible to have two or more necessary and sufficient pathway components, that do not have any overlap, but describe the same process. These processes are then typically discovered to be earlier (upstream) or later in the pathway (downstream).

Modern genetics has advanced our ability to study *in situ*

Reductionist approaches are powerful and particularly useful in generating potential molecular targets for further investigation. However, these models are limited and there is always a concern that what is observed in the laboratory may not occur in living organisms. Modern genetics has led to a wide variety of approaches that can help answer this question. For instance, transgenic animals (typically mice) can be created with a particular gene removed (*knocked out*) from their genomic DNA. The case for the function of that particular protein is greatly strengthened if the expected loss of function is seen in the phenotype of the transgenic animal.

Often knockout animals have such functional impairments that they are not viable. The limitations of global deletion approaches can be overcome by using a *site-specific recombinase* (SSR). Rather than deleting a gene from all of the cells in an organism, SSR technologies allow deletion in only certain cells or tissues. The most popular SSR system is the *Cre-Lox* strategy, which relies on the activity of *Cyclic Recombinase* (Cre). Isolated from a class of viruses that infect bacteria, Cre recognizes a targeted gene or gene fragment and deletes it. By creating an artificial genetic construct in which Cre is placed downstream of a particular promoter and delivering the construct to an animal, expression of Cre (and hence gene deletion) is limited to only those cells in which the promoter is activated. Promoters are available that restrict Cre to specific tissues and cells. Inducible promoters are even available that can activate Cre expression at the time of the investigator's choosing by administration of an otherwise innocuous drug. By studying the loss of function in an animal with a missing protein, the function of that protein can sometimes be inferred. Perhaps the most convincing support for a proposed function is a *rescue* experiment in which reintroducing the deleted gene restores the lost function (the analog of necessary and sufficient experiments in organisms instead of in cell culture). These tools and others have sparked a revolution in genetics and allow translation and validation of insights made in simple systems.

Bioinformatics allows us to use vast amounts of genomic data

A dramatic advance in the capacity of computer technology is one of the defining characteristics of the last decade. The Information Revolution has led to advances in the computing power available to scientists and their ability to compute and to transmit vast amounts of information. In biology, the sequencing of the human genome has led to vast quantities of data as a result of the large-scale, high-throughput biotechnologies. Sequencing the genome of entire organisms, once thought impractical, is now relatively routine (the current price for sequencing an entire individual human is $200,000, down from the millions the genome project cost. The *exome*, the part of the genome formed by exons, can be sequenced for less than that $1,000.). The complete genetic sequences of hundreds of organisms are now available to anyone online.

Microarray technology allows investigators to determine which genes and mRNA sequences a biological sample (cells, tissues, or even an organism) is expressing at any given time. Modern mass spectrometers allow the identification of hundreds to thousands of the proteins produced by a cell over time. This deluge of information and the computer power to analyze it has led to the formation of a new field, *bioinformatics*.

In terms of genetic information or *genomics*, the ability to search a genome database for a specific sequence has greatly simplified primer/probe design. Pieces of sequence can be added or deleted, allowing the creation of new gain- or loss-of-functional mutations to dissect how specific proteins work. Searching for sequence homology allows proteins to be identified that might have similar structure and function. Functional annotation allows rapid identification of what role a particular gene might play in organisms or regulatory and signaling pathways. A similar advance in high-throughput protein sequencing is the foundation of *proteomics*, a field that is only now beginning to yield benefits. As a field, bioinformatics is extremely new. No longer merely a tool supporting traditional approaches, it has grown into a paradigm in which unique hypotheses can be advanced and tested. Although it is not possible to predict the future, large-scale, information-based approaches are likely to continue to produce important insights and tools.

Systems biology is integration rather than reduction

Bioinformatics has led to the development of another new field in modern biology. Systems biology espouses a philosophy of integration rather than reduction. At the beginning of the twenty-first century, the systems approach began incorporating the interaction of a biological network in all of its inherent complexity, rather than reducing it to its simplest isolated components. It is expected that biological networks, because of their highly interconnected nature, can be simulated, and emergent behaviors can be identified. Such properties would be indicative of the network itself and could not be understood in terms of the behavior of any single component. The databases and information manipulation tools from bioinformatics are significant enabling technologies for systems biology. Systems analysis has been applied to gene expression patterns from DNA microarrays, protein expression as determined by mass spectrometry, analysis of all of the small molecule metabolites in a cell or tissue, and simulation of the kinetics of kinase/phosphatase signaling cascades and regulatory networks. Traditional reductionist tools have also been scaled up to the systems approach with systematic exposure of cells to libraries of chemical blockers and arrays of siRNAs. Although examples of emergent behavior are still few and therapeutics nonexistent, it is clear that much of cell biology can only be understood in terms of complex network behavior.

Biomechanics and mechanobiology are integrative

Finally, it is important to acknowledge the differences in approaches between biologists and engineers. Traditionally, the former tend to use detection methods that are empirical, coupled with statistical testing. Mechanistic approaches are important, in terms of elucidating regulatory pathways. Repetition and rigorous protocols are highly valued because of the degree of variability found among samples. By contrast, engineering approaches tend to be mathematical. They value long-established physical principles that can be applied analytically or simulated. To study biomechanics/mechanobiology, an appreciation for both approaches is desirable. Not all aspects of mechanobiology are easily modeled and not all aspects of biomechanics can be experimentally measured, but one key to approaching this sort of research is identification of where the two areas are close and attempting to bridge the gap between them.

Key Concepts

- Mechanobiology deals with the biological response of cells to mechanical stimuli.
- Structurally, cell components are influenced by polymers. Proteins are amino acid polymers. DNA and RNA are nucleic acid polymers. Polysaccharides are sugar polymers. Lipids are fatty acid polymers.
- The central dogma of cell biology states that proteins are translated from mRNA, which in turn is transcribed from DNA.
- Cell organelles play various roles in cell physiology, but several components are also involved in making or modifying proteins.
- Receptors are proteins that span the cell membrane and receive external signals and engage signaling pathways inside the cell.

- Second messengers transfer information within the cell quickly, but are broad in effect. Protein signaling cascades are more specific, but act slower.
- Microscopy is used to examine cells visually. Many fluorescence techniques exist to help probe cell behavior.
- Gels electrophoresis is used to separate molecules for analysis of various biological polymers.
- Biological experiments tend to be reductionist, mechanistic, and hypothesis-driven. Engineering approaches are more quantitative but complement biological approaches.

Problems

1. Describe a scenario whereby stretching a cell will lead to changes in transcription of some gene, with calcium signaling as an intermediary. Your response does not have to be based on a real pathway, but should involve the material discussed in this chapter.

2. You are examining a cell that was stained for a component of the cytoskeleton called actin. You see a single fiber at a location but suspect there may be two fibers instead. The objective you are using is a 40×, NA 1.3 oil immersion. What is the closest distance between two actin fibers at which you can still distinguish them?

3. A particular gene is known to be activated upon mechanical stimulation. When you assay the mRNA concentration 1 hour after applying stimuli, you do not see any changes. List possible reasons for this observation.

4. Some signaling pathways will involve multiply-linked proteins that either engage or disengage in response to some stimuli. You are interested in examining whether

two proteins, X and Y, are engaged at a particular time. How would you experimentally determine this, given that you do not have access to a microscope sensitive enough to see the proteins themselves?

5. Here is an abstract from a 1997 article in *Science* (275, 1308–1311): "The small guanosine triphosphatase (GTPase) Rho is implicated in the formation of stress fibers and focal adhesions in fibroblasts stimulated by extracellular signals such as lysophosphatidic acid (LPA). Rho-kinase is activated by Rho and may mediate some biological effects of Rho. Microinjection of the catalytic domain of Rho-kinase into serum-starved Swiss 3T3 cells induced the formation of stress fibers and focal adhesions, whereas microinjection of the inactive catalytic domain, the Rho-binding domain, or the pleckstrin-homology domain inhibited the LPA-induced formation of stress fibers and focal adhesions. Thus, Rho-kinase appears to mediate signals from Rho and to induce the formation of stress fibers and focal adhesions."

To help you better understand the abstract, please note the following: Swiss 3T3 cells are a type of fibroblast from mice; microinjection is using a special microsyringe to inject selected chemicals into cells directly, rather like a shot from a conventional syringe. Do not worry about the pleckstrin-homology domain (which means the local protein sequence is nearly identical to that of a protein called pleckstrin). Do not worry about what LPA actually does, except for the function outlined in the abstract.

(a) Briefly explain the hypothesis of the authors and how they tested the hypothesis.

(b) If you had a technique for measuring how stiff a cell is, and you compared LPA-treated cells versus untreated cells, which would I expect to be stiffer?

(c) Why did the authors serum-starve the cells?

(d) If you had unlimited resources, what other controls would you add to what the authors did to further test their hypothesis?

6. If you wanted to examine the expression of a particular protein, there are several molecular readouts you can use. Suppose you predict that in response to stretching a cell for an hour, the protein SSP (stretch-sensitive protein) will be more abundant in the cell.

(a) How can you experimentally validate your prediction?

(b) Suppose you did the experiment from (a) and found that SSP increases. What would you think if a colleague told you that she also stretched the cell but the SSP mRNA levels decreased? Does it matter when she assayed the cells?

7. A gene mutates so that the generated protein has a single amino acid substitution somewhere in the middle whereby an Asp is replaced with a Glu. (You may consult sources to determine what these side chains actually are.)

(a) What difference in protein structure, if any, do you expect to result from this substitution? That is, do you expect the protein to fold exactly the same way? Why or why not?

(b) Next you grab the ends of the protein and try to pull it apart. You find that the protein is "stiffer" (i.e., harder to extend) with the Glu mutation. Could this affect the signaling capacity of the protein? If so, how?

8. We are interested in the interactions of two proteins, JP (junctional protein) and JPAP (JP-associating protein). We know these proteins are linked at the cell boundaries between adjacent, adhering cells. We also know that JPAP will dissociate (i.e., separate from) JP when the cell is mechanically stimulated. We do not know whether JP leaves the cell boundaries when JPAP dissociates from it. Describe an experiment using fluorescent-conjugated JP and JPAP to determine whether JP leaves the cell boundaries in response to mechanical stimuli.

9. For the following experimental results involving protein N, determine whether N phosphorylation is necessary, sufficient, or whether one cannot tell, the down-regulation of gene given the findings below. Note that we do not know where in this pathway N acts, but under normal conditions mechanical stimuli will also increase N phosphorylation.

mechanical stimuli → calcium increase → gene 1 up-regulation → gene 2 down-regulation

(a) Under mechanical stimulation N phosphorylation was inhibited by a chemical. Gene 1 failed to up-regulate, but gene 2 was down-regulated.

(b) Without mechanical stimuli, phoshorylated N was induced by injecting phosphorylated N directly into cells. Gene 2 was down-regulated.

(c) Under mechanical stimuli, calcium changes were suppressed by a chemical. N exhibited increased phosphorylation, but gene 2 was not down-regulated.

10. Discuss the following statement. "To produce one molecule of each possible kind of polypeptide chain, 300 amino acids in length, would require more atoms than exist in the universe." Given the size of the universe, do you suppose this statement could possibly be correct? Since counting atoms is a tricky business, consider the problem from the standpoint of mass. The mass of the observable universe is about 10^{80} g, give or take an order of magnitude or so. Assuming that the average mass of an amino acid is 110 Da, what would be the mass of one molecule of each possible kind of polypeptide chain 300 amino acids in length? Is this greater than the mass of the universe?

11. It is often said that protein complexes are made from subunits (that is, individually synthesized proteins) rather than as one long protein because the former is more likely to give a correct final structure.

(a) Assuming that the protein synthesis machinery incorporates one incorrect amino acid for each 10,000 it inserts, calculate the fraction of bacterial ribosomes that would be assembled correctly if the proteins were synthesized as one large protein versus built from individual proteins. For the sake of calculation, assume that the ribosome is composed of 50 proteins, each 200 amino acids in length, and that the subunits—correct and incorrect—are assembled with equal likelihood into the complete ribosome. [The probability that a polypeptide will be made correctly, P_c, equals the fraction correct for each operation, f_c, raised to a power equal to the number of operations, n, $P_c = (f_c)n$. For an error rate of 1/10,000, $f_c = 0.9999$.]

(b) Is the assumption that correct and incorrect subunit assembly is equally likely true? Why or why not? How would a change in that assumption affect the calculation in part (a)?

12. If a sample of human DNA contains 20% cysteine (C) on a molar basis, what are the molar percentages of A, G, and T?

13. All small intracellular mediators (second messengers) are water soluble and diffuse freely through the cytosol (indicate true or false, and explain why).

14. Cells communicate in ways that resemble human communication. Decide which of the following forms of human communication are analogous to autocrine, paracrine, endocrine, and synaptic signaling by cells and briefly say why.

    (a) A telephone conversation
    (b) Talking to people at a cocktail party
    (c) A radio announcement
    (d) Talking to yourself

15. Why do signaling responses that involve changes in proteins already present in the cell occur in milliseconds to seconds, whereas responses that require changes in gene expression require minutes to hours?

16. You want to amplify DNA between the two stretches of sequence in the figure. Of the listed primers, choose the pair that will allow you to amplify the DNA by PCR.

DNA to be amplified

```
5'-GACCTGTGGAAGC ———————— CATACGGGATTGA-3'
3'-CTGGACACCTTCG ———————— GTATGCCCTAACT-5'
```

Primers

```
(1) 5'-GACCTGTCCAAGC-3'    (5) 5'-CATACGGGATTGA-3'
(2) 5'-CTGGACACCTTCG-3'    (6) 5'-GTATGCCCTAACT-3'
(3) 5'-CGAAGGTGTCCAG-3'    (7) 5'-TGTTAGGGCATAC-3'
(4) 5'-GCTTCCACAGGTC-3'    (8) 5'-TCAATCCCGTATG-3'
```

17. A typical mammalian cell is about 1000 μm^3 in volume. The concentration of protein within a typical cell is 200 mg/ml. Using western blotting, you can detect 10 ng of a specific protein from loading 100 μg of total protein. The specific protein you are interested in has a molecular weight of 100 kDa.

    (a) How many cells do you need to harvest to be able to load 100 μg of total protein?
    (b) How many copies of a given protein per cell are required to detect the western band?

18. After treating cells with a chemical mutagen, you isolate two mutants. One carries alanine and the other carries methionine at a site in the protein that normally carries valine. After treating these two mutants again with the mutagen, you isolate mutants from each that now carry threonine at the site of the original valine. Assuming that all mutations involved single nucleotide changes, deduce the codons that are used for valine, methionine, threonine, and alanine at the affected site. Would you expect to be able to find valine-to-threonine mutations in one step?

Annotated References

Alberts B, Johnson A, Lewis J et al. (2008) Molecular Biology of the Cell, 5th ed. Garland Science. *A comprehensive text on molecular biology that covers in great detail the process of transcription and translation, and the structure and function of cells. This text is a useful reference not only for Chapter 2, but for many parts of the rest of the book.*

Lodish H, Berk A, Kaiser CA et al. (2007) Molecular Cell Biology, 6th ed. W.H. Freeman. *Another good textbook on molecular biology. Actually, there are several texts and you may choose the one that is most appropriate for your level of interest.*

Mullis KB & Faloona FA (1987) Specific synthesis of DNA in vitro via a polymerase-catalyzed chain reaction. *Methods Enzymol.* 155, 335–350. *This is the original article that describes the PCR reaction, although there are previous publications reporting its use. The technique was rapidly adopted.*

Periasamy A (2001) Methods in Cellular Imaging. Oxford University Press. *A good, broad introductory reference to imaging techniques used in cell biology.*

Watson JD (2001) The Double Helix: A Personal Account of the Discovery of the Structure of DNA. Touchstone. *An autobiographical explanation of the discovery of the double-helical structure of DNA by one of the Nobel Prize–winning team.*

CHAPTER 3

Solid Mechanics Primer

In the field of cell mechanics, there is more than one way to understand the mechanical behavior of cells. Our review of basic solid mechanics begins with rigid-body mechanics and goes on to consider small deformations in solid bodies. We discuss simple loading configurations, including axial, torsion, and bending configurations. From there we discuss some aspects of large deformation mechanics, although a complete treatment is beyond the scope of this text.

3.1 RIGID-BODY MECHANICS AND FREE-BODY DIAGRAMS

What is a "rigid" body?

In this section, we confine ourselves to purely mechanical changes in state to a given body. We exclude such things as chemical, radioactive, or other changes to overall mass. Mechanical changes can be described as a combination of translation, rotation, and deformation to the body. A surprising amount can be learned by ignoring the deformation and only considering translation and rotation. By considering only the balance of forces on a body, we are able to determine its acceleration and vice versa.

First, we consider all bodies to be rigid—that is, undeformable. This assumption is only good for examining force balances and approximating distributions; under close examination it will lead to ludicrous conditions. We will see later that stress is the force applied to a given surface divided by the surface area. If a perfectly rigid sphere hits a perfectly rigid flat wall, the stress at the point of impact is infinity because there is zero area in the single point of contact. For now, we ignore these issues and focus on the broader problem of determining force balance.

One of the most powerful, but underused, tools is a free-body diagram

Free-body diagrams are representations of an object, a part of an object, or a collection of objects for which all external forces are described. They are derived from the fact that Newton's laws of motion apply not only to complete structures, but also to all components or sub-structures we care to define. The fundamental process in a free-body diagram analysis is to (1) identify a component (or components) of interest by surrounding it with an imaginary boundary, (2) imagine removing this portion of a larger structure or system, and (3) identify all forces across our imaginary boundary that must be acting to replace the external surroundings. The forces that act across our imaginary boundary may be known fixed forces or they may be unknown internal forces. Therefore, the key is knowing where to place the boundary so that the desired forces can be obtained. As we will see, in deformational mechanics the fundamental equations of equilibrium in solid mechanics, or the Navier–Stokes equation in fluid mechanics, are the

Figure 3.1 A crawling cell can pull itself forward. The cell is crawling from left to right. It has produced an extension called a pseudopod and is pulling itself forward with it.

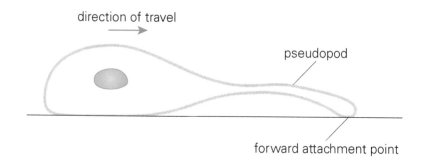

direction of travel

pseudopod

forward attachment point

result of free-body diagram analysis. For now, let us ignore deformation and focus on problems in which the forces are simple and discrete.

Identifying the forces is the first step in drawing a free-body diagram

It is important to identify all of the forces applied to the isolated component. Explicitly drawing the imaginary boundary needed to isolate the structure of interest from the larger system will ensure that no mistakes are made. Consider a cell that is slowly crawling away from a wall that has left a cellular extension behind, pulling the cell back toward the wall (Figure 3.1). Assume that the motion is slow enough that it can be ignored (velocity = constant = 0).

The free-body diagram of the cell can be constructed by first drawing the imaginary boundary that would isolate the structure of interest (the cell). Then we redraw the cell in isolation, but representing the forces acting across the boundary that would need to be applied to keep the cell in place (Figure 3.2).

A frequent error in drawing the free-body diagram of the cell is to locate the force at the extension tip pointing toward the cell—it is wrongly imagined that the tension in the extension is pulling inward toward the cell. If you remember that the forces in a free-body diagram are those that act across the boundary needed to replace the external environment, you will never go wrong. The force being drawn is the force of the wall on the cell, which is the reaction force in the opposite direction and of equal magnitude. Because the cell has a constant velocity (of nearly zero), there must be a force that counterbalances it (Newton's first or second law, net change in acceleration is zero), and this is at the base of the cell. The arrows down and up represent the weight of the cell and the reaction force from the ground. These are both external forces, as gravitation is the force exerted by the planet on the cell, and the reaction force is the force exerted by the substrate on the cell.

Influences are identified by applying the equations of motion

Once the free-body diagram is complete, one may then apply Newton's second law [$\Sigma F = 0$, or $\Sigma F = ma$ if there is acceleration) and moment balance [$\Sigma M = 0$, or $\Sigma M = I\alpha$ where I is the (mass, or first) moment of inertia and α is the angular

Figure 3.2 The imaginary boundary around the cell (A) and the resulting free-body diagram (B). The arrow inside the cell represents its weight.

(A)   (B)

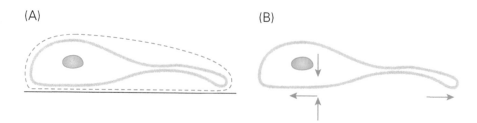

acceleration if there is rotation]. For cases in which accelerations are permitted, we may use the inertial vector and the moment of inertial torque on the free-body diagram to have all influences displayed. On the scale of a typical cell, however, inertial forces are generally small and therefore can generally be ignored.

Free-body diagrams can be drawn for parts of objects

In the example above, the forces between the cell and substrate are easy to identify because they are clearly separate things. However, free-body diagrams can also be very useful for determining forces *through* something, even when there is no obvious separation or boundary. Remember, we can draw our imaginary boundary anywhere we like, including right through a solid structure, to identify the forces acting internally. Consider the mechanics of a polymer network, which is considered in more detail later (Figure 3.3).

Say you wish to know the forces in the individual polymers as a function of the overall force on the network. We can again do this by constructing a free-body diagram. Start by drawing our imaginary boundary (Figure 3.4).

We selected this boundary because it takes advantage of the symmetry of the problem. In addition, it cuts through the center polymer and we would like to determine the force going through that element of the network. Next, we isolate the portion of the structure within the boundary and replace the boundary with the forces that must act across it to maintain equilibrium (Figure 3.5).

The force balance and moment balance can now be applied. However, there are some important things to note. First, we could have used the left part of the network to draw the free-body diagram. This would have produced the same results. Second, we have drawn all of the horizontal forces pointing to the left. A quick inspection of the network reveals that the center polymer is under compression. Therefore, it might have made more sense to draw this arrow pointing to the right. Actually, it does not make a difference—drawn as it is, the analysis produces a negative value for this force, indicating that the force is indeed applied from left to right.

For this free body, we now apply the force and moment balance. Notice that we know the downward force on the lower pin. It is simply half the applied force (by symmetry we know that the other half of the applied force is supported by the left-hand side of the network). We can also see that force in the middle polymer is twice the horizontal component of force in the obliquely oriented polymers. However, we cannot determine the magnitude of these forces. We say that it is statically indeterminate. Nevertheless, we can create another free-body diagram—there is nothing to prevent more than one free-body solution or analysis for a given structure. Specifically, it is informative to isolate each of the pins (Figure 3.6), which tells us that the vertical and horizontal forces at each pin must sum to zero. With this information, and some simple geometry, we can solve the problem relatively easily.

3.2 MECHANICS OF DEFORMABLE BODIES

Rigid-body mechanics is not very useful for analyzing deformable bodies

Although we can learn a great deal about the mechanical behavior of structures by considering their elements to be rigid, this approach is limited. It cannot tell us about the distribution of forces within a structure. To do this we need to consider deformation. When we first consider that objects that appear solid and hard can actually deform, it can be difficult to understand intuitively. However, in reality, even the most rigid material undergoes deformations when exposed to loading.

In the mechanics of rigid bodies, we were concerned with characterizing the forces and displacements of a structure. Because we idealized the behavior of the

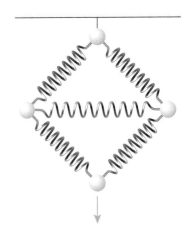

Figure 3.3 Schematic of a polymer network. Individual polymers are represented as springs connected by freely rotating hinges or pins. We would like to determine the internal forces on the polymers as a function of the total force on the network.

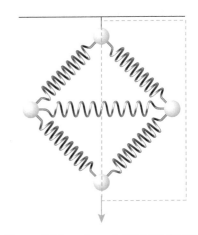

Figure 3.4 Imaginary boundary drawn through the polymer network.

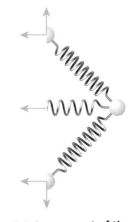

Figure 3.5 Component of the network isolated from Figure 3.4. There are forces across the boundary at the upper and lower pins as well as the middle polymer.

Figure 3.6 Free-body diagram for the lower pin (A), middle pin (B), and upper pin (C). Notice that we have reversed the direction of the force in the middle polymer from Figure 3.5 to better reflect its compressive nature.

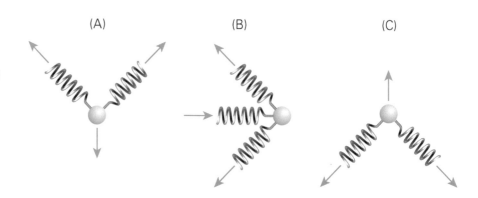

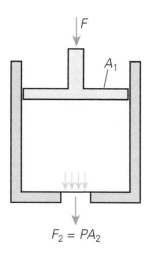

$$F_2 = PA_2$$

Figure 3.7 A container filled with a fluid has a piston applied to its upper surface of area A_1. The piston force F is supported by the substance, meaning that the contents are placed under pressure $P = F/A_1$. A small region of the bottom wall of the container with area $A_2 < A_1$ will experience a smaller force $F_2 = PA_2 = F(A_2/A_1)$ acting on it.

individual components as rigid, we did not need to be concerned with the mechanical properties of the materials. When we now relax this assumption and consider material deformations, we take an important step forward in developing our mechanical understanding. However, we will find that the tools of forces and deformations alone are insufficient to describe material deformation. This is because two distinct things contribute to the mechanical behavior of a structure: the properties of the material from which it is formed and its shape. Consider a cantilevered beam such as a swimming pool diving board. The stiffness at the end of the board can be increased by using a stiffer material or by making a thicker board. Therefore, forces and deformations need to be scaled in some way to the shape of structure to which they are applied. Now, our first order of business is to define our scaled forces (stress) and our scaled displacement (strain).

Mechanical stress is analogous to pressure

When it is initially encountered, the concept of stress can be quite baffling. It is not necessarily intuitive to imagine forces "flowing" and being distributed throughout a structure. Conceptually, stress is the abstract idea of force being distributed over an area. We already have an intuitive understanding of distributed forces when we consider pressure. Imagine a simple vessel filled with a fluid, such as a gas or liquid. A force F is then applied to a piston that pressurizes the contents (Figure 3.7).

We can determine the pressure of the contents as the force F divided by the area of the piston A_1, $P = F/A_1$. Furthermore, if we imagine a small cutout or window in the bottom of the vessel, a force would be exerted on the area by the pressurized fluid. The distributed pressure across the face of the piston adds up to produce a net force on the cutout. And we know that this force is proportional to the pressure and to the area of the piston, $F_2 = PA_2$. When the pressure is larger, the force is larger, and when the area is larger, the force is also larger. This is all very straightforward and intuitive.

What is critical to realize is that we can imagine a distributed force due to pressure anywhere within the fluid. For example, we can imagine a plane within the fluid of area A_3. The force across this area is $F_3 = PA_3$ and is exerted by the fluid on itself, regardless of the orientation of the plane. So, the idea of pressure is not limited only to those locations where there are pistons and cutouts. The pressure is exerted everywhere within and is defined as the force per unit area exerted on any small imaginary cut-plane regardless of position or orientation, so long as the plane is wholly immersed in the fluid.

Normal stress is perpendicular to the area of interest

Let us try to apply our analogy with pressure to a solid structure and see how far it can take us. Imagine a simple cylindrical column supporting a load at its top. In the interior of the column the force resulting from the load is evenly distributed through the cross section, assuming the column is uniform. If we move far away from the end, the force becomes evenly distributed throughout the cross section. As with pressure, we can imagine a cut-plane through the column. Consider a cut-plane that is perpendicular to the long axis of the column (Figure 3.8). Because the force across this area is distributed evenly, we can apply an analogy to pressure and define *stress*, σ, to be the force per unit area

$$\sigma = \frac{F}{A}.$$ (3.1)

This stress is considered a *normal* stress, because the force is acting perpendicularly across the area. Historically, tensile stresses, as illustrated in Figure 3.8, are assumed to be positive by convention, and compressive stresses are negative. Unfortunately for the analogy with pressure, the opposite sign convention is used for pressure as stress.

Strain represents the normalized change in length of an object to load

Just as we defined stress to be a force scaled by the area it is acting across, we can also define *strain* as a change in the length of an object, scaled by its original length. Consider again the column under tension from Figure 3.8. Let the initial length of the column be denoted as L. When the load is applied, the column extends by a small amount (ΔL) and the total length of the column is now $L + \Delta L$ (Figure 3.9). Here, the strain is defined as

$$\varepsilon = \frac{\Delta L}{L}.$$ (3.2)

A strain associated with normal stress (that is, along the same axis as the applied force) is called normal strain. Because strain is defined as the ratio of two lengths, it is inherently dimensionless. It is sometimes expressed as either percent strain ($\Delta L/L \times 100$) or microstrain ($\Delta L/L \times 1,000,000$). Also note that the same sign convention for stress holds for strain (tensile positive, compressive negative).

The stress–strain plot for a material reveals information about its stiffness

We have now defined stress and strain in our simple uniaxial column example. Before we go on to more complicated examples, let us see how they relate to each other to help us understand the behavior of a material. Our column is loaded with axial tension and we record the deformation of the tip as a function of the load we apply. Assume that the column is uniform in its material properties (homogeneous), that it behaves as an elastic material, and that its material behavior is the same in all directions (isotropic). Although the two quantities we directly measure are the tip displacement and the force, we calculate the stress ($\sigma = F/A$) and strain ($\varepsilon = \Delta L/L$) and plot them versus each other (Figure 3.10).

For a broad range of sizes and shapes of columns, this plot will exhibit the same overall shape. For many materials, this plot is a simple line. Such materials are known as linear (or linear elastic) materials. The slope of this line is known as the Young's modulus of the material. Typically this proportionality constant is given by E in the linear relationship $\sigma = E\varepsilon$, known as Hooke's law (physics students may recognize its similarity to the one-dimensional Hooke's law $F = kx$ from modeling a spring). As the load increases, a stress is eventually reached, known

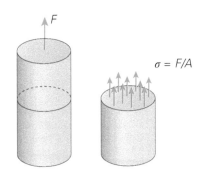

$\sigma = F/A$

Figure 3.8 Column loaded with a distributed force *F* at both ends experiences internal normal stress *F/A* at any cut-plane oriented perpendicular to the long axis of the cylinder.

Nota Bene

Tensile stresses are positive. The convention that tensile stresses are positive comes about because much of the material-testing literature is focused on tensile testing.

Nota Bene

Distinguishing material and structural behaviors. The column is a good example of why scaling can be so critical. Imagine that the column was twice as long initially: the amount by which it extends under the same load would be twice as large. However, the stress in the material would not have changed. This larger displacement at the tip would be due entirely to the shape of the column, in this case its length. This is an example of the influence of geometry on mechanical behavior and is sometimes called the structural behavior to distinguish it from the influence of the material properties.

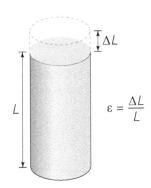

ΔL

L

$\varepsilon = \dfrac{\Delta L}{L}$

Figure 3.9 Same column from Figure 3.8 with the same load (load not shown) will lengthen along the long axis of the cylinder. The ratio of the change in length (ΔL) to the original length (L) of the column is the strain.

Nota Bene

Who was Hooke? The relationship between stress and strain for linear elastic materials, $\sigma = E\varepsilon$, is known as Hooke's law after Robert Hooke, the seventeenth-century physicist. Actually, Hooke studied linear springs in terms of force and displacement ($F = kX$). Hooke never would have used the equation that bears his name. Nonetheless, his contribution to understanding elastic behavior was very significant. Coincidentally, for the subject of cell mechanics, he is also known as the father of microscopy and coined the term *cell*.

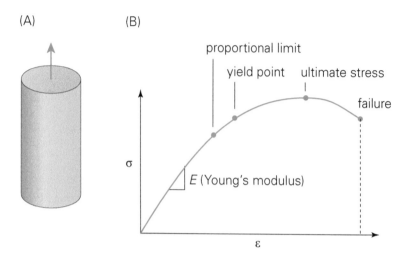

Figure 3.10 **For the loading configuration (A), a stress–strain relationship can be obtained (B).** For small strains, the relationship can be a straight line; this is the linear region with the slope representing the stiffness, or elastic modulus. As the load increases, the material will pass through the proportional limit (extent of linear region), yield point (where deformation is not recovered upon unloading), ultimate stress (maximum stress), and failure.

Nota Bene

Who was Young? Young's modulus is named after the eighteenth-century British physician Thomas Young. He conducted research in light and vision, circulation, mechanics, and even helped translate some of the Rosetta Stone. He extended the work of Hooke, who characterized elastic behavior in terms of force and displacement ($F = kX$) into normalized stress and strain, whereby, the proportionality constant k—known as the stiffness and a function of both material and geometry—becomes the Young's modulus, E, and is a true material property. It is also sometimes referred to as the elastic modulus or modulus of elasticity. However, it is only one of several possible elastic moduli, such as the bulk modulus or shear modulus.

as the *proportional limit* where it no longer behaves linearly. The stress at failure is the *ultimate stress* or *strength* of the material. Independence from geometry is a major benefit of using stress and strain rather than force and displacement.

Stress and pressure are not the same thing, because stress has directionality

One key distinction between stress and pressure has to do with orientation. As we mentioned previously, the pressure at any point in a fluid does not depend on the orientation. For a given point inside a fluid, the vertical and horizontal pressures at that location are always the same. By contrast, stress depends strongly on orientation. In our axially loaded column example, the stress in the vertical direction is F/A, yet the stress in the horizontal direction is zero, because there is no load in

Example 3.1: Cytoskeletal proteins in extension

At this point in our discussion we can already create a simple model for cytoskeletal proteins. We can treat them as simple rods and we can ask how stiff they are. We are interested not in the stiffness as a material property *per se*; rather, what is relevant is the axial stiffness of a polymer as a structure to give us some intuition of the mechanical behavior of the cell's components. This stiffness would be influenced by both the material behavior and geometry of the polymer. *EA* is the relevant measure of structural stiffness (per unit length) in this case Table 3.1.

Table 3.1 Cytoskeletal axial stiffnesses

	R	A	E	EA
Microtubule	12.5 nm	492 nm^2	1.9×10^9 N/m^2	0.934×10^{-6} N
Intermediate filament	5.0 nm	79 nm^2	2×10^9 N/m^2	0.16×10^{-6} N
Actin filament	3.5 nm	38 nm^2	1.9×10^9 N/m^2	0.23×10^{-9} N

R, radius; *A*, cross-sectional area; *E*, Young's modulus.

the horizontal direction. The reason behind this difference is the ability of the molecules in a fluid to flow under load and reorient the internal force distribution. Molecules in a solid, on the other hand, are more strongly bound to their neighbors and are not free to redistribute under load. In mathematical terms, pressure is a scalar quantity, depending (as do forces, displacements, stresses, and strains) on the location within the material where the quantity is measured. Because forces and displacements are vectors (these quantities are described by a magnitude and a direction), directionality is built into them. Stress and strain, which depend on the direction of forces and displacements, therefore exhibit directionality-like force and displacement. However, stress and strain are even more complicated, because they depend not only on the direction of forces and displacements, but also on the direction of the area (or displacement) to which they apply. To specify a component of stress, we must know the direction of the forces relative to the orientation of the surface (the surface direction or orientation is given by the vector normal to its local area). Stress and strain are mathematical quantities known as *tensors*.

Shear stress describes stress when forces and areas are perpendicular to each other

Let us return to the column-loading example. Initially, the forces in the cut-plane were perpendicular to the plane, resulting in *normal* stresses. What sort of stress results if we load the column perpendicular to the long axis (that is, parallel to the plane)? Consider again the imaginary cut-plane some distance away from the region of load application. As before, the force is distributed evenly through the cross section, but now the small distributed forces are lined up parallel to the plane rather than perpendicular to it (Figure 3.11). We term this type of stress *shear stress*,

$$\tau = \frac{F}{A}. \tag{3.3}$$

In general, the cut-plane through the column can be in any orientation we desire. If we select a plane that is oriented obliquely to the axis of the column, the distributed forces across the surface will be a mixture of parallel and perpendicularly oriented forces. Therefore, in general stress is a mixture of normal stress and shear stress (Figure 3.12).

Shear strain measures deformation resulting from shear stress

In addition to shear stress, we need to define the complementary deformation measure, shear strain. In the case of normal strain, we divided the vertical displacement at the tip (representing the length change) by the column length. We can similarly define the shear strain to be the transverse displacement divided by the original vertical length, or $\gamma = \delta/L$. Because this is a result of shear stress, the direction of displacement is perpendicular to the direction of the original length (Figure 3.13). Like normal strain, shear strain is a dimensionless quantity and is typically expressed in percent strain or microstrain.

To illustrate shear strain, consider a rectangular body. Assume the block is fixed at the bottom and exposed to a shear force at the top. Finally, assume that points along the top surface of the block displace by an amount δ as a result of the application of the shear force, sometimes referred to as "pure shear." The shear strain is the ratio of δ to L or, $\gamma = \delta/L$. The shear strain can be related to the amount by which the right angle at the base of the block is reduced, which we denote θ, with $\delta/L = \tan\theta$. If $\delta \ll L$, $\tan\theta \approx \theta$ when θ is measured in radians. As with Hooke's law, we can then

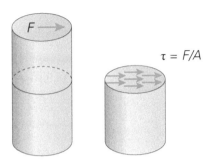

Figure 3.11 Shear stress occurs when the loading force is perpendicular to the area across which the stress acts.

$\tau = F/A$

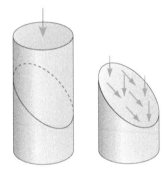

Figure 3.12 A general cut-plane through a column with a normal load will experience both shear and normal stresses (since the vector components can be decomposed into components parallel and perpendicular to the surface). This is true even if the load is a mixture of shearing and normal forces.

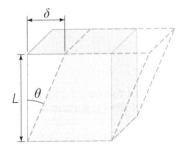

Figure 3.13 Shear strain is measured by dividing the lateral displacement (δ) by the original normal length (*L*).

define the linear relationship between shear stress and shear strain with the shear modulus, G,

$$\tau = G\gamma = G\theta \tag{3.4}$$

for small θ.

Torsion in the thin-walled cylinder can be modeled with shear stress relations

Let us work through an example using shear stresses and strains. We are given a thin-walled cylinder that is clamped at one end and a moment is applied at the other end (Figure 3.14).

Let M be the torsional moment applied to the end of the shell of radius R. In response to this torque, the shell will twist along its length. The overall deformation is related to the internal strain, and the relationship between the displacements and strain is known as the *kinematics* of the problem. We can measure the amount of this twist by the angle of twist at the end, θ. For the same strain in the material, a longer shell will exhibit a larger angle of twist compared with that of a shorter shell. So we relate the twist angle to the angle by which a line drawn on the surface of the shell parallel to the long axis of the shell rotates (which is simply the shear strain, γ). We start by considering how much a point on the surface displaces, denoted δ,

$$\delta = \theta R, \tag{3.5}$$

as a function of the displacement along the circumference of the shell. In terms of shear strain,

$$\gamma = \frac{\delta}{L} = \frac{\theta R}{L}, \tag{3.6}$$

where L is the length of the rod.

Next, we relate the applied moment to the angle of twist. Similar to the axial loading case, the resultant load (or moment) is the net resultant of the stress acting on the end of the shell. The moment $M = \tau AR$, where R is the radius of the shell (we ignore the thickness as being small compared with the radius R) and τ is the shear stress on the end surface of the shell which has area A. The area $A = 2\pi Rt$, where t is the rod thickness (again, $t \ll R$). So,

$$M = \tau 2\pi R^2 t, \tag{3.7}$$

which can be rearranged to obtain the shear stress

$$\tau = \frac{M}{2\pi R^2 t}. \tag{3.8}$$

Then, by using our material model, $\tau = G\gamma$ from Equations 3.4 and 3.6

$$\theta = \frac{\gamma L}{R} = \frac{ML}{2G\pi R^3 t}. \tag{3.9}$$

Figure 3.14 Torsion of a thin-walled cylindrical shell results in a twisting of the shell. We can model this by cutting open the shell and treating this object as a plane rectangle.

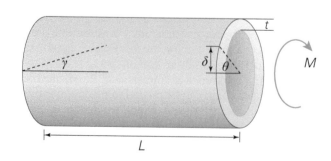

Torsion of a solid cylinder can be modeled as a torsion of a series of shells of increasing radius

Now consider a solid cylindrical rod. We need to think about how the mechanics changes at different points throughout the thickness rod. Indeed, we can solve the solid rod problem by dividing it up into several thin rods and superimposing those solutions (Figure 3.15).

Starting with the kinematics, let us determine how the strain changes through the thickness. As we determined for the thin-walled shell, at a given angle of twist, shear strain (and shear stress) varies linearly with radius R (Equations 3.5 and 3.6), where we now replace the constant shell radius R with the variable radius R, and turn the shear into functions of R:

$$\gamma(R) = \frac{\theta}{L} R. \tag{3.10}$$

For a small element of thickness, $t = dR \ll R$, the shear stress in that thin-walled tube is

$$\tau(R) = G\gamma = \frac{G\theta R}{L}. \tag{3.11}$$

Multiplying by the cross-sectional area of the shell (along which the applied torque) gives the contribution to the net moment of the shell,

$$dM = \tau 2\pi R^2 dR = \frac{2G\theta\pi R^3 dR}{L}. \tag{3.12}$$

To find the total moment for the entire rod, we integrate the contribution for each shell,

$$M = \int_0^R dM = \int_0^R \frac{2G\theta\pi R^3}{L} dR = \frac{G\theta\pi R^4}{2L}. \tag{3.13}$$

Equation 3.13 can also be expressed in terms of the *polar moment of inertia, J*,

$$J = \int_A R^2 dA = \int_0^R 2\pi R^3 dR, \tag{3.14}$$

which is a measure of a given cross-sectional geometry to resist torsion. This allows us to express Equation 3.13 in a more compact form,

$$M = \frac{GJ\theta}{L}. \tag{3.15}$$

Notice that this analysis only works for cylindrical-shaped rods. If a rod is not a cylinder, if it has a square or rectangular cross section, our kinematic assumption breaks down. Plane sections will not deform by simply sliding over one another, but they will also become deformed (no longer remain planar). This deformation along the axis of the rod is known as *warping* and occurs even for small deformations.

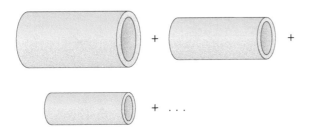

Figure 3.15 A solid rod can be modeled as an integral over thin shells of varying radii. Each shell has a radius of r and a thickness of dr.

Example 3.2: Cytoskeletal proteins in torsion

We wish to determine the rough structural stiffness of typical cytoskeletal proteins in torsion. Although it is not possible to compare axial structural stiffness with torsional stiffness directly, it is still helpful to make a general order-of-magnitude comparison to develop our intuition at this level. The parameter that captures both the material behavior and the structural behavior is GJ (Table 3.2). The effective shear modulus is not readily available experimentally, but we can get within

an order of magnitude by simply taking E for G. We also make the assumption that the polymers are solid cylinders. This is not the case for microtubules but it is still a reasonable order-of-magnitude approximation. We leave it to you to determine the size of this error. From this simple calculation, we can expect that the forces due to axial deformation are much more important than those due to torsion because of their very low relative magnitude.

Table 3.2 Cytoskeletal torsional stiffnesses

	R	J	G	GJ
Microtubule	12.5 nm	38,400 nm⁴	1.9×10^9 N/m²	73×10^{-24} Nm²
Intermediate filament	5.0 nm	980 nm⁴	2×10^9 N/m²	1.8×10^{-24} Nm²
Actin filament	3.5 nm	230 nm⁴	1.9×10^9 N/m²	0.45×10^{-24} Nm²

Kinematics, equilibrium, and constitutive equations are the foundation of solid mechanics

At this stage in our review of mechanics, let us work through a more challenging problem. Beam bending is illustrative, because we can solve the problem with only the background and definitions we have developed thus far. Yet it also requires the three fundamental relationships of solid mechanics: kinematics, equilibrium, and constitutive behavior. Although we have not stated it explicitly, we have been using these equations in the simple examples so far. Indeed, all problems in solid mechanics require these three components, and, with the addition of boundary conditions, each must be included to find a solution. It is helpful to consider each of these in turn in this simple situation before we develop them in the general continuum mechanics formulation.

Kinematics in a beam are the strain–displacement relationship

In every mechanics problem there needs to be a relationship between displacement and strain. In our simple axial and torsion examples, this was straightforward. In the case of a beam, it is a bit more complex. Consider a beam that is loaded by a bending moment such that it takes on a curved shape (Figure 3.16).

We make three simplifying assumptions about the deformation that characterizes the kinematics of the problem. First, we assume that any y–z-plane in the undeformed beam will remain a plane in the deformed situation. In other words,

Nota Bene

Cytoskeletal polymers often experience small strains. Beams following these simplifications conform to *Euler–Bernoulli* beam theory. It is particularly useful for slender structures in which shear deformation is quite small compared with displacements due to flexion. The long slender structural polymers of the cells satisfy these assumptions quite well.

Figure 3.16 Schematic of a beam exposed to bending moments.

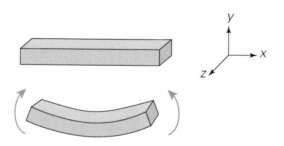

y–z-planes rotate, but they do not deform to lose their planar shape. This assumption is sometimes called the "plane sections remain plane" assumption. Next, we assume that the plane sections only rotate and do not slide relative to each other. In other words, we are going to assume that we can ignore the effects of shear. Finally, we assume that as the beam deforms, the y–z-planes rotate in such a way as to remain perpendicular to any imaginary line in the beam that was originally oriented in the x-direction. These three assumptions serve to define fully the deformation of every point in the beam. Furthermore, the deformed state is fully defined by knowing the displacement of the central axis of the beam only. The beam becomes a simple one-dimensional problem of determining the displacement in the y-direction, which we denote $w(x)$. Next we need to determine the strain in the beam as a function of the displacement $w(x)$.

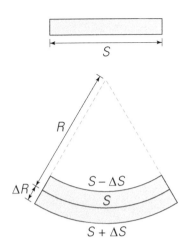

Figure 3.17 Small section of the beam *S* long, ± Δ*R* thick.

To accomplish this, we need to determine the amount of stretch or strain that lines initially oriented parallel to the x-axis experience as the beam deforms. Consider a very small portion of the deformed beam (Figure 3.17). Further, consider three lines on the beam; the top surface, the bottom surface, and the *central axis* of the beam. If the section is small enough, the deformed shape of each of these lines can be approximated as sections of a circle (this is known as the *osculating* circle and can be uniquely defined at every point in a curve as the best-fitting circle at that point). Let the radius of the deformed central axis be R, the top surface $R - \Delta R$, and the bottom surface $R + \Delta R$. Next, assume that the initial length of a line along the central axis in our section of beam is S. In the deformed configuration, the central axis is neither lengthened nor shortened. In other words, the strain at the central axis is zero and is therefore often referred to as the neutral axis. Because the ratio of circumference to radius is constant for similar circular sections,

$$\frac{S + \Delta S}{R + \Delta R} = \frac{S}{R}. \tag{3.16}$$

Equation 3.16 can be rearranged to show

$$\frac{\Delta S}{S} = \frac{\Delta R}{R}. \tag{3.17}$$

Notice that the quantity $\Delta S / S$ is simply the change in length of one of our imaginary lines normalized by its initial length. That is exactly our definition of strain. Also notice that ΔR is the distance to the imaginary line in the y-direction measured from the neutral axis. If we position the origin of our coordinate system so that it lies on the neutral axis, we have a simple expression for the strain in the beam,

$$\varepsilon(y) = \frac{y}{R}. \tag{3.18}$$

So we have used the term "central axis" without really defining it. This was an intentional oversight to simplify the explanation. It would have been more appropriate to refer to the neutral axis. For cross sections that are symmetric about the central axis, the two coincide, but this need not be the case in general.

Now, notice that the radius, R, is not a constant; it is a function of the displacement $w(x)$. To formulate equations that we can solve, we must make this dependency explicit. To do this, notice that the quantity $1/R$ is also the local *curvature* of the beam typically denoted as κ. In calculus, we learned that the local slope of a curve is given by its first derivative, but the local curvature is given by the second derivative. Therefore strain is the product of the second derivative of the displacement, and the distance from the neutral axis becomes

$$\varepsilon(y) = \frac{y}{R} = y\kappa = y\frac{\mathrm{d}^2 w}{\mathrm{d}x^2}. \tag{3.19}$$

Equilibrium in a beam is the stress–moment relationship

The next step in deriving the beam-bending equations is to make use of equilibrium to relate the moment at a section within the beam to the stress at each point within the beam. Let us think intuitively about the distributed forces or stresses within our beam (Figure 3.18). At the inner side, the beam will be compressed to a shorter total length, whereas at the outer side, the beam will be stretched. The stress in a beam varies linearly through the cross section. At the inner side, the stress is negative or compressive while at the outer side, it is positive or tensile. At the neutral axis, the stress is zero. The stress resultant of this linearly varying stress is the bending moment M.

Consider a plane that cuts through an arbitrary point in the beam at a location x. For a small strip across the face in the y–z-plane of thickness dy and width h, the contribution to the total moment by the stress exerted on the strip is simply the total force times the moment arm. The moment arm can be calculated from any arbitrary point; however, in this example, the most logical place is the origin of the coordinate system that we located at the neutral axis. Also, note that we must define the sign of the moment, which is arbitrary. You can define a positive moment to be one that causes the beam to bend with its inner surface facing up (as we have done here). It is equally acceptable to define it in the other sense, but just make sure you keep track of which way you have defined it, or you will end up with a minus sign that will not go away. It is a good idea to make a note of the convention you are using on any problem you work out. In our case remember that the stress is positive (tensile) on the lower surface and negative (compressive) on the upper surface. We now know the moment associated with our strip is d$M = -y\sigma(y)h$dy. The total moment is obtained by integrating the contributions of all the strips that make up the beam or

$$M = \int_{-t/2}^{t/2} \mathrm{d}M = \int_{-t/2}^{t/2} y\sigma(y)h\mathrm{d}y. \tag{3.20}$$

Figure 3.18 The resultant moment on the end of the beam is the integral of the stresses through the thickness of the beam.

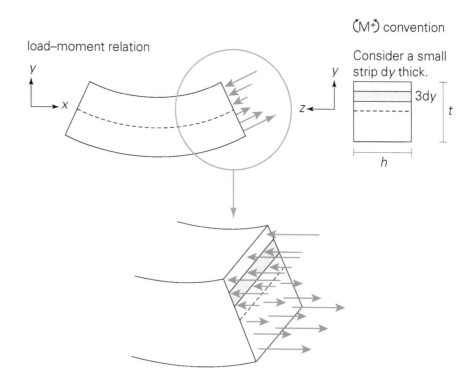

load–moment relation

$(M\overset{+}{\frown})$ convention

Consider a small strip dy thick.

The constitutive equation is the stress–strain relationship

The last part of our development is the constitutive equation. This is the equation that relates stress to strain. It is the fundamental description of how a material behaves. For this example, assume a simple linear elastic material behavior. Hooke's law gives us

$$\sigma(y) = E\varepsilon(y). \tag{3.21}$$

Remarkably, with the kinematics, equilibrium, and constitutive equations, we have completed the theoretical development for this problem. All that remains is to combine what we know. Substituting the stress–strain relationship (Equation 3.21) into the equilibrium equation (Equation 3.20) yields

$$M = E \int_{-t/2}^{t/2} y\varepsilon(y)h\mathrm{d}y. \tag{3.22}$$

Next, using the strain–displacement (Equation 3.19), we have

$$M = E \int_{-t/2}^{t/2} y^2 \frac{\mathrm{d}^2 w}{\mathrm{d}x^2} h\mathrm{d}y. \tag{3.23}$$

Because w does not depend on y, it can come out of the integral. The end result is a governing relationship between the moment and the curvature,

$$M = E \int_{-t/2}^{t/2} y^2 h\mathrm{d}y \frac{\mathrm{d}^2 w}{\mathrm{d}x^2} = EI \frac{\mathrm{d}^2 w}{\mathrm{d}x^2}, \tag{3.24}$$

where

$$I = \int_{-t/2}^{t/2} y^2 h\mathrm{d}y. \tag{3.25}$$

The second moment of inertia is a measure of bending resistance

Note that the integral term I in Equation 3.25 is a constant function of the cross-sectional geometry only. It is known as the *second moment of area*, the *second moment of inertia,* or *area moment of inertia*. It is a measure of a cross section's geometric contribution to a beam's resistance to bending. Also notice that unlike for the polar moment of inertia, J, for the second moment of inertia we must specify the direction in which the moment is taken. For this reason a subscript is sometimes added,

$$I_y = \int_{-t/2}^{t/2} y^2 h\mathrm{d}y \quad \text{and} \quad I_x = \int_{-t/2}^{t/2} x^2 h\mathrm{d}x. \tag{3.26}$$

The term that relates curvature to moment, EI, is a measure of the combined effects of material stiffness and geometry to resist bending. To reflect this, it is known as the *flexural rigidity*.

Example 3.3: Cytoskeletal proteins in bending

Simple bending occurs frequently in the world of the cell. One of the key elements that gives the cell its structure is the cytoskeleton. There are three components that make up the cytoskeleton: microtubules, intermediate filaments, and the actin cytoskeleton. We can estimate the flexural rigidity of a polymer of each of these components from its Young's modulus and radius (Table 3.3). Note that, again, we have neglected to account for the hollow shape of the microtubule.

As with torsion, notice the very low relative magnitude of bending forces compared with axial forces. Cytoskeletal polymers are relatively flexible in bending and torsion and relatively stiff axially. In a way, they are similar to strings or ropes, hard to stretch, but easy to bend and twist. Also notice that of the three, microtubules are the most able to resist bending (and torsion). Actin is the most string-like, and intermediate filaments are in between. As seen in future analyses, very flexible polymers, like actin, can have their axial behavior affected by their flexibility due to the thermal behavior of the molecules surrounding them.

Table 3.3 Cytoskeletal bending stiffnesses

	R	I	E	EI
Microtubule	12.5 nm	19,175 nm^4	1.9×10^9 N/m^2	364×10^{-25} Nm^2
Intermediate filament	5.0 nm	491 nm^4	2×10^9 N/m^2	10×10^{-25} Nm^2
Actin filament	3.5 nm	118 nm^4	1.9×10^9 N/m^2	2×10^{-25} Nm^2

The cantilevered beam can be solved from the general beam equations

Now that we have all the tools in hand, let us solve an example beam-bending problem. One classic problem is the *cantilevered beam*. In this problem, the beam is clamped at one end and has a single force applied to the other end (Figure 3.19).

As we have derived, the governing equation for the transverse displacement of the beam is

$$\frac{d^2w}{dx^2} = \frac{M(x)}{EI}. \tag{3.27}$$

Notice that M is a function of x and E and I are constants. Given a beam of length L, the moment produced by the force on the end is simply the force multiplied by the moment arm, $M(x) = (x - L)F$. That takes care of the right side, but what about the boundary conditions on the left side? Clearly, the displacement must be zero, $w(0) = 0$. But it is also clamped, so the slope must also be zero, $dw(0)/dx = 0$. We need to solve

$$\frac{d^2w}{dx^2} = \frac{(x - L)F}{EI} \quad w(0) = \frac{dw(0)}{dx} = 0. \tag{3.28}$$

Figure 3.19 The cantilevered beam is clamped at one end and a vertical force on the other.

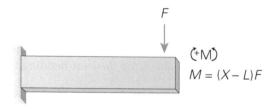

Integrating (Equation 3.28) twice yields

$$w = \frac{F}{EI}\left(\frac{x^3}{6} - L\frac{x^2}{2} + C_1 x + C_2\right).$$

(3.29)

Using our boundary conditions:

$$w(0) = 0 \Rightarrow C_2 = 0$$

$$\frac{dw(0)}{dx} = 0 \Rightarrow C_1 = 0.$$

(3.30)

Therefore, the final solution is

$$w = \frac{F}{EI}\left(\frac{x^3}{6} - L\frac{x^2}{2}\right).$$

(3.31)

Buckling loads can be determined from the beam equations

Another important problem in cytoskeletal mechanics that we can solve with the beam equations is that of buckling. Consider a beam that is loaded axially (Figure 3.20).

This is different from the axially loaded rod, because we are considering the deformation of the beam in the y-direction (not the x-direction) as a function of the axial loading at the tip. Consider a small element of the deformed beam (Figure 3.21).

Notice from this free-body diagram that we have replaced the portion of the beam to the right. To compensate for what we have removed, we need to add an equivalent force and moment, F_s and M_s. Equilibrium requires that the force on the section, F_s, is simply equal to the force on the tip F. Because this force is not applied along the same line as F, it produces a moment, $F_s y$, that must be countered in order for the free body to be in equilibrium. This moment acts across the cut-plane and must be $M_s = -Fy$. Our beam equation is now

$$\frac{d^2w}{dx^2} = \frac{M(x)}{EI} = -\frac{Fw}{EI}.$$

(3.32)

This is a little more challenging to solve than Equation 3.28. We need a function that returns itself when differentiated twice. This suggests

$$w(x) = C_1 \sin(kx) + C_2 \cos(kx).$$

(3.33)

To satisfy the boundary condition, $y(0) = 0$, $C_2 = 0$, and

$$\frac{d^2w}{dx^2} = -\frac{Fw}{EI} = -C_1 k^2 \sin(kx).$$

(3.34)

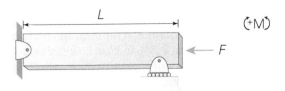

$$w(0) = w(L) = 0$$

Figure 3.20 Beam exposed to an axial load.

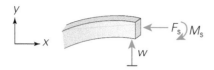

Figure 3.21 Free-body diagram for a small section of the beam shown in Figure 3.20.

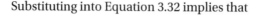

Substituting into Equation 3.32 implies that

$$C_1\left(k^2 - \frac{F}{EI}\right)\sin(kx) = 0.$$

(3.35)

One solution to Equation 3.35 is $k = 0$ or $y(x) = 0$. This trivial solution corresponds to the beam remaining in a straight configuration. We will come back to this possibility in a moment. Alternatively, if we assume that k is not zero, we can see that k must be

$$k = \pm\sqrt{\frac{F}{EI}}.$$

(3.36)

To satisfy our second boundary condition, $y(L) = 0$,

$$C_1\sin\left(L\sqrt{\frac{F}{EI}}\right) = 0 \quad \text{or} \quad L\sqrt{\frac{F}{EI}} = n\pi.$$

(3.37)

Each of the solutions corresponding to different values of the integer, n, is a different *mode* of bending (Figure 3.22).

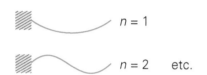

Figure 3.22 Deformation patterns corresponding to the first two bending modes of the beam.

Notice that none of the solutions to Equation 3.36 place a constraint on C_1. The bending mode that corresponds to the lowest possible force is the $n = 1$ mode. The force that corresponds to this mode is

$$F_b = \frac{\pi^2 EI}{L^2},$$

(3.38)

which is known as the *Euler buckling load*. If the applied force is $< F_b$, the beam remains straight and $y(x) = 0$ is the solution. It can also be shown that this solution is stable. In other words, if the beam is deflected slightly away from $y(x) = 0$, it will return to the $y(x) = 0$ configuration. On the other hand, if $F > F_b$, the $y(x) = 0$ solution is not stable, and the beam will buckle, taking on a deformed configuration. This simple analysis to determine the buckling load can be used to find the EI of a biopolymer by measuring the longest length it is able to attain.

Transverse strains occur with axial loading

When we conducted our analysis of axial deformation, we neglected one important thing. When we stretch a rod axially, not all the deformation is in the axial direction. There is also a reduction in the width of the rod in the transverse direction. This effect is quantified by *Poisson's ratio* (Siméon Poisson, 1781–1840), the ratio of the axial and transverse strains, and is typically denoted by ν which is defined as $\nu = \varepsilon_t/\varepsilon_a$.

The general continuum equations can be developed from our simple examples

Now that we have worked a few problems and developed a bit of intuition for mechanics, we are ready to state the general form of these equations. Of course, we are making several simplifying assumptions. Specifically, the type of continuum mechanics we are going to apply is useful for linear elastic materials undergoing small deformations. This is a bit limiting, because the mechanical behavior of the cell is far from a linear–elastic–infinitesimal structure. Also, what is being presented here is not meant to be a complete treatment of this topic. Indeed, elastic small-deformation theory is typically taught as a year-long subject in mechanical engineering departments. The linear elastic infinitesimal theory of plates and shells is typically another entire year-long subject.

We aim to provide you some familiarity with these tools, but the presentation is far from complete.

As we described before, there are three parts to a problem in continuum mechanics: kinematics, equilibrium, and constitutive behavior. So, let us take each one in turn.

Equilibrium implies conditions on stress

In our discussion of stress and equilibrium, let us first introduce a notation system for stress. In our examples, we defined stress conceptually to be the distributed force per unit area through a given test plane. In the example, the location of the test plane was obvious. Now we want to create a general definition and notation system associated with our x-, y-, z-coordinate system. To specify the orientation of the test plane, we will make use of the normal vector to the plane.

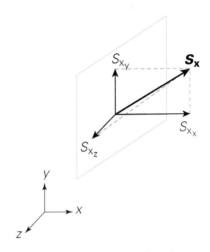

Figure 3.23 Forces on an imaginary cut-plane in the y–z-plane.

Next consider the resultant force denoted by $\mathbf{S}_x$ (Figure 3.23). The x refers to the cut-plane we selected, which has a surface normal aligned in the x-direction. Like any vector, $\mathbf{S}_x$ can be broken into its three components, $S_{x_x}, S_{x_y}, S_{x_z}$. So, for this face, we have three potential stresses, one associated with each of the components of $\mathbf{S}_x$. To express the components of stress, we need a double subscript notation system with the first referring to the direction of the cut-plane normal, and the second referring to the direction of the internal force acting through that plane. Now we have everything we need to define the stress components in this coordinate system. For a cut-plane oriented perpendicular to the x-direction,

$$\sigma_{xx} = \lim_{A \to 0} \frac{S_{x_x}}{A}. \tag{3.39}$$

Similarly, for the other components for the x-cut-plane

$$\tau_{xy} = \lim_{A \to 0} \frac{S_{x_y}}{A} \quad \text{and} \quad \tau_{xz} = \lim_{A \to 0} \frac{S_{x_z}}{A}. \tag{3.40}$$

And for the y-cut-plane

$$\tau_{yx} = \lim_{A \to 0} \frac{S_{y_x}}{A}, \quad \sigma_{yy} = \lim_{A \to 0} \frac{S_{y_y}}{A} \quad \text{and} \quad \tau_{yz} = \lim_{A \to 0} \frac{S_{y_z}}{A} \tag{3.41}$$

and for the z-cut-plane

$$\tau_{zx} = \lim_{A \to 0} \frac{S_{z_x}}{A}, \quad \tau_{zy} = \lim_{A \to 0} \frac{S_{z_y}}{A} \quad \text{and} \quad \sigma_{zz} = \lim_{A \to 0} \frac{S_{z_z}}{A}. \tag{3.42}$$

Now that we have a consistent notation system, let us see what equilibrium can tell us about stress. Remember that the equilibrium condition is a condition on forces. Specifically, for a nonaccelerating body, the forces must sum to zero. Imagine a small piece of material within a general solid body, and consider the forces on the infinitesimal element as shown (Figure 3.24). We assume that the dimensions of the element are dx, dy, and dz respectively. Each face has one normal force and two shear forces. However, because the element is vanishingly small, we can approximate the force on the faces far from the origin as a function of the force on the closer face and its derivative. Taking the first term in a Taylor series,

$$S_{x_x}(x + dx) = S_{x_x}(x) + \frac{dS_{x_x}(x)}{dx}dx. \tag{3.43}$$

Figure 3.24 All the forces on a small element of volume oriented with the coordinate axes for convenience.

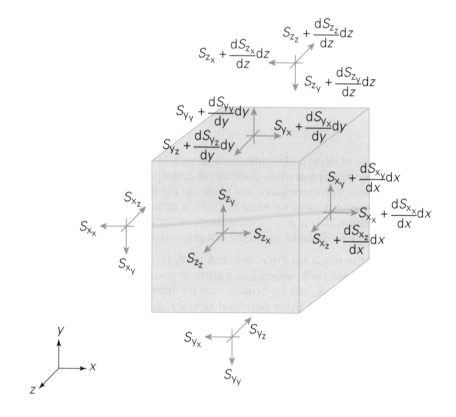

Now we simply sum the forces in the x-, y-, and z-directions in turn:

$$\sum f_x = 0 \Rightarrow \left(-S_{x_x} + S_{x_x} + \frac{dS_{x_x}}{dx}dx\right) + \left(-S_{y_x} + S_{y_x} + \frac{dS_{y_x}}{dy}dy\right) + \left(-S_{z_x} + S_{z_x} + \frac{dS_{z_x}}{dz}dz\right)$$

$$\sum f_y = 0 \Rightarrow \left(-S_{x_y} + S_{x_y} + \frac{dS_{x_y}}{dx}dx\right) + \left(-S_{y_y} + S_{y_y} + \frac{dS_{y_y}}{dy}dy\right) + \left(-S_{z_y} + S_{z_y} + \frac{dS_{z_y}}{dz}dz\right)$$

$$\sum f_z = 0 \Rightarrow \left(-S_{x_z} + S_{x_z} + \frac{dS_{x_z}}{dx}dx\right) + \left(-S_{y_z} + S_{y_z} + \frac{dS_{y_z}}{dy}dy\right) + \left(-S_{z_z} + S_{z_z} + \frac{dS_{z_z}}{dz}dz\right).$$

$$(3.44)$$

Notice that the first two terms in each quantity in parentheses cancel. Next, we can divide each row by the infinitesimal volume $(dx\,dy\,dz)$ and simplify

$$\frac{\left(\dfrac{dS_{x_x}}{dx}\right)}{dydz} + \frac{\left(\dfrac{dS_{y_x}}{dy}\right)}{dxdz} + \frac{\left(\dfrac{dS_{z_x}}{dz}\right)}{dxdy} = 0$$

$$\frac{\left(\dfrac{dS_{x_y}}{dx}\right)}{dydz} + \frac{\left(\dfrac{dS_{y_y}}{dy}\right)}{dxdz} + \frac{\left(\dfrac{dS_{z_y}}{dz}\right)}{dxdy} = 0 \qquad (3.45)$$

$$\frac{\left(\dfrac{dS_{x_z}}{dx}\right)}{dydz} + \frac{\left(\dfrac{dS_{y_z}}{dy}\right)}{dxdz} + \frac{\left(\dfrac{dS_{z_z}}{dz}\right)}{dxdy} = 0.$$

Now, notice that in each term we have a differential area. In each case, this differential area does not depend on the derivative in the numerator. Therefore, we can write

$$\frac{d}{dx}\left(\frac{S_{x_x}}{dydz}\right) + \frac{d}{dy}\left(\frac{S_{y_x}}{dxdz}\right) + \frac{d}{dz}\left(\frac{S_{z_x}}{dydz}\right) = 0$$

$$\frac{d}{dx}\left(\frac{S_{x_y}}{dydz}\right) + \frac{d}{dy}\left(\frac{S_{y_y}}{dxdz}\right) + \frac{d}{dz}\left(\frac{S_{z_y}}{dydz}\right) = 0 \qquad (3.46)$$

$$\frac{d}{dx}\left(\frac{S_{x_z}}{dydz}\right) + \frac{d}{dy}\left(\frac{S_{y_z}}{dxdz}\right) + \frac{d}{dz}\left(\frac{S_{z_z}}{dydz}\right) = 0.$$

The differential area for each term in parentheses is the area normal to the respective force vector component. Therefore, each of these terms is simply our definition of stress. The equilibrium equations then take the following remarkably simple form:

$$\frac{d\sigma_{xx}}{dx} + \frac{d\sigma_{yx}}{dy} + \frac{d\sigma_{zx}}{dz} = 0$$

$$\frac{d\sigma_{xy}}{dx} + \frac{d\sigma_{yy}}{dy} + \frac{d\sigma_{zy}}{dz} = 0 \qquad (3.47)$$

$$\frac{d\sigma_{xz}}{dx} + \frac{d\sigma_{yz}}{dy} + \frac{d\sigma_{zz}}{dz} = 0.$$

Example 3.4: Symmetry of stress

Our free-body analysis of the resultant forces on an infinitesimal element can tell us one other important fact about stress. Notice that the equilibrium equation is the result of requiring the forces to sum to zero. What about the moments? Remember that in a body that is not accelerating, the moments about any arbitrary axis must also sum to zero. In our example, calculate the moments about an axis passing through the center of the element in the x-direction (Figure 3.25).

Summing moments implies

$$\sum M_x = 0 \Rightarrow 2S_{y_z} + 2S_{z_y} = 0$$

or $S_{y_z} = S_{z_y}$. In terms of stress

$$\iint \sigma_{zy} dxdy = \iint \sigma_{yz} dxdy \quad \text{or} \quad \sigma_{zy} = \sigma_{yz}.$$

Likewise for the y- and z-axes, we obtain

$$\sigma_{xy} = \sigma_{yx} \quad \text{and} \quad \sigma_{zx} = \sigma_{xz}.$$

This important property of stress is that it must be symmetric for any nonaccelerating body. Because

there is no additional information in the σ_{yx}, σ_{zx}, and σ_{zy} terms they are typically replaced by σ_{xy}, σ_{xz}, and σ_{yz}, respectively.

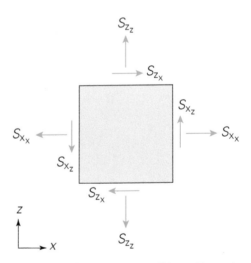

Figure 3.25 The forces on a small two-dimensional element.

Kinematics relate strain to displacement

What is strain? The equations that relate strain and displacement are the kinematic equations and serve as the formal definition of strain. Unlike in our simple examples above, these are general equations that characterize the deformation of normal strain:

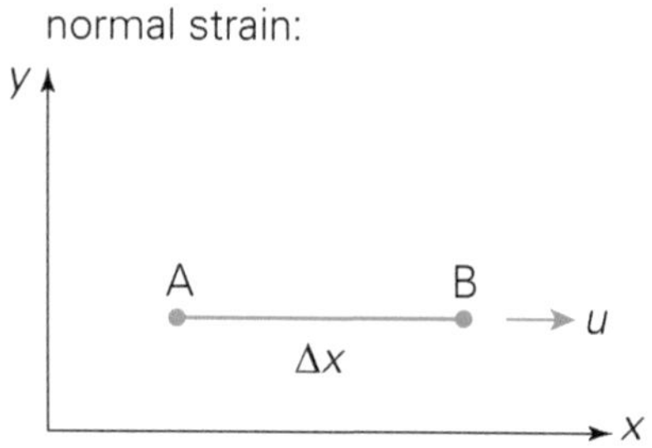

Figure 3.26 An imaginary line in the x-direction embedded in the material being analyzed.

a physical body. We begin our discussion by assuming we have specified a deformable body in a coordinate system specified x, y, and z. There is a deformation at each point in the body given as u, v, w, the displacements in the x-, y-, and z-directions respectively.

Let us start by defining normal strains. Imagine a general body undergoing a small deformation. We can define a test line within the body in the undeformed condition and ask what happens to it during the deformation (Figure 3.26). We are going to consider every possible deformation and orientation of the test line in turn, but, for now, we assume that the test line is oriented along the x-direction. The ends of the test line are denoted by A and B, and are able to displace independently.

Now we need to determine how much the test line is elongated. In general, this will be a function of u, v, and w. However, if the deformations are "small" the extension of the test line is dominated by u. The strain quantity we are defining here is the so-called *infinitesimal* or *small deformation* strain. Technically, it is known as the *Cauchy* strain. The extension of the line is given by $u_A - u_B$ and the average strain in the test line is simply the change in length over the original length $(u_A - u_B)/(x_A - x_B) = \Delta u / \Delta x$. The strain at any given point can now be defined as the average strain in the test line as the length of the line shrinks to zero:

$$\varepsilon = \lim_{\Delta x \to 0} \frac{\Delta u}{\Delta x}, \tag{3.48}$$

which is the definition of the derivative. To specify that we are referring to the displacement in the x-direction of a test line originally oriented along the x-axis, strain components are typically denoted with two subscripts,

$$\varepsilon_{xx} = \frac{du}{dx}. \tag{3.49}$$

Similarly in the y- and z-directions,

$$\varepsilon_{yy} = \frac{dv}{dy}, $$
$$\varepsilon_{zz} = \frac{dw}{dz}. \tag{3.50}$$

Now we consider the shear strain. Perhaps the most logical thing to do would be to define them as similar to normal strains. We could define

$$\varepsilon_{xy} = \frac{dv}{dx}. \tag{3.51}$$

However, there is a problem with this approach. Strain defined in this way is not symmetric, because in general,

$$\frac{dv}{dx} \neq \frac{du}{dy}. \tag{3.52}$$

It will simplify things greatly to define strain as symmetric, and this can be easily achieved by taking our definition of shear strain to be

$$\varepsilon_{xy} = \frac{1}{2}\left(\frac{dv}{dx} + \frac{du}{dy} \right), \tag{3.53}$$

which preserves the symmetry condition, because $\varepsilon_{xy} = \varepsilon_{yx}$, $\varepsilon_{xz} = \varepsilon_{zx}$, and $\varepsilon_{yz} = \varepsilon_{zy}$.

To clearly distinguish normal and shear strains, the symbol γ is sometimes used for the components of the shear strain. The notation γ refers to the engineering strain defined in our pure shear and torsion examples. It differs from the continuum strain ε by a factor of two, that is, $\gamma_{xy} = 2\varepsilon_{xy}$, $\gamma_{xz} = 2\varepsilon_{xz}$, and $\gamma_{yz} = 2\varepsilon_{yz}$.

The constitutive equation or stress–strain relation characterizes the material behavior

How are strain and stress related? As we have noted, the equations that relate stress and strain are the constitutive equations. These are the equations that capture the behavior of the material. They will change depending on the material being considered. Earlier in this chapter, we introduced Hooke's law in the one-dimensional case, $\sigma = E\varepsilon$. Let us see if we can generalize Hooke's law to describe the material behavior of a three-dimensional, isotropic, linearly xy elastic solid. In fact, we have already described three parts of Hooke's law in simple example cases. We described the uniaxial behavior $\sigma = E\varepsilon$, the transverse contraction due to Poisson's ratio $\varepsilon_p = -\nu\varepsilon_a$, and the shear behavior $\tau = G\gamma$. Now, let us see if we can figure out the general case. If we are given a set of six stresses (three normal stresses and three shear stresses) can we figure out the set of six strains? In other words, what are the 36 (6 × 6) coefficients that must multiply the stresses to get the strains. Note, it is easier to determine the coefficients that multiply the stresses to get the strains, because the situations in which the material's properties were defined were given with several components of stress being zero. Because we are describing a linear material, we can simply apply each of these behaviors in our new notation system and add them up. We know from our example and definition of Young's modulus, when we hold all other stresses to be zero and apply a stress in a normal direction, the coefficient multiplying the strain in the same direction is $1/E$. This gives us three of our coefficients.

Now, what can Poisson's ratio tell us? If the stress is applied in the x-direction, it tells us that the normal strain in the y- and z-directions are $-\nu$ times the strain in the x-direction. However, we already know that strain is $1/E$ times the applied stress. This gives us six more coefficients. For shear strain, we know that for an applied pure shear stress, the normal strains and the shear strains in the other directions are all zero. This gives us 24 more coefficients that must be zero. Finally, we know from the relationship between shear strain and shear stress $(1/G)$ the final three coefficients we need. We can write the general form of the equations:

$$\varepsilon_{xx} = \frac{1}{E}\left[\sigma_{xx} - \nu\sigma_{yy} - \nu\sigma_{zz}\right], \quad \gamma_{xy} = (1/G)\tau_{xy}$$

$$\varepsilon_{yy} = \frac{1}{E}\left[-\nu\sigma_{xx} + \sigma_{yy} - \nu\sigma_{zz}\right], \quad \gamma_{xz} = (1/G)\tau_{xz} \qquad (3.54)$$

$$\varepsilon_{zz} = \frac{1}{E}\left[-\nu\sigma_{xx} - \nu\sigma_{yy} + \sigma_{zz}\right], \quad \gamma_{yz} = (1/G)\tau_{yz}.$$

Notice that there are three material constants (E, ν, G) in Equation 3.54, though only two of these are independent constants. The shear modulus can be expressed in terms of Young's modulus and Poisson's ratio (the proof of this is left to you), $G = E/2(1 + \nu)$.

Therefore,

$$\varepsilon_{xx} = \frac{1}{E}\left[\sigma_{xx} - \nu\sigma_{yy} - \nu\sigma_{zz}\right], \quad \tau_{xy} = \frac{2(1 + \nu)}{E}\gamma_{xy}$$

$$\varepsilon_{yy} = \frac{1}{E}\left[-\nu\sigma_{xx} + \sigma_{yy} - \nu\sigma_{zz}\right], \quad \tau_{xz} = \frac{2(1 + \nu)}{E}\gamma_{xz} \qquad (3.55)$$

$$\varepsilon_{zz} = \frac{1}{E}\left[-\nu\sigma_{xx} - \nu\sigma_{yy} + \sigma_{zz}\right], \quad \tau_{yz} = \frac{2(1 + \nu)}{E}\gamma_{yz}.$$

These equations can be inverted to yield the more traditional form of stress as a function of strain:

$$\sigma_{xx} = \frac{E}{(1+v)(1-2v)}\left[(1-v)\varepsilon_{xx} + v\varepsilon_{yy} + v\varepsilon_{zz}\right] \quad \tau_{xy} = \frac{E}{2(1+v)}\gamma_{xy}$$

$$\sigma_{yy} = \frac{E}{(1+v)(1-2v)}\left[v\varepsilon_{xx} + (1-v)\varepsilon_{yy} + v\varepsilon_{zz}\right] \quad \tau_{xz} = \frac{E}{2(1+v)}\gamma_{xz} \qquad (3.56)$$

$$\sigma_{zz} = \frac{E}{(1+v)(1-2v)}\left[v\varepsilon_{xx} + v\varepsilon_{yy} + (1-v)\varepsilon_{zz}\right] \quad \tau_{yz} = \frac{E}{2(1+v)}\gamma_{yz}.$$

Vector notation is a compact way to express equations in continuum mechanics

Writing out in detail and manipulating the continuum equations of solid mechanics can become quite cumbersome, therefore many compact forms of notation have been introduced. One very powerful and extensively used system is known as *Voigt notation* or *vector notation*. In this approach, the components of stress and strain are organized into vectors,

$$\boldsymbol{\sigma} = \begin{Bmatrix} \sigma_{xx} \\ \sigma_{yy} \\ \sigma_{zz} \\ \tau_{xy} \\ \tau_{xz} \\ \tau_{yz} \end{Bmatrix} \quad \text{and} \quad \boldsymbol{\varepsilon} = \begin{Bmatrix} \varepsilon_{xx} \\ \varepsilon_{yy} \\ \varepsilon_{zz} \\ \gamma_{xy} \\ \gamma_{xz} \\ \gamma_{yz} \end{Bmatrix}. \qquad (3.57)$$

Using this notation the stress–strain relationship Equation (3.55) is simply

$$\begin{Bmatrix} \sigma_{xx} \\ \sigma_{yy} \\ \sigma_{zz} \\ \tau_{xy} \\ \tau_{xz} \\ \tau_{yz} \end{Bmatrix} = \frac{E}{(1+v)(1-2v)}$$

$$\begin{bmatrix} (1-v) & v & v & 0 & 0 & 0 \\ v & (1-v) & v & 0 & 0 & 0 \\ v & v & (1-v) & 0 & 0 & 0 \\ 0 & 0 & 0 & \dfrac{(1-2v)}{2} & 0 & 0 \\ 0 & 0 & 0 & 0 & \dfrac{(1-2v)}{2} & 0 \\ 0 & 0 & 0 & 0 & 0 & \dfrac{(1-2v)}{2} \end{bmatrix} \begin{Bmatrix} \varepsilon_{xx} \\ \varepsilon_{yy} \\ \varepsilon_{zz} \\ \gamma_{xy} \\ \gamma_{xz} \\ \gamma_{yz} \end{Bmatrix} \quad \text{or} \quad \boldsymbol{\sigma} = \boldsymbol{C}\boldsymbol{\varepsilon}$$

$$(3.58)$$

and

$$
\begin{Bmatrix} \varepsilon_{xx} \\ \varepsilon_{yy} \\ \varepsilon_{zz} \\ \gamma_{xy} \\ \gamma_{xz} \\ \gamma_{yz} \end{Bmatrix} =
\begin{bmatrix}
\dfrac{1}{E} & \dfrac{-v}{E} & \dfrac{-v}{E} & 0 & 0 & 0 \\[2mm]
\dfrac{-v}{E} & \dfrac{1}{E} & \dfrac{-v}{E} & 0 & 0 & 0 \\[2mm]
\dfrac{-v}{E} & \dfrac{-v}{E} & \dfrac{1}{E} & 0 & 0 & 0 \\[2mm]
0 & 0 & 0 & \dfrac{2(1+v)}{E} & 0 & 0 \\[2mm]
0 & 0 & 0 & 0 & \dfrac{2(1+v)}{E} & 0 \\[2mm]
0 & 0 & 0 & 0 & 0 & \dfrac{2(1+v)}{E}
\end{bmatrix}
\begin{Bmatrix} \sigma_{xx} \\ \sigma_{yy} \\ \sigma_{zz} \\ \tau_{xy} \\ \tau_{xz} \\ \tau_{yz} \end{Bmatrix}
\quad \text{or} \quad \boldsymbol{\varepsilon} = \boldsymbol{D\sigma}
$$

$$(3.59)$$

Nota Bene: Lamé constants allow a compact form of Hooke's law

Equation 3.58 can be expressed in a more compact form.

$$
\begin{Bmatrix} \sigma_{xx} \\ \sigma_{yy} \\ \sigma_{zz} \\ \tau_{xy} \\ \tau_{xz} \\ \tau_{yz} \end{Bmatrix} =
\begin{bmatrix}
\lambda + 2\mu & \lambda & \lambda & 0 & 0 & 0 \\
\lambda & \lambda + 2\mu & \lambda & 0 & 0 & 0 \\
\lambda & \lambda & \lambda + 2\mu & 0 & 0 & 0 \\
0 & 0 & 0 & 2\mu & 0 & 0 \\
0 & 0 & 0 & 0 & 2\mu & 0 \\
0 & 0 & 0 & 0 & 0 & 2\mu
\end{bmatrix}
\begin{Bmatrix} \varepsilon_{xx} \\ \varepsilon_{yy} \\ \varepsilon_{zz} \\ \varepsilon_{xy} \\ \varepsilon_{xz} \\ \varepsilon_{yz} \end{Bmatrix},
$$

where

$$\lambda = \frac{Ev}{(1+v)(1-2v)}$$

and

$$\mu = \frac{E}{2(1+v)},$$

note that μ and λ are known as the *Lamé constants*.

Advanced Material: Coordinate rotations

It is important to note that the vector notation we use is a notational system only. We arrange the components of stress and strain in a vector to make them easy to manipulate. However, stress and strain are not mathematically vectors. As you remember, a vector is a quantity that has both magnitude and direction, like force, deformation, or velocity. An implication of this is that a vector maintains its magnitude and direction from one coordinate system to another. This implies that the components of a vector must behave in a very specific way in terms of how they relate to one another when expressed in different coordinate systems. Assume that we have two (orthogonal) coordinate systems, one specified by the vectors $\boldsymbol{x}$, $\boldsymbol{y}$, and $\boldsymbol{z}$ and another specified by another set of vectors $\boldsymbol{x'}$, $\boldsymbol{y'}$, and $\boldsymbol{z'}$ ($\boldsymbol{x}$, $\boldsymbol{y}$, $\boldsymbol{z}$, and $\boldsymbol{x'}$, $\boldsymbol{y'}$, and $\boldsymbol{z'}$ are known as basis vectors). Also, define the angle between any of these vectors to be $\theta_{xx'}$, $\theta_{xy'}$, etc. A rotation matrix $\boldsymbol{Q}$ can be defined such that

$$
\boldsymbol{Q} = \begin{bmatrix}
\cos(\theta_{xx'}) & \cos(\theta_{xy'}) & \cos(\theta_{xz'}) \\
\cos(\theta_{yx'}) & \cos(\theta_{yy'}) & \cos(\theta_{yz'}) \\
\cos(\theta_{zx'}) & \cos(\theta_{zy'}) & \cos(\theta_{zz'})
\end{bmatrix}.
$$

Then any vector can be expressed in the new coordinate system by multiplying the vector expressed in the old coordinate system by Q. If P is a vector in the x,y,z-system then P' in the x',y',z'-system is given by $P' = QP$. Obeying this coordinate transformation rule is the formal definition of a vector. However, our stress and strain "vectors" do not obey this rule. In fact, stress and strain are more general mathematical quantities known as *tensors*. The appropriate transformation rule for tensors of this type requires the terms to be written in matrix notation. For example,

$$\sigma = \begin{bmatrix} \sigma_{xx} & \sigma_{xy} & \sigma_{xz} \\ \sigma_{yx} & \sigma_{yy} & \sigma_{yz} \\ \sigma_{zx} & \sigma_{zy} & \sigma_{zz} \end{bmatrix} \quad \text{and} \quad \sigma' = Q^T\sigma Q.$$

In this text we will not be using tensors or tensor mathematics. However, some of the quantities are in fact tensors, like stress and strain. In Section 3.3 we define some new quantities and refer to them by their names from continuum mechanics, so that you can follow up with additional reading. However, we avoid the use of any tensor mathematics, and for our purposes they can be manipulated like matrices.

Stress and strain can be expressed as matrices

Using vector notation for stress and strain makes it easier to express the three-dimensional Hooke's law as a matrix equation. Notice that each component of stress and strain has two subscripts referring to coordinate directions. Therefore, it might make more sense to depict stress and strain as matrices,

$$\sigma = \begin{bmatrix} \sigma_{xx} & \sigma_{xy} & \sigma_{xz} \\ \sigma_{yx} & \sigma_{yy} & \sigma_{yz} \\ \sigma_{zx} & \sigma_{zy} & \sigma_{zz} \end{bmatrix} \quad \text{and} \quad \varepsilon = \begin{bmatrix} \varepsilon_{xx} & \varepsilon_{xy} & \varepsilon_{xz} \\ \varepsilon_{yx} & \varepsilon_{yy} & \varepsilon_{yz} \\ \varepsilon_{zx} & \varepsilon_{zy} & \varepsilon_{zz} \end{bmatrix}. \tag{3.60}$$

Known as *matrix notation*, this approach is useful for mathematical manipulations of stress and strain and proving new relationships. Bear in mind that the same amount of information is contained in each notation (here, there are nine components each for stress and strain, but three of them are redundant because of symmetry). To illustrate one advantage of using this notation, we next derive what are called the principal stresses and strains.

In the principal directions shear stress is zero

Recall from linear algebra that a matrix that is symmetric and positive definite has an associated eigenvector problem. Specifically, there is a vector that, when multiplied by the matrix in question, yields the same vector multiplied by a scalar,

$$Av = \varphi v, \tag{3.61}$$

where v is the *eigenvector*, and φ is the *eigenvalue*. In general, for a 3×3 matrix, there are three eigenvectors/values. We also know from linear algebra that the principal stresses are solutions of the so-called *characteristic equation*

$$|A - \varphi I| = \det \begin{bmatrix} A_{11} - \varphi & A_{12} & A_{13} \\ A_{21} & A_{22} - \varphi & A_{23} \\ A_{31} & A_{32} & A_{33} - \varphi \end{bmatrix}, \tag{3.62}$$

Nota Bene

Where does the word "eigenvalue" come from? Eigen is a German word that means "own," as in "inherent to."

where I is the identity matrix (one of the diagonal zeros elsewhere). This is actually a third-order polynomial equation. The three eigenvalues of stress are known as principal stresses, and we will denote them φ_1, φ_2, and φ_3. Likewise, the principal strains are denoted, ε_1, ε_2, and ε_3. For isotropic materials, the eigenvectors for the stress and strain are aligned with each other and are known as the *principal directions* $\mathbf{v}_1$, $\mathbf{v}_2$, and $\mathbf{v}_3$. We call these principal stresses and strains because they are *invariant* with respect to the coordinate system. That is, regardless of how you arrange the coordinate axes, the principal stresses and strains for a given deformation will always be the same. Similarly, the principal directions will not change. If you were to grab a square-shaped object at opposite corners and pull, the material will deform into a kite-shape. One of the principal directions is along the long diagonal of the kite. This remains true regardless of which coordinate system is used.

The principal directions possess some particularly useful properties. They are mutually perpendicular or *orthogonal* to each other. This means that they can serve as a coordinate system, and it is possible to rotate our coordinate system from our original x-, y-, and z-axes into v_1, v_2, and v_3. When we do this, the stress and strain matrices also change to match the new coordinate system, with the result that the off-diagonal components of the stress matrix vanish, and the diagonal components are the principal stresses. In other words, using the principal directions as our coordinate system, σ becomes

$$\sigma = \begin{bmatrix} \sigma_1 & 0 & 0 \\ 0 & \sigma_2 & 0 \\ 0 & 0 & \sigma_3 \end{bmatrix}. \tag{3.63}$$

This remarkable property means that no matter how complex the state of stress, there is always a coordinate system in which all the shear stresses vanish and only normal stresses remain. Indeed, specifying the principal stresses and principal directions is sufficient to fully define the state of stress. The same holds true for strain.

Example 3.5: Principal strains

Suppose you are analyzing the deformation of a flexible membrane on which you plan to seed cells to study their reaction to substrate stretch. You make three marks that have initial coordinates (0, 0), (1, 0), and (0, 1). After you apply the deformation, they move to (0, 0), (1.015, 0.005), and (0.005, 1.015), respectively. What would the strains be? What are the principal strains and principal directions?

We start by determining our strains. We have placed our marks purposely to make this very easy. We have one test line ranging from (0, 0) to (1, 0) and the other from (0, 0) to (0, 1). Both of them are originally of length one. Consider the normal strains first. Our original line in the x-direction extends by 1.5%, so $\varepsilon_{xx} = 0.015$. It is also displaced by 0.5% in the y-direction, so $\varepsilon_{xy} = 0.005$. Likewise, $\varepsilon_{yy} = 0.015$ and $\varepsilon_{yx} = 0.005$.

$$\varepsilon = \begin{bmatrix} 0.015 & 0.005 \\ 0.005 & 0.015 \end{bmatrix}. \tag{3.64}$$

To find the principal strains, we need to find the eigenvalues of ε.

$$\varepsilon v_a = \varepsilon_a v_a \tag{3.65}$$

is the characteristic equation relating our strain ε, the principal strains ε_a, and the principal directions v_a. Noticing that we can rearrange the equation,

$$\left(\varepsilon v_a - I\varepsilon_a v_a \right) = 0. \tag{3.66}$$

We can then write,

$$\left(\varepsilon - I\varepsilon_a \right) v_a = 0. \tag{3.67}$$

This must be true for non-zero v_a (our eigenvectors) and therefore the quantity in parentheses must be non-invertible. That means the determinant of $(\varepsilon - I\varepsilon_a) = 0$, or

$$\begin{vmatrix} 0.015 - \varepsilon_a & 0.005 \\ 0.005 & 0.015 - \varepsilon_a \end{vmatrix} = 0. \tag{3.68}$$

Some algebra leads to

$$0.000225 - 0.03\varepsilon_a + \varepsilon_a^2 - 0.000025$$
$$= \varepsilon_a^2 - 0.03\varepsilon_a + 0.0002 = 0. \tag{3.69}$$

The two roots of this equation are $\varepsilon_1 = 0.02$ and $\varepsilon_2 = 0.01$. Physically this means that the two principal strains are

2% and 1%. What directions are these strains in? To answer this we need the principal directions. We simply substitute back into the characteristic equation

$$\boldsymbol{\varepsilon}\boldsymbol{v}_a = \varepsilon_a\boldsymbol{v}_a$$

for each ε_a in turn. We find that the eigenvectors $(1, 1)$ (corresponding to ε_1) and $(1, -1)$ (corresponding to ε_2). Note that when you are solving for the eigenvectors, any eigenvector can be multiplied by a scalar, and the resulting vector is also an eigenvector. So, you will either need to constrain the length or to set a component to an arbitrary value (we use ones for clarity). (Alternatively, you can construct eigenvectors of unit length.)

Finally, notice that the eigenvectors are oriented 45 degrees from our original coordinate system. We can define a new coordinate system rotated by 45 degrees counterclockwise such that

$$x' = \frac{1}{\sqrt{2}}\begin{Bmatrix} 1 \\ 1 \end{Bmatrix}, \ y' = \frac{1}{\sqrt{2}}\begin{Bmatrix} 1 \\ -1 \end{Bmatrix}. \tag{3.70}$$

The rotation matrix (described in Advanced Material, Coordinate rotations p. 73) between the two coordinate systems would be

$$Q = \begin{bmatrix} \cos\left(\dfrac{-\pi}{4}\right) & \cos\left(\dfrac{-3\pi}{4}\right) \\ \cos\left(\dfrac{\pi}{4}\right) & \cos\left(\dfrac{-\pi}{4}\right) \end{bmatrix} = \begin{bmatrix} \dfrac{\sqrt{2}}{2} & -\dfrac{\sqrt{2}}{2} \\ \dfrac{\sqrt{2}}{2} & \dfrac{\sqrt{2}}{2} \end{bmatrix}. \tag{3.71}$$

In this new coordinate system

$$\boldsymbol{\varepsilon} = \begin{bmatrix} 0.02 & 0.0 \\ 0.0 & 0.01 \end{bmatrix}. \tag{3.72}$$

An illustration of the principal strains can be obtained with the strain ellipse (Figure 3.27). It also shows that the maximum and minimum normal strains occur in the principal directions.

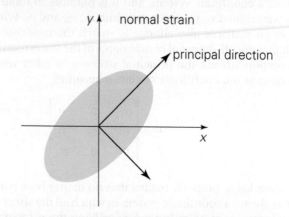

Figure 3.27 The strain ellipse is a depiction of the normal strain as a function of direction. In this example, the larger of the two principal strains is in the (1,1) direction and the smaller is in the (1, −1).

3.3 LARGE DEFORMATION MECHANICS

In mechanics, if the deformations we are considering are too large to be approximated as infinitesimal, they fall in the realm of *large deformation* or *finite deformation* mechanics. Cellular deformations can be quite large (exceeding 5–10%). We are going to discuss how such deformation can be quantified (kinematics), forgoing discussion of the other two components of mechanics—equilibrium (stress) and constitutive models.

The deformation gradient tensor describes large deformations

Consider a general object in its initial undeformed state (Figure 3.28). Every point in the object can be described by a vector A. The object undergoes a deformation such that it takes on a new configuration. In the new configuration it is now described by a vector a.

This is a general description and can describe any deformation as long as certain simple requirements are met, such as no holes or cracks and the deformation being smooth. Consider a small line in the undeformed configuration, dA, which

Nota Bene

A note about notation. Typically, the undeformed vector is denoted by (capital) **X** and the deformed vector by (lowercase) **x**. We have changed notation here to distinguish "x" from the x-coordinate.

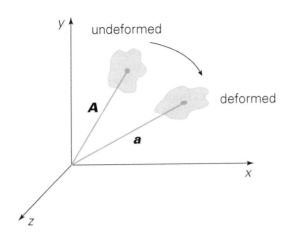

Figure 3.28 A general object can be defined in its undeformed configuration by a set of vectors A and in its deformed configuration by a set of vectors a.

becomes a small line, $\mathrm{d}a$, in the deformed configuration. We can relate these two small elements by approximating the deformation with a Taylor series expansion and neglecting higher-order terms with the matrix equation:

$$\mathrm{d}a \approx F\mathrm{d}A, \tag{3.73}$$

where F is known as the deformation gradient and is given by

$$F = \frac{\partial a}{\partial A} = \begin{bmatrix} \dfrac{\partial a_x}{\partial A_x} & \dfrac{\partial a_x}{\partial A_y} & \dfrac{\partial a_x}{\partial A_z} \\[2mm] \dfrac{\partial a_y}{\partial A_x} & \dfrac{\partial a_y}{\partial A_y} & \dfrac{\partial a_y}{\partial A_z} \\[2mm] \dfrac{\partial a_z}{\partial A_x} & \dfrac{\partial a_z}{\partial A_y} & \dfrac{\partial a_z}{\partial A_z} \end{bmatrix}. \tag{3.74}$$

F is an incredibly useful quantity and forms the basis of almost all of large deformation kinematics. Here we develop a few particularly relevant applications.

Stretch is another geometrical measure of deformation

Although strain is a natural way to quantify deformation, it is far from the only one. Indeed several other deformation measures exist depending on the application. Another popular measure of deformation is *stretch*. Conceptually, stretch is simple. If something is pulled twice as long, we say it has a stretch of two. If it does not change, it has a stretch of one. Stretch is quantified by the *stretch ratio*, λ, the ratio of the deformed length divided by the original length, in contrast to strain, which is the change in length over original length. In our infinitesimal deformation notation

$$\varepsilon = \frac{\Delta L}{L}, \ \lambda = \frac{L + \Delta L}{L}. \tag{3.75}$$

In our large deformation notation

$$\lambda = \frac{|\mathrm{d}a|}{|\mathrm{d}A|}. \tag{3.76}$$

Nota Bene

Vernacular of large deformation mechanics. Much of the notation and terminology of large deformation mechanics was introduced by Clifford Truesdell and Walter Noll (who had been Truesdell's graduate student) in the classic 1965 text *The Non-Linear Field Theories of Mechanics*. More recently, Noll has suggested that the term deformation gradient, which they introduced, may be misleading and should be replaced with *transplacement gradient*, which was adopted by Truesdell and Noll in subsequent work. However, deformation gradient is now so deeply embedded in the literature that it seems unlikely to replaced.

Unlike strain, which can be normal or shear, the stretch ratio is always a measure of the normal deformation of a prescribed differential line element. The stretch ratio is related to the *extension ratio*, given by $\lambda - 1$, which is similar, but not identical, to our concept of strain. In particular, stretch is measured along a line segment of interest and follows the line segment as it rotates and translates. In Example 3.5, the test line that originally goes from $(0, 0)$ to $(1, 0)$ is deformed to a new line segment from $(0, 0)$ to $(1.015, 0.005)$. We decomposed the strain of this segment into the ε_{xx} and ε_{xy} components. For stretch, we do not perform this decomposition. Instead, we work with the entire vector and create F to describe how the line segment changes. For this particular deformation, the stretch ends up being just above 1.015, so there is only a small difference between stretch and strain measures. However, for larger deformations, the two measures can be divergent. If instead of $(1.015, 0.005)$ the test line ends up at $(1.5, 0.5)$ then ε_{xx} is 0.5 and $\varepsilon_{xy} = 0.5$, but stretch is 1.58 and the extension ratio is 0.58. So although many of the mathematical manipulations will be similar (such as determining principal directions), these quantities are distinct.

Stretch ratios can be defined for any particular line segment. One might choose dA or da to be aligned with one of the coordinate axes. One particularly useful approach is to use line segments that are aligned with the principal directions of stretch. Neglecting rigid-body motions, the deformation in the principal directions is entirely normal (that is, there is no shear). The stretches are known as the principal stretches $(\lambda_1, \lambda_2, \lambda_3)$. If the coordinate system is aligned with the principal directions, F takes a particular form with the principal stretches on the diagonals and zeros elsewhere,

$$F = \frac{\partial a}{\partial A} = \begin{bmatrix} \lambda_1 & 0 & 0 \\ 0 & \lambda_2 & 0 \\ 0 & 0 & \lambda_3 \end{bmatrix}. \tag{3.77}$$

The principal stretches can capture the full deformation, and formulations of mechanical behavior in terms of principal stretches are particularly common in large deformation constitutive modeling because of their simplicity.

Large deformation strain can be defined in terms of the deformation gradient

The concept of strain is a measure of how much distortion an object experiences. In large deformation mechanics, there are actually several quantities that are acceptable measures of strain. However, we will only define one, the Green-Lagrange strain, because it can be thought of as the small deformation strain with an additional term. In terms of F, The Green–Lagrange strain, E, is defined as

$$E = \frac{1}{2}\left(F^{\mathrm{T}}F - I\right). \tag{3.78}$$

Nota Bene

Right Cauchy–Green deformation tensor. In continuum mechanics, C is called the *right Cauchy–Green deformation tensor*. It is called "right" not because it is correct, but because it is related to the deformation matrix of a polar decomposition when the deformation matrix is on the right of the rotation matrix.

The term $F^{\mathrm{T}}F$ is a special quantity denoted as $C = F^{\mathrm{T}}F$.

It is special because it is related to local stretch. In terms of our small test lines, it can be shown that $da^2 = (dA)C(dA)$. It also has interesting properties: its trace (sum of the diagonals) is equal to the sum of the squares of the principal stretches, and its determinant is equal to the product of the squares of the principal stretches. On an intuitive level, the Green-Lagrange strain is a measure of how much the local stretches, C, differ from one. Written out in detail in terms of deformation the Green-Lagrange strain is

$$E_{xx} = \quad \frac{du}{dx} \quad +\frac{1}{2}\left(\left(\frac{du}{dx}\right)^2 + \left(\frac{dv}{dx}\right)^2 + \left(\frac{dw}{dx}\right)^2\right)$$

$$E_{yy} = \quad \frac{dv}{dy} \quad +\frac{1}{2}\left(\left(\frac{du}{dy}\right)^2 + \left(\frac{dv}{dy}\right)^2 + \left(\frac{dw}{dy}\right)^2\right)$$

$$E_{zz} = \quad \frac{dw}{dz} \quad +\frac{1}{2}\left(\left(\frac{du}{dz}\right)^2 + \left(\frac{dv}{dz}\right)^2 + \left(\frac{dw}{dz}\right)^2\right).$$

$$E_{xy} = E_{yx} \quad \frac{1}{2}\left(\frac{du}{dy} + \frac{dv}{dx}\right) \quad +\frac{1}{2}\left(\frac{du}{dx}\frac{du}{dy} + \frac{dv}{dx}\frac{dv}{dy} + \frac{dw}{dx}\frac{dw}{dy}\right)$$

$$E_{xz} = E_{zx} \quad \frac{1}{2}\left(\frac{du}{dz} + \frac{dw}{dx}\right) \quad +\frac{1}{2}\left(\frac{du}{dx}\frac{du}{dz} + \frac{dv}{dx}\frac{dv}{dz} + \frac{dw}{dx}\frac{dw}{dz}\right)$$

$$E_{yz} = E_{yx} \quad \frac{1}{2}\left(\frac{dv}{dz} + \frac{dw}{dy}\right) \quad +\frac{1}{2}\left(\frac{du}{dy}\frac{du}{dz} + \frac{dv}{dy}\frac{dv}{dz} + \frac{dw}{dy}\frac{dw}{dz}\right).$$

(3.79)

Notice that each component of strain contains a first term corresponding to small deformation strains and an additional higher-order term that captures nonlinear effects of large deformation. With linear homogeneous deformations, the Green–Lagrange strain and the small strain measures are the same, because these higher-order terms vanish.

Example 3.6: Large deformation principal strains

Consider our stretching example from Example 3.5, but this time we apply much larger stretches. Again, our three markers have initial coordinates $(0, 0)$, $(1, 0)$, and $(0, 1)$. After the deformation is applied, the markers are at $(0, 0)$, $(2.5, 0.5)$, and $(0.5, 2.5)$, respectively. Determine the principal strains and principal directions of the deformation.

This is similar to our previous principal strain example, except that the deformations are not very small. Our deformation may be represented by Figure 3.29 (note that in the previous example, the changes would be so small as to be invisible if drawn). The black lines represent the undeformed state and the blue lines represent the deformed state. You can see that our test lines are quite a bit different, so we need to use the finite deformation approach.

Let us build our F first. From the data, we have the positions of two lines in both the undeformed and deformed configurations,

$$dA_1 = \begin{Bmatrix} 1 \\ 0 \end{Bmatrix},\ dA_2 = \begin{Bmatrix} 0 \\ 1 \end{Bmatrix},\ da_1 = \begin{Bmatrix} 2.5 \\ 0.5 \end{Bmatrix},\ da_2 = \begin{Bmatrix} 0.5 \\ 2.5 \end{Bmatrix} \quad (3.80)$$

We also know that $da = F dA$, so

$$\begin{Bmatrix} 2.5 \\ 0.5 \end{Bmatrix} = \begin{bmatrix} F_{11} & F_{12} \\ F_{21} & F_{22} \end{bmatrix} \begin{Bmatrix} 1 \\ 0 \end{Bmatrix},\ \begin{Bmatrix} 0.5 \\ 2.5 \end{Bmatrix} = \begin{bmatrix} F_{11} & F_{12} \\ F_{21} & F_{22} \end{bmatrix} \begin{Bmatrix} 0 \\ 1 \end{Bmatrix}$$

(3.81)

$$\Rightarrow F = \begin{bmatrix} 2.5 & 0.5 \\ 0.5 & 2.5 \end{bmatrix}.$$

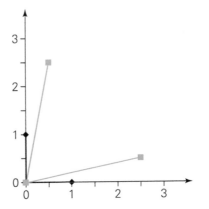

Figure 3.29 Two lines in the original configuration (black) and deformed configuration (blue).

With **F** in hand, we can compute the Green–Lagrange strain, **E**,

$$E = \frac{1}{2}\left(F^{\mathrm{T}}F - I\right) = \begin{bmatrix} 2.75 & 1.25 \\ 1.25 & 2.75 \end{bmatrix}. \tag{3.82}$$

To find the principal strains, we need to find the eigenvalues of **E**. Using the same approach as in Example 3.5, we obtain principal strains of 1.5 and 4.0 and principal directions of (1, –1) and (1, 1). The largest of the two

strains is aligned 45 degrees to the original coordinate system. In that direction the strain is 400% and in the perpendicular direction it would be 150%. If we were to rotate our coordinate system 45 degrees counterclockwise, the new **E** would be

$$E = \begin{bmatrix} 4.0 & 0 \\ 0 & 1.5 \end{bmatrix}. \tag{3.83}$$

Notice that the shear strains are zero.

The deformation gradient can be decomposed into rotation and stretch components

F can be expressed by *polar decomposition* as

$$F = RU, \tag{3.84}$$

where **R** is what is an *orthonormal* matrix. That is, it has special properties such that it is *orthogonal*, meaning that all of its rows or columns are pairwise orthogonal (dot products are one if self-multiplied, zero otherwise) and it is *normal*, or its determinant is 1. The properties of **R** are exactly those required for it to be a rotation matrix like **Q** discussed in the Advanced Materials on coordinate rotation (Section 3.2). **R** represents a rigid-body rotation of the object. This part of **F** does not lead to distortions or strain, but is an integral part of measurements made if the principal stretch directions are not known ahead of time. The remaining term, **U**, captures the distortion of the object.

U sits on the right side of the decomposition of **F**. You can alternatively decompose **F** into **VR**, where the stretch component **V** sits on the left side. **U** and **V** are not the same, because the order of matrix multiplication is important (it is not commutative), although the rotations are the same.

In Figure 3.30, we see that a square can be rotated by 45 degrees counterclockwise first, then stretched to expand one side to twice the original length. Alternatively, we can stretch the square to twice the original length, and then rotate it by 45 degrees counterclockwise. The rotations are identical. The material stretch is the same, but the stretch tensors are different, because in the rotate-first case, the square is being stretched along a 45-degree line, whereas the stretch-first case stretches the square along the horizontal axis.

Note that if we include translation in this analysis, we will have a general mathematical transformation for any mechanical deformation—a general deformation can be expressed as a single translation, a single rotation, and a single distortion.

Figure 3.30 Deformation of an object can be decomposed into a stretch and rotation, or a rotation and stretch. Whereas rotation is the same in both cases, the stretch is not.

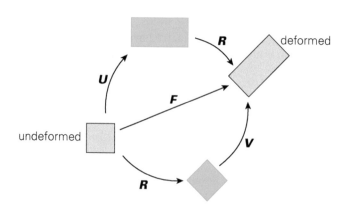

Example 3.7: Rotation can be extracted from F

In our previous examples, we mostly focused on deformations without really understanding what physical process was being represented by F. Now that we know that F can be decomposed, let us see how R and U result in a given deformation. Consider a square that gets deformed as in Figure 3.31, where the black lines are the undeformed configuration and the blue lines are the deformed configuration.

Say we measure the deformation such that A deforms to a as

$$A_1 = \begin{Bmatrix} 1 \\ 0 \end{Bmatrix}, \quad A_2 = \begin{Bmatrix} 0 \\ 1 \end{Bmatrix}, \quad a_1 = \begin{Bmatrix} 1/2 \\ -\sqrt{3}/2 \end{Bmatrix}, \quad a_2 = \begin{Bmatrix} \sqrt{3} \\ 1 \end{Bmatrix}. \quad (3.85)$$

From this we can construct F as in Example 3.6,

$$F = \begin{bmatrix} 1/2 & \sqrt{3} \\ -\sqrt{3}/2 & 1 \end{bmatrix}. \quad (3.86)$$

In some cases, we can extract rotation by finding U first. To obtain U, we note that $C = F^T F$ is also U^2, because R^T is the inverse of R (this characteristic of R is true for rotation matrices in general; you may show it for yourself using the Q in Advanced Material, Coordinate rotations p. 75). If we're lucky, we can then obtain U. Therefore,

$$F^T F = \begin{bmatrix} 1 & 0 \\ 0 & 4 \end{bmatrix}, \quad (3.87)$$

from which U is easily derived (noting that stretches cannot be negative):

$$U = \begin{bmatrix} 1 & 0 \\ 0 & 2 \end{bmatrix}. \quad (3.88)$$

This means that we can now solve for R, given that $F = RU$,

$$F = \begin{bmatrix} 1/2 & \sqrt{3} \\ -\sqrt{3}/2 & 1 \end{bmatrix} = \begin{bmatrix} \cos\theta & -\sin\theta \\ \sin\theta & \cos\theta \end{bmatrix} \begin{bmatrix} 1 & 0 \\ 0 & 2 \end{bmatrix}. \quad (3.89)$$

We obtain $\cos\theta = 1/2$ and θ is either 60 or –60 degrees ($\pm\pi/3$ radians). The sin term is negative, meaning that θ is –60 degrees. Note that $\cos\theta_i z = 0$ for $i = x, x', y$ or y' because the z-axis is not moved in this two-dimensional problem. Also note that $\cos\theta_{xy'} = \sin\theta$ because the x- and y-axes are 90 degrees apart and the only rotation is about the z-axis.

If we were to use the previous approach of taking eigenvalues, we would obtain the principal stretches as 1 and 2. We also would obtain the principal stretch directions, based on U, as 0 and 90 degrees. But it is important to note that these principal stretch directions are not in the coordinate system we started out with—these are for U, where the rotation is already removed. Example 3.6 gives us the principal directions in the coordinate system we started out with, because there's no rotation in that deformation.

So, how much is the rotation? Because we derived R, we know that the object being deformed is rotated clockwise by 60 degrees. So now we know how the deformation occurs; we stretch the square so it doubles in length in one direction and remains unstretched in the other, and then rotate it by 60 degrees clockwise. The principal directions are along –60 degrees and 30 degrees (0 and 90 degrees, rotated clockwise by 60 degrees).

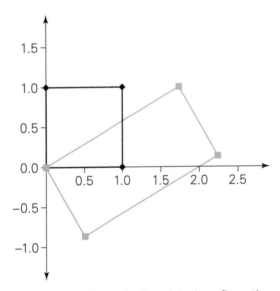

Figure 3.31 Two boxes in the original configuration (black) and deformed configuration (blue).

3.4 STRUCTURAL ELEMENTS ARE DEFINED BY THEIR SHAPE AND LOADING MODE

These equations of solid continuum mechanics are rather general and, as you might agree, somewhat complicated. However, many structural elements are much more simple. They have been simplified through assumptions of the dimensionality of the element and the dimensionality of the loading that it is exposed to. In our review of solid mechanics, we have already employed several examples that we used to develop our intuition (the axial rod, torsional rod, and beam). We have also seen how these simplifying assumptions have led to specialized governing equations between force and displacement. These tend to be much simpler to solve than the general continuum equations. Here is a list of *structural elements,* each with its set of assumptions and the order of the resulting governing equations:

Truss: One-dimensional straight element exposed to axial loading. Axial deformation only. Second order.

Beam: One-dimensional straight element exposed to transverse loading. Bending deformation only. Fourth order.

Wall: Two-dimensional flat structure exposed to in-plane loading and experiencing in-plane deformation. Second order.

Plate: Two-dimensional flat structure exposed to transverse loading. Bending deformation only. Fourth order.

Membrane: Two-dimensional curved structure exposed to in-plane loading. Deformation is only considered within the plane of the membrane. Second order.

Shell: Two-dimensional curved structure exposed to transverse loading. Bending deformation only. Fourth order.

Key Concepts

- The forces and moments acting on a body that is not accelerating must sum to zero. The forces and moments acting across any closed boundary must also sum to zero. This gives rise to the free-body diagram.
- For deformable bodies, stress is a normalized measure of force analogous to pressure.
- Strain is a normalized deformation.
- Stress and strain can be normal or shear.
- Continuum mechanics gives rise to three fundamental relationships: kinematics, constitutive behavior, and equilibrium.
- Euler–Bernoulli beam theory describes small transverse deformations of slender bodies and accounts only for lengthwise stresses.

- Cytoskeletal proteins are relatively stiff in axial tension, but flexible in torsion and bending.
- A column will collapse when it is exposed to axial compression that exceeds the Euler buckling load.
- In large deformation mechanics, the strains are not assumed to be infinitesimal.
- The deformation gradient is the fundamental large deformation kinematic measure, and captures a body's rotation and deformation.
- The Green–Lagrange strain is a common large deformation strain measure, although others exist as well.

Problems

1. Let V_1 be the volume of a hollow cylinder of outer radius R_0 and inner radius R_i (do not use the thin-shell approximation) consisting of material with shear modulus G. This cylinder is fixed at one end; a moment of M is applied to the other end, resulting in an angular twist of θ. Let V_2 be the volume of a solid cylinder of radius r_0 consisting of material with shear modulus G. Under the same loading conditions as the hollow cylinder, the angular twist is $\theta/2$. Derive an expression for r_i in terms of the other parameters provided.

2. A cylinder is a composite of two materials, with shear moduli G_1 and G_2. The cylinder is composed of material with modulus G_1 from $0 \leq R \leq R_i$ and G_2 from $R_i \leq R \leq R_0$. Derive the net effective shear modulus G for the composite material for torsion.

3. If an object of arbitrary shape is completely submerged in water without touching any other surface (such as the bottom floor), is there any pressure on the surface of the object? Is there net pressure on the object (that

is, if you integrate the pressure over the entire surface of the object, is there a net force in any direction)? Justify your response.

4. You are given two materials, one with elastic modulus E_1 called Mat A and the other with elastic modulus $E_2 = 10E_1$ called Mat B. The materials are shaped into cubes of 1 m³ and loaded with 1 kN of force, evenly across the top surface (and supported by an infinitely rigid bottom surface). Estimate the strain (in terms of E_1) for the following:

   (a) The cube is made entirely of Mat A.

   (b) The cube is made entirely of Mat B.

   (c) The cube is made of 50% Mat A and 50% Mat B, with the different materials layered horizontally (interface is parallel to the load).

   (d) The cube is made of 50% Mat A and 50% Mat B, with the different materials layered vertically (interface is perpendicular to the load).

   (e) The cube is made up of 50% Mat A and 50% Mat B, with the different materials arranged in a "patchwork" pattern, composed of 1 cm cubes placed in alternating patterns (like a chessboard, but in both directions).

5. A toy balances on the edge of a table or on a small platform (see below). These toys typically look like they are about to fall over, but because of the presence of a counterweight, they stay balanced. Even if they are nudged, they just rock back and forth. Draw a free-body diagram for such a system at rest. All relevant information should be provided in your response, including location of the forces and relative magnitudes.

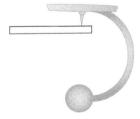

6. You want to build a table with a single support column underneath it. Assume that the tabletop is symmetrically balanced and centered on the column. The design constraints are that the height of the table fixed, and the column requires a rectangular cross section of constant perimeter, for which you can alter the aspect ratio (the ratio of the lengths of adjacent sides).

   (a) If the sides of the rectangle are a and b, determine the relationship between a and b such that the stress (magnitude) in the column is minimized. Briefly justify your answer.

   (b) Given that $a + b = 20$ cm, and the tabletop has a mass of 10 kg and the column material has

$E = 10$ GPa (about the stiffness of wood), how much strain do you expect in the column when the table is assembled under the minimal stress condition derived in (a)? Neglect the weight of the column itself.

   (c) The strain in part (b) is quite small, so you may neglect it for this part. Someone sitting at the table is tilting their chair back and resting on the edge of the tabletop. You model this as a block of mass 70 kg with an arrangement as shown in the illustration below. Assume that there is no friction between the block and tabletop. If the shear modulus of the column is 5 GPa, find the total displacement of the tabletop in the lateral (horizontal) direction. Neglect the deformation of the tabletop. Does the orientation of the column matter?

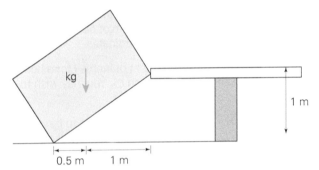

7. Starting over with setup similar to Problem 6, you instead wish to maximize the resistance of the bar to being twisted, as this bar will be used to support a table, and it is quite irritating using a table whose top twists back and forth. Here, you wish to use a cylinder to support the table. This cylinder is symmetrically centered with the table and is 1 m tall (ignore any changes in height during torque application).

You are using plastic ($G = 100$ MPa) to build the column. The design constraint is that you want a maximum angular twist of 0.01 radians under a torque of 100 Nm.

   (a) You can select from a set of plastic tubes that have a wall thickness of 1 cm but with any outer radius you want. Find the smallest outer radius, to the nearest centimeter, that satisfies the design constraint. Hint 1: use MATLAB/Excel/calculator to solve the inequality. Hint 2: do not round down if the rounded solution does not satisfy the design requirement.

   (b) You now can select from a set of solid plastic cylinders of the same material, with any radius you want. Again, find the smallest radius (to the nearest centimeter) that satisfies the design constraint, and calculate the ratio of areas of the suitable solid to the hollow cylinder you found in (a). What does the ratio tell you about where the torsion load is being most resisted?

8. Assuming a model of a cell as a thin-walled sphere enclosing a pressurized fluid, derive the relationship between the stress in the membrane and the cytoplasm pressure.

9. In calculating *GJ* for cytoskeletal proteins (Table 3.2), we ignored the fact that microtubules are hollow. Recalculate *G* and *GJ* for a hollow microtubule, assuming that the inner radius is 11 nm. How large was our error? Was our order-of-magnitude estimate acceptable?

10. The other error we made in Table 3.2 was that we could estimate the shear modulus *G* with the Young's modulus *E*. Assume an effective Poisson's ratio of 0.3. How large was the error? Was our order-of-magnitude estimate acceptable? What is the result if you assume a Poisson's ratio of 0.0 or 0.5?

11. Determine the second moment of inertia *I* for a cylindrical cross section of a beam of diameter *D*. You may wish to use cylindrical coordinates.

12. Determine the second moment of inertia *I* for a hollow cylinder, diameter *D*, and lumen D_0. You may wish to use cylindrical coordinates.

13. Why is it advantageous for a microtubule to be hollow? Using your result from the preceding question, determine the mass ratio and flexural rigidity ratio for a microtubule with inner and outer radii 11.5 nm and 14 nm respectively, compared with a solid microtubule with the same outer radius. What is the most efficient use of proteins to gain bending rigidity (attaining the most rigidity with the least mass): one solid microtubule or several hollow ones?

14. Show that $J = I_x + I_y$.

15. Glass micropipettes in bending are often used as cell and molecular force transducers. The tip defection acts as a linear spring, obeying Hooke's law. Determine the spring constant of the tip for a glass rod of radius 0.25 μm, length 100 μm, and Young's modulus 70 GPa. The spring constant can be decreased further by increasing the length or decreasing the radius. What is the effect on the spring constant of doubling the length? What is the effect on the spring constant of halving the radius?

16. Imagine that a cantilevered beam is loaded until the point of failure by a single transverse load at the tip. Assume that the beam is made of a material that has a tensile failure stress that is lower than its compressive failure stress. Where in the beam will failure initiate? If the failure stress of the material is σ_f, what is the failure load at the tip?

17. Solve the beam equation for a beam that is free to rotate at the supports and is loaded with a single point load at mid-span.

18. We solved for the buckling load assuming that the base of the beam is free to rotate. What is the buckling load for the clamped configuration? This can be done with symmetry arguments. Do not solve the beam equation.

19. Solve the beam equation for a beam supporting a distributed pressure of *Q* N/m that is free to rotate at the supports.

20. Often we are interested in the energy stored in a structure because of its deformation or *strain energy*. One approach is to integrate the *strain energy density* (strain energy per unit volume) over the structure,

$$dw = \frac{1}{2}\sigma\varepsilon$$

where the product of stress and strain is taken component-by-component. Taking this approach, show that for a beam

$$dw = \frac{1}{2}E(y\kappa)^2 = \frac{1}{2}E\left(y\frac{d^2w}{dw^2}\right)^2 = \frac{1}{2}E\left(\frac{y}{R}\right)^2.$$

21. Provide a one-sentence description for each of the three basic relationships in continuum mechanics: kinematics, constitutive behaviour, and equilibrium.

22. Using the result from Problem 20, show that the strain energy of a beam that is bent in a hairpin (180 degrees) is $\dfrac{E\pi I_y}{2R}$.

23. Assume a state of stress $\sigma_x = \sigma$, $\sigma_y = -\sigma$, and $\sigma_z = 0$. Show that this is a state of pure shear. Use this information to derive the shear modulus in terms of the Young's modulus and Poisson's ratio.

24. Consider a membrane that is in a bi-axial state of stress such that $\sigma_{xx} + \sigma_{yy} = \sigma$ and $\sigma_{zz} = 0$. We can define the bi-axial strain to be $\varepsilon_b = \varepsilon_{xx} + \varepsilon_{yy}$. Determine the bi-axial modulus $E_b = \sigma/\varepsilon_b$ in terms of *E* and ν. Also, determine how ε_b is related to the change in area over the original area for a differential portion of the membrane.

25. Consider Examples 3.5 and 3.6. You will notice that the displacements were selected in a specific way—the large displacement example is 100 times the small displacement example. However, **E** is not simply 100 times ε. Why is that? What happens if you neglect the second-order terms in **E**?

Annotated References

Fung YC (1977) A First Course in Continuum Mechanics. Prentice-Hall. *An outstanding introduction to infinitesimal fluid and solid continuum mechanics that includes a very lucid introduction to tensor analysis.*

Malvern LE (1969) Introduction to the Mechanics of a Continuous Medium. Prentice-Hall. *A comprehensive and classic text on large deformation mechanics. Mathematically intense and rigorous.*

Timoshenko SP & Goodier JN (1934) Theory of Elasticity. McGraw-Hill. *A classic text in small deformation elasticity. Includes easy-to-follow treatments of beams, plates, torsion, etc.*

Truesdell C & Noll W (1919) The Non-Linear Field Theories of Mechanics. Springer. *A seminal book that first presented a standard unified set of concepts and notations for large deformation mechanics. Currently out of print, but the third edition incorporates corrections from the late Professor Truesdell's personal copy. The history and rationale for updating the terminology is outlined on Noll's (currently at Carnegie Mellon University) website (http://www.math.cmu.edu/~wn0g/noll).*

CHAPTER 4

Fluid Mechanics Primer

The mechanics of fluids can play a critical role in maintaining normal cellular physiology and in mediating pathological processes. Many physiological processes depend on the presence of extracellular fluid flow to transport nutrients and waste products to and from various locations. In addition, mechanical loads imparted by the fluid in the form of pressure and fluid shear stress can function as potent regulatory signals to cells. Because cells are largely composed of fluid, it comes as no surprise that within cells, fluid mechanics can influence a variety of processes, such as those associated with cell motion or intracellular transport. Given that fluid mechanics is essential to understanding many aspects of cellular mechanobiology, in this chapter, we give a brief introduction to fluid mechanics principles. As in Chapters 2 and 3, this primer cannot cover the field in its entirety. Instead, our goal is to provide a basic understanding of fluid mechanics principles that will allow readers to have a better grasp of topics associated with fluid statics and dynamics in subsequent chapters. It will facilitate the understanding of the process of mechanotransduction in cells regulated by fluid-derived forces (covered in Section 11.1), as well as some experimental methods in cell mechanics (covered in Section 6.3). In this chapter, we cover the basic fluid statics and dynamics, the Navier–Stokes equations, the distinction between Newtonian and non-Newtonian fluids, and rheological analysis. We conclude the chapter with a treatment of dimensional analysis.

4.1 FLUID STATICS

We begin our discussion of fluid mechanics with a basic introduction to fluid statics. Fluid statics, which is sometimes referred to as hydrostatics, is a subdiscipline within fluid mechanics that deals with the mechanics of fluids at rest. In being at rest, the fluid is assumed to take the shape of the container that it is in (this ability is what is often used to defined a substance as being a fluid rather than a solid). In this section, we will see that although the fluid is at rest, it exerts forces on surfaces it contacts. We will also see that these forces, which are derived from the gravitational forces acting on the fluid, can be understood from the fluid pressure.

Hydrostatic pressure results from gravitational forces

To introduce the concept of hydrostatic pressure, consider a cylindrical glass container filled with water. Because the water has mass, it exerts a force on the bottom of the glass from gravity. Assume that the container has cross-sectional area A, the height of the water within the container is h, and the density of water is constant and equal to ρ (Figure 4.1). The total gravitational force exerted on the bottom of the container is

$$F = \rho g A h, \tag{4.1}$$

Figure 4.1 A glass of water experiences hydrostatic pressure on its bottom from the weight of the water. The height of the water column within the glass is *h*, and the cross-sectional area of the glass is *A*. The density of water is ρ.

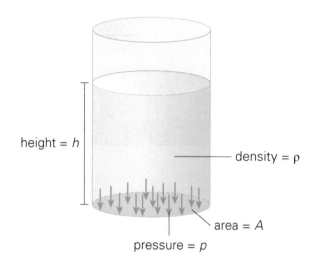

Nota Bene

Incompressible fluids. In deriving Equation 4.1, we assumed that the density of the water was constant, in other words the density of the fluid is not altered even when subjected to large pressures. Such fluids are said to be incompressible. Most fluids (including water) are generally assumed to be incompressible under most relevant circumstances. However, there are cases in which pressures are extremely large and compressibility may need to be accounted for, even for fluids that are normally considered incompressible. For example, die casting involves the injection of molten metal into a die at extremely high pressures. This pressure causes the liquid metal to compress, and this compressibility must be taken into account for in the design of the die.

---

where g is the acceleration due to gravity. In this case, *F* acts in the direction normal to the bottom of the container and in a downward direction. As we discussed in Section 3.2, pressure is the force per unit area in the direction normal to the surface on which it is acting. Given this definition, we can express Equation 4.1 in terms of pressure,

$$p = F/A = \rho g h. \tag{4.2}$$

Such a pressure is referred to as *hydrostatic pressure*, because it is the pressure exerted by a fluid at equilibrium, resulting here from gravitational forces. Equation 4.2 states that the hydrostatic pressure acting on the bottom of the container is proportional to the height of the fluid. More generally, the hydrostatic pressure experienced by a cell or organism in contact with a fluid is dependent on the height of the fluid above it. Cells and organisms may be exposed to hydrostatic pressures that vary quite a bit. For instance, aquatic organisms may experience different hydrostatic pressures as they swim at different depths. If one were to hang upside down by one's feet (Figure 4.2), vascular cells in the feet and skull would experience a decrease and increase in hydrostatic pressure, respectively.

Figure 4.2 Schematic of a man hanging upside down during inversion therapy, which has been used in the past by patients seeking relief from back pain. Upon hanging upside down, vascular cells in the feet and skull experience a reversal in hydrostatic pressure due to the change in fluid height above the cells.

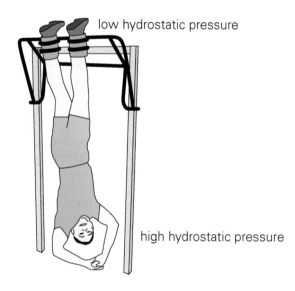

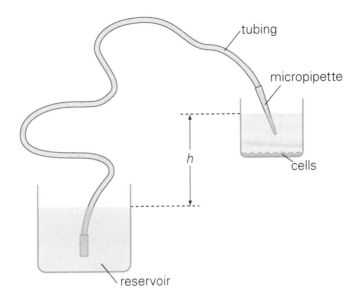

Figure 4.3 Schematic depicting an apparatus for micropipette aspiration. The negative pressure at the micropipette tip only depends on the height difference between the surface of the fluid in the reservoir and the tip of the micropipette, and is independent of the specific path the tube traverses.

Hydrostatic pressure is isotropic

In the last section, we found that the water in a cylindrical container will exert pressure on the bottom of the container. But what about the sides? Consider an infinitesimally small cube of fluid within the container. Because the fluid is at rest, it follows from equilibrium that the pressure on all sides of the cube must be equal. Therefore, the pressure within a fluid at rest is *isotropic*. If we assign a coordinate system to our container where the z-axis is perpendicular to the container bottom, and the x–y plane is parallel to it, and we were to traverse to any point within a given x–y plane, the pressure would be identical, no matter the path taken or distance traveled. An example of this principle can be observed in a typical experimental setup for micropipette aspiration, which we introduced in Section 1.3. In such a setup, one seeks to generate a negative pressure at the tip of a micropipette by coupling a micropipette and a reservoir with fluid-filled tubing (Figure 4.3). In this instance, the pressure only depends on the height difference between the surface of the fluid in the reservoir and the tip of the micropipette, and is independent of the path the tube traverses in the x- and y-directions.

The isotropic nature of hydrostatic pressure also suggests that, within our fluid-filled container, the water will exert force against the sides of the container in addition to exerting force against the bottom. Consider a small fluid element in contact with the side of the container (Figure 4.4). The fluid element experiences a force from the glass that is equal in magnitude to the force exerted by the surrounding water. If the glass were suddenly removed, the fluid element would experience an imbalance in force, and would flow, collapsing into a puddle.

Nota Bene

Absolute pressure versus gauge pressure. In many cases, pressures are measured relative to a convenient baseline that is taken to be zero. Often this baseline is the *atmospheric* pressure (the pressure due to the weight of the air in the atmosphere). These types of pressure are called *gauge*, or *gage*, pressures. Blood pressure is a gauge pressure. In contrast, *absolute* pressure is relative to a vacuum.

Figure 4.4 The fluid element adjacent to the container side is subjected to forces from the glass and the surrounding fluid. In this case, the glass exerts a force that resists the net force imparted on the element by the surrounding forces which keeps it from accelerating. If the glass were suddenly removed, the fluid would flow, and the column of fluid would collapse into a puddle.

Resultant forces arising from hydrostatic pressure can be calculated through integration

In the last section, we saw that hydrostatic pressure is isotropic, and that our fluid-filled container experiences forces from the fluid on its bottom and sides. Because pressure is depth-dependent, the pressure exerted by the fluid on the container sides will vary spatially: low near the top of the water surface and high near the bottom of the container. We might ask, how do we calculate the resultant forces acting on the sides of the container if the pressures are depth-dependent? Integration is required, as is shown in the following example.

Example 4.1: Force of fluids being held by a wall

Consider the case in which we wish to construct an open-top, rectangular-shaped culture vessel, or bioreactor, that will be filled completely to the top with culture media. The walls of the vessel are oriented vertically and have height h and width w. Calculate the force exerted by the media on the walls of the bioreactor.

The pressure acting on a small fluid element is $p = \rho g z$, where z is the vertical distance between the surface of the water and the fluid element, assuming the pressure at the surface is zero (gauge pressure) indicating we are using gauge pressure. For an infinitesimally thin strip along the wall of the vessel of height dz and width w located at depth z (Figure 4.5), the resultant force is

$$dF = \rho g z dA,$$

where

$$dA = w dz.$$

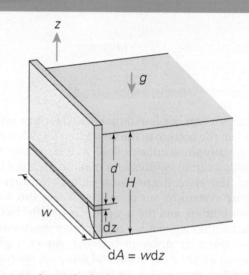

Figure 4.5 Schematic depicting a single wall of a fluid-filled cell culture vessel.

The total force can be found as

$$F = \int dF = \int_{z=0}^{h} \rho g z w dz = \frac{\rho g w h^2}{2}.$$

4.2 NEWTONIAN FLUIDS

We now turn our attention from fluid statics to fluid dynamics. We previously stated that a fluid is classified as a substance that adapts to the shape of its container. A more precise definition can be stated as follows: a fluid is a substance that responds to shear stress with continual deformation. The specific way in which deformation occurs under shear can vary between different types of fluids. A *Newtonian fluid* is a substance that responds to shear stress by deforming at a shear rate proportional to the applied shear stress. To illustrate, consider a thin layer of fluid between two plates exposed to a constant shear stress (Figure 4.6). Such a shear stress can be imposed by sandwiching fluid between two large parallel plates, and moving the top plate at a constant velocity V_0, while the bottom plate is kept stationary. This generates a flow that is assumed to be *steady*, in other words the velocity field is not changing

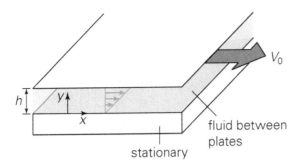

with time. Between the plates, the fluid velocity varies linearly in the y-direction, with zero velocity at $y = 0$, and a velocity of V_0 at $y = h$. This characterization of the fluid velocity is called a *flow profile*. In the case of a linear flow profile, a constant shear stress is applied to the upper plate to maintain this profile. For a Newtonian fluid, this shear stress is

$$\tau = \mu \frac{\partial u}{\partial y}, \tag{4.3}$$

where τ is the shear stress, u is the fluid velocity and y is a direction perpendicular to the direction of u. The derivative of u with respect to y is called the *shear rate* or *velocity gradient*. The constant μ represents the *dynamic viscosity*, and has units kg/ms. We will see later in Section 4.3 that Equation 4.3 comes from a more general set of constitutive relations for Newtonian fluids.

Given Equation 4.3, we can determine the magnitude of the shear stress in our chamber as

$$\tau = \mu \frac{V_0}{h}. \tag{4.4}$$

Upon inspection of Equation 4.4, we see that how fast the fluid deforms in response to a given shear stress is determined by the fluid viscosity. A highly viscous fluid such as honey will undergo a much slower velocity for the same-sized gap, h, in response to a given shear stress, compared with a fluid with a lower viscosity, such as water. Alternatively, it takes a lot more shear stress to move the upper plate with a desired velocity, V_0, for a fluid like honey.

Fluids obey mass conservation

Fluids, just like solids, obey mass conservation. For incompressible fluids, mass conservation states that the volumetric flow rate into a fixed volume must equal the volumetric flow rate out, or

$$V_{in} A_{in} = V_{out} A_{out}, \tag{4.5}$$

where V_{in} is the average velocity of all the fluid entering, A_{in} is the cross-sectional area normal to the flow entry, and V_{out} and A_{out} are similarly defined for fluid exiting (Figure 4.7). The product $(V)(A)$ is referred to as the volumetric flow rate. The reason why volumetric flow rate must be conserved is that for incompressible fluids, where the density is assumed constant, volume is directly proportional to mass. Conservation of mass is one of the most useful concepts in fluid mechanics. It explains why a pipe, a narrowing of the diameter in a pipe leads to increased velocity of the fluid flowing within. As we will see, this relation will be critical in allowing us to solve the Navier–Stokes equations in Section 4.3.

Nota Bene

No-slip condition. In our example of the linear flow profile, the fluid contacting the plates themselves has the same velocity as the plates. This condition is called the *no-slip condition*, which stipulates that fluid immediately adjacent to a solid surface cannot slip along that surface, and will travel with the same velocity as the surface itself. An everyday example of the no-slip condition is that dust on fan blades tends to stay in place no matter how fast the fan spins.

Nota Bene

Streams and rivers. An everyday example of conservation of mass is the water speed of streams and rivers. In deep regions of a channel (which have a larger cross-sectional area), the water tends to flow slowly, while shallow regions tend to be moving quickly.

Figure 4.7 In some fixed volume, the rate that fluid mass enters the volume must be matched by the fluid mass leaving the volume. Under incompressible conditions, this can be expressed mathematically as the velocity times the area must be constant. In this figure, fluid leaving the volume must be moving faster than the fluid entering the volume, because the exit area, A_{out} is smaller than the inlet area A_{in}.

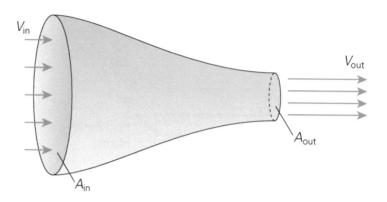

Convection versus diffusion. A fluid that is in motion is able to carry things with it as a consequence of that motion by *convection*. Typically this would be either a dissolved solute or heat. Convection is an important mechanism of mixing in turbulent flow. In contrast to convection, *diffusion* is the motion of particles in a fluid that occurs independent of the bulk velocity.

Fluid flows can be laminar or turbulent

Before we present some analytical solutions, it is necessary to discuss the concepts of laminar and turbulent flow. In general, turbulent flow can be roughly characterized as chaotic and irregular and involving some degree of stochasticity and/or randomness. Laminar flow is any flow in which turbulence is not exhibited. You may already have some intuition for the differences between such flows based on everyday occurrences. Pouring a thick fluid, such as honey or certain oils, into a dish generally results in a flow that is laminar. Opening the faucet slightly, so that the water is clear, will also generally result in laminar flow (assuming you do not have an aerator). Opening the faucet more will result in a chaotic flow, with the water churning and mixing—a characteristic of turbulent flow (Figure 4.8).

More formally, laminar flows are flows without any internal convective mixing, in which fluid elements travel in well-defined "lines." These lines can be visualized by injecting a small amount of dye into the flow. The flow can be injected with dye at many spots simultaneously to visualize the flow field. For laminar flows, the lines of dye will remain cohesive. Because there is no active mixing, such flow visualization generally will yield many parallel strands, from which one can imagine layers of, without crossing from one layer to the next. This pattern of layers is the source of the name "laminar."

By measuring certain aspects of the flow and the geometry through which it is occurring, we can distinguish between laminar and turbulent flow with a dimensionless number called the Reynolds number. Abbreviated *Re*,

$$Re = \frac{\rho VL}{\mu} \tag{4.6}$$

Figure 4.8 Laminar versus turbulent flows. Laminar flow tends to be steady and smooth, with unbroken streamlines (which may be curved or straight and may change with time). Turbulent flows will exhibit mixing and disrupted streamlines and are time-dependent.

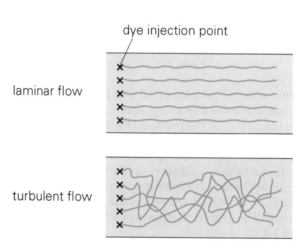

where ρ and μ are the fluid density and dynamic viscosity, respectively, *V* is what is called a *characteristic velocity*, and *L* is some length scale. For flow in tubes and pipes, *L* is generally the diameter, and *V* is generally the average velocity (although the radius and peak velocity may also be used).

The Reynolds number measures the relative magnitudes of inertial versus viscous forces in a given flow and is widely considered one of the most important quantities in fluid mechanics. In general, a Reynolds number greater than one indicates that inertial forces, in other words how much momentum the fluid has, dominates the flow. Such inertial forces tend to drive mixing, and high Reynolds numbers are associated with turbulent flow. Reynolds numbers of less than 1 indicate that viscous forces dominate. Note that at *Re* = 1, we have a balance between viscous and inertial terms, and this is sometimes referred to as the *transition region*, where the flow can start developing instabilities, but is not yet fully turbulent.

Many laminar flows can be solved analytically

The ability to distinguish between laminar and turbulent is of great importance when attempting to solve for a flow profile. Turbulent flows can generally not be solved analytically; however, many laminar flows can. In this section, we solve for pressure-driven flow between two parallel plates using a simple free-body analysis of a differential fluid element. Consider a left-to-right, pressure-driven flow between parallel plates. We assume the fluid is Newtonian and incompressible and that the flow is steady and laminar. We also assume that the flow is *fully developed*, in other words the flow is far from any inlet or outlet, so that the flow profile is not changing in the flow direction. Finally, we assume that the flow is exclusively in the *x*-direction, with zero velocity in the *y*- and *z*-directions. Now, consider a small fluid element of volume (d*x*d*y*d*z*) within the flow and construct a free-body diagram for the fluid element as in Figure 4.9.

In the *x*-direction, the fluid element is subjected to shear forces acting on the top and bottom surfaces, as well as forces due to pressure exerted by the surrounding fluid. Summing these forces, we obtain the expression

$$P\mathrm{d}y\mathrm{d}z - \left(P + \frac{\partial P}{\partial x}\mathrm{d}x\right)\mathrm{d}y\mathrm{d}z + \left(\tau \frac{\partial \tau}{\partial y}\mathrm{d}y\right)\mathrm{d}x\mathrm{d}z - \tau\mathrm{d}x\mathrm{d}z = 0 \qquad (4.7)$$

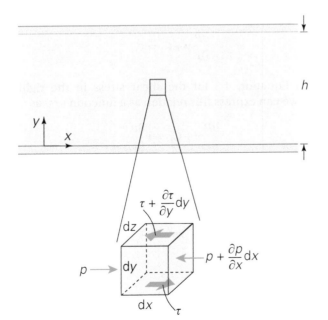

Nota Bene

Transition to turbulence. Though *Re* = 1 is often considered a transition point at which flows begin to exhibit turbulence, for steady pipe flows, *Re* can rise as high as 2000 or so before such a transition occurs. The reason for this is that in pipe flows, the fluid is highly constrained, moving in a single dimension. In this instance, the changes in inertial terms tend to be very small and mixing is more a function of surface roughness.

Nota Bene

Laminar flow at the cellular level. Most flows at the cellular level are laminar by virtue of the small length scales and low velocities prevalent in the cellular environment. For instance, a typical capillary has a diameter of 10 μm and a blood velocity on the order in 0.1 mm/s. If we assume the blood has a density of approximately 1000 kg/m³ and a viscosity of 0.001 kg/ms, plugging these numbers into Equation 4.6, we obtain a value of *Re* = 0.001. This indicates that viscous forces dominate the flow, suggesting that turbulence is unlikely to occur.

Figure 4.9 Force balance in pressure-driven flow. A pressure-driven flow can be solved by considering a small infinitesimal fluid element (a differential element) within the fluid. There are shear stresses acting on the top and bottom surfaces and pressures acting on the left and right surfaces. The remaining two surfaces experience no shear stress, and the pressures on them are equal and opposite.

which can be simplified to

$$-\frac{\partial P}{\partial x}dxdydz + \frac{\partial \tau}{\partial y}dydxdz = 0. \tag{4.8}$$

Dividing Equation 4.8 by the volume of the fluid element (dxdydz) yields

$$-\frac{\partial P}{\partial x} + \frac{\partial \tau}{\partial y} = 0. \tag{4.9}$$

Next, we examine the spatial dependence of the pressure and shear gradient in the above expression. Consider the dependence of the shear gradient, $\partial \tau / \partial y$, on x, y, and z. First, because the flow is fully developed and not changing in the x-direction, shear stress is also not changing in the x-direction. Because we assume no flow in the z-direction, this implies that the shear gradient can only be a function of y,

$$\frac{\partial \tau}{\partial y} = f(y). \tag{4.10}$$

Now consider the pressure gradient. Because the flow velocity is assumed to be zero in the y- and z-directions, this implies that pressure is not a function of y and z, and the pressure gradient can only be a function of x,

$$\frac{\partial P}{\partial x} = g(x). \tag{4.11}$$

However, Equation 4.9 states that the difference of Equations 4.10 and 4.11 must equal zero,

$$-f(y) + g(x) = 0. \tag{4.12}$$

The condition in Equation 4.12 can only be met if $f(y)$ and $g(x)$ are both equal to some constant, or

$$f(y) = g(x) = \text{constant}. \tag{4.13}$$

To determine this constant, one can solve for $\partial \tau / \partial y$ in Equation 4.9 and integrate with respect to y to obtain

$$\frac{\partial P}{\partial x}y + C_1 = \tau. \tag{4.14}$$

Substituting in Equation 4.3 for the shear stress in the right-hand side of Equation 4.14, we can express this relation as a function of u as

$$\frac{\partial P}{\partial x}y + C_1 = \mu\frac{\partial u}{\partial y}, \tag{4.15}$$

where u is the velocity in the x-direction. Dividing Equation 4.15 by viscosity and integrating with respect to y once more, we obtain the velocity profile

$$\frac{1}{2\mu}\frac{\partial P}{\partial x}y^2 + C_1 y + C_2 = u(y), \tag{4.16}$$

where C_1 and C_2 are unknown constants.

To solve for these constants, we use the boundary conditions. In particular, we rely on the no-slip characteristic of fluids—that the fluid velocity at a solid boundary is exactly equal to the velocity of the solid boundary. The top and bottom plates are stationary and $u(y = 0) = 0$, and $u(y = h) = 0$. This gives us the final flow profile,

$$\frac{1}{2\mu}\frac{\partial P}{\partial x}(y^2 - hy) = u(y). \tag{4.17}$$

Equation 4.17 describes a parabolic velocity profile, which is commonly encountered in a variety of situations involving fluid flow. We will encounter this profile again in our discussion of flow chambers in Section 6.3. In addition, it is straightforward to show—using a similar approach—that within a pipe of circular cross section, a parabolic profile is also obtained.

Many biological fluids can exhibit non-Newtonian behavior

In the last section we found that a Newtonian fluid will exhibit a parabolic velocity profile when subjected to pressure-driven flow between parallel plates. This solution depended on the incorporation of the constitutive equation for Newtonian fluids, Equation 4.3, which related shear stress and shear rate (or velocity gradient) through a constant, viscosity. Not all fluids follow the linear relation between shear rate (or velocity gradient) and shear stress that characterize Newtonian fluids. The fluid here is considered *non-Newtonian*. For many biological fluid flows, viscosity is not constant, but a function of other parameters such as shear rate.

One non-Newtonian fluid is the Bingham plastic or Bingham fluid, which has been used to model the flow of biological fluids such as blood and mucus that exhibit solidlike behavior at low stresses but fluidlike behavior at large stresses. For this fluid there is a critical value of shear stress, τ_0, below which the shear rate is zero, such that

$$\frac{\partial u}{\partial y} = 0 \quad \text{for } \tau < \tau_0$$

$$\frac{\partial u}{\partial y} = \frac{\tau - \tau_0}{\mu} \quad \text{for } \tau \geq \tau_0. \tag{4.18}$$

If the shear stress is below τ_0, the fluid does not deform or flow. To demonstrate this, consider our parallel plate example in the previous section, except we substitute the Newtonian fluid with a Bingham fluid. Assume that the peak level of shear stress computed using Equations 4.13 and 4.17 is less than τ_0. Here, $\partial u/\partial y = 0$ and u must be constant. However, we know that the velocity is zero at the plate surfaces by the no-slip condition and $u = 0$ everywhere. If this level of shear is greater than or equal to τ_0, then the fluid can flow. A classic example of the application of the Bingham fluid model is to model paint sticking to a wall—if the shear forces arising from gravity are not sufficiently large, the fluid stays motionless until it dries out.

Another non-Newtonian fluid is a power-law fluid, which can be modeled using the relationship

$$\tau = \beta\left(\frac{\partial u}{\partial y}\right)^\alpha, \tag{4.19}$$

where β is a constant viscous factor (not viscosity) and α is a constant scaling exponent. Equation 4.19 can be recast as

$$\tau = \mu_{\text{eff}}\frac{\partial u}{\partial y}, \tag{4.20}$$

Nota Bene

Bernoulli's equation. Consider a steady and laminar flow in which we inject dye, leaving a line of dye that is tangent everywhere to the velocity of the particle. Formally, we call this line a streamline. As a particle moves along a streamline, it will speed up or slow down due to the different forces acting on it, such as those due to changes in fluid pressure and from gravity. If the flow is inviscid, in other words the fluid is assumed to have zero viscosity, then we can relate these parameters using a relationship called Bernoulli's equation,

$$P + \frac{1}{2}\rho V^2 + \rho g z = \text{constant}$$

In the above expression, P is the local pressure, ρ is the fluid density, V is the local fluid velocity, g is the gravitational constant, and z is the local elevation. We can see that along a streamline, for a given height, as pressure increases, velocity decreases. A great way to experience this phenomenon is to hold two pieces of paper close together and gently blow between them. They will tend to move together due to the increased velocity of air, which results in a decrease in pressure between the two pieces of paper.

Figure 4.10 Flow profiles for (A) Newtonian, (B) Bingham plastic, and (C) shear-thinning fluids. The Newtonian fluid exhibits a parabolic profile. Both the Bingham plastic and the shear-thinning fluid have blunted profiles, reminiscent of having cells concentrated in the middle, and are thus used to model blood flow in certain circumstances. The Bingham plastic has a sharply flat profile in the middle, where shear is lower than the critical shear.

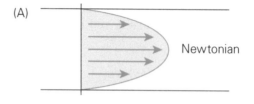

(A) Newtonian

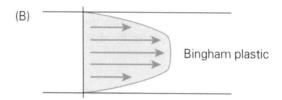

(B) Bingham plastic

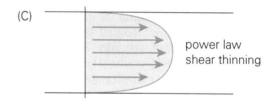

(C) power law shear thinning

where

$$\mu_{\text{eff}} = \beta \left(\frac{\partial u}{\partial y} \right)^{\alpha-1}.$$

(4.21)

In this formulation, the effective viscosity of the fluid is shear rate–dependent, and that dependency changes whether the scaling exponent is greater or less than one. When α is greater than one, the fluid undergoes shear-thickening, in other words the effective viscosity increases as the shear rate increases. There are not too many everyday examples of shear-thickening solutions, but a watery mixture of cornstarch can act as one—so much so that even though you can stir a vat of it, you can also walk across its surface if you step quickly enough.

If the exponent is less than one, the fluid undergoes shear-thinning, where the viscosity decreases as the shear rate increases. An example of a shear-thinning fluid is ketchup, which shows its non-Newtonian behavior when one is trying to get it out of a glass bottle. Initially, getting the ketchup to flow is difficult, but once it starts flowing, it tends to flow easily. So, stirring or shaking the bottle of ketchup will shear the fluid and temporarily decrease its viscosity and allow it to flow faster. Blood is a well-known biological fluid that tends to exhibit shear-thinning behavior. One reason that blood exhibits shear-thinning is that with flow the red blood cells tend to become oriented with one another, thereby reducing viscocity. When the scaling exponent is equal to one, we recover Newtonian fluid behavior (Figure 4.10).

4.3 THE NAVIER–STOKES EQUATIONS

In Section 4.2, we derived an expression for flow with simple geometry assuming the flow was laminar, steady, and fully developed. Such conditions may not be applicable *in vivo*, where typical flows may be unsteady, spatially heterogeneous, turbulent, and occurring over complicated geometries. We now derive a set of equations called the Navier–Stokes equations that provide a mathematical characterization of general fluid flows.

The Navier–Stokes equations are a general set of equations that describe the motion of fluids. They are incredibly important and are used describe a variety of phenomena such as air flow around an airplane wing, water flow in pipes, and currents in the ocean. With regard to cellular mechanobiology, they have been critical in shedding light on the types of mechanical forces that cells are exposed to *in vivo*. For example, they have been used to predict fluid shear stresses and pressure profiles for cells subjected to blood or interstitial fluid flow. In addition, they are useful for modeling flow profiles for cells subjected to fluid flow *in vitro*, such as within flow chambers and bioreactors.

Derivation of the Navier–Stokes equations begins with Newton's second law

In this section, we will derive the Navier–Stokes equations for incompressible, time-dependent flow. We begin our derivation with an application of Newton's second law of motion. This is analogous to our application of the equilibrium condition in Section 3.2 (recall that the key components of continuum mechanics are equilibrium, constitutive, and kinematic relationships); however, this case is broader because we do not assume accelerations are zero. Our strategy is to consider an infinitesimally small fluid element (similar to our approach in deriving our definitions of stress), determine the external forces acting on it, and then use these forces to apply Newton's second law of motion.

Consider a flowing fluid defined by the velocity vector $\boldsymbol{u}(x, y, z, t)$, where x, y, z are spatial coordinates, and t is time. The velocity vector has three components, $u, v,$ and w, which give the velocities in the x-, y-, and z-directions respectively

$$\boldsymbol{u}(x, y, z, t) = \begin{bmatrix} u(x, y, z, t) \\ v(x, y, z, t) \\ w(x, y, z, t) \end{bmatrix}. \tag{4.22}$$

We now consider a small rectangular-shaped fluid element moving with velocity $\boldsymbol{u}(x, y, z, t)$. The fluid element is assumed to be aligned with the coordinate axes, centered at point (x, y, z), and assumed to have dimensions $\Delta x, \Delta y,$ and Δz, in the x-, y-, and z-directions, respectively. The external forces are assumed to arise from one of two sources; stresses imparted by the surrounding fluid acting on the faces of the control volume, and body forces (such as that due to gravity).

First, consider the forces that act on the faces of the control volume. For simplicity, we will begin with the forces in the x-direction. Because the fluid element has six faces, there are six forces we must consider, one on each face: $F_x(x - dx/2, y, z)$, $F_x(x + dx/2, y, z)$, $F_x(x, y - dy/2, z)$, $F_x(x, y + dy/2, z)$, $F_x(x, y, z - dz/2)$, and $F_x(x, y, z + dz/2)$ (Figure 4.11). These forces can be computed as the product of the stress acting on the face and the area of the face over which it is acting. In the x-direction,

$$F_x\left(x - \frac{\Delta x}{2}, y, z\right) = \sigma_{xx}\left(x - \frac{\Delta x}{2}, y, z\right)\Delta y \Delta z$$

$$F_x\left(x + \frac{\Delta x}{2}, y, z\right) = \sigma_{xx}\left(x + \frac{\Delta x}{2}, y, z\right)\Delta y \Delta z$$

$$F_x\left(x, y - \frac{\Delta y}{2}, z\right) = \sigma_{xy}\left(x, y - \frac{\Delta y}{2}, z\right)\Delta x \Delta z$$

$$F_x\left(x, y + \frac{\Delta y}{2}, z\right) = \sigma_{xy}\left(x, y + \frac{\Delta y}{2}, z\right)\Delta x \Delta z \tag{4.23}$$

$$F_x\left(x, y, z - \frac{\Delta z}{2}\right) = \sigma_{xz}\left(x, y, z - \frac{\Delta z}{2}\right)\Delta x \Delta y$$

$$F_x\left(x, y, z + \frac{\Delta z}{2}\right) = \sigma_{xz}\left(x, y, z + \frac{\Delta z}{2}\right)\Delta x \Delta y.$$

As in Section 3.2, we express each of the forces as a Taylor series expansion about the center of the fluid element $(0, 0, 0)$, and ignore second-order and higher terms

$$F_x\left(x - \frac{\Delta x}{2}, y, z\right) \approx \left(\sigma_{xx}(x, y, z) - \frac{\partial \sigma_{xx}}{\partial y}\frac{\Delta x}{2}\right)\Delta y \Delta z$$

$$F_x\left(x + \frac{\Delta x}{2}, y, z\right) \approx \left(\sigma_{xx}(x, y, z) + \frac{\partial \sigma_{xx}}{\partial y}\frac{\Delta x}{2}\right)\Delta y \Delta z$$

$$F_x\left(x, y - \frac{\Delta y}{2}, z\right) \approx \left(\sigma_{xy}(x, y, z) - \frac{\partial \sigma_{xy}}{\partial y}\frac{\Delta y}{2}\right)\Delta x \Delta z$$

$$F_x\left(x, y + \frac{\Delta y}{2}, z\right) \approx \left(\sigma_{xy}(x, y, z) + \frac{\partial \sigma_{xy}}{\partial y}\frac{\Delta y}{2}\right)\Delta x \Delta z \tag{4.24}$$

$$F_x\left(x, y, z - \frac{\Delta z}{2}\right) \approx \left(\sigma_{xz}(x, y, z) - \frac{\partial \sigma_{xz}}{\partial y}\frac{\Delta z}{2}\right)\Delta x \Delta y$$

$$F_x\left(x, y, z + \frac{\Delta z}{2}\right) \approx \left(\sigma_{xz}(x, y, z) + \frac{\partial \sigma_{xz}}{\partial y}\frac{\Delta z}{2}\right)\Delta x \Delta y.$$

If we let F_x^{ext} be the sum of the forces in the x-direction,

$$F_x^{\text{ext}} = -F_x\left(x - \frac{\Delta x}{2}, y, z\right) + F_x\left(x + \frac{\Delta x}{2}, y, z\right) - F_x\left(x, y - \frac{\Delta y}{2}, z\right)$$

$$+ F_x\left(x, y + \frac{\Delta y}{2}, z\right) - F_x\left(x, y, z - \frac{\Delta z}{2}\right) + F_x\left(x, y, z + \frac{\Delta z}{2}\right) \tag{4.25}$$

Figure 4.11 Schematic depicting external forces acting on faces of the fluid element that is immersed in the fluid and moving at the same local velocity. In other words, the boundaries follow the motion of the fluid. The surface forces arise from stresses imparted on the control volume by the surrounding fluid.

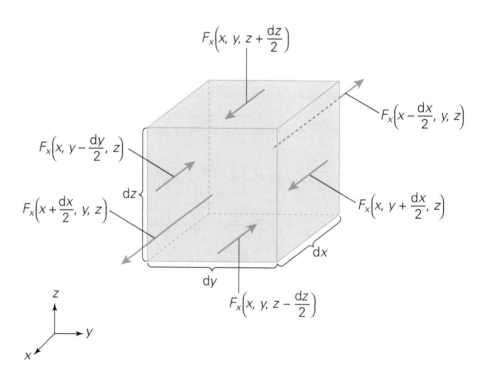

and following the substitution of the expressions in Equation 4.24 into 4.25,

$$F_x^{\text{ext}} = \left(\frac{\partial \sigma_{xx}}{\partial x} + \frac{\partial \sigma_{yx}}{\partial y} + \frac{\partial \sigma_{zx}}{\partial z} \right) \Delta x \Delta y \Delta z. \tag{4.26}$$

The volume is also subject to a body force. If f_x is the body force in the x-direction per unit mass, and ρ is the mass density of the fluid, then the total body force is the product of f_x and the mass of the fluid in the control volume,

$$F_x^{\text{body}} = f_x \rho \Delta x \Delta y \Delta z. \tag{4.27}$$

Now that we have expressions for the external forces acting on the control volume, we are able to invoke Newton's second law ($\mathbf{F} = m\mathbf{a}$). Our fluid element is moving in the fluid with velocity $\mathbf{u}(x, y, z, t)$. Because $\mathbf{u}$ is dependent on both time and space, the velocity of the fluid element can be changing due to two distinct phenomena; changes in the flow velocity with time, or changes in the flow field with space. Consider the case in which the flow is spatially uniform ($\mathbf{u}$ is identical at all points in space), but increasing with time. The fluid element is accelerating solely due to the increase in velocity with time. Now consider the case in which $\mathbf{u}$ does not change with time (it is steady), but the flow field is spatially heterogeneous. The fluid element will undergo changes in velocity due to spatial alterations in the flow profile. As an example, refer to Figure 4.7—there, the flow can be steady (time-invariant), but as you move from the entrance to the exit, the fluid must accelerate to satisfy mass conservation. This spatial change in velocity is sometimes referred to as *convective acceleration*.

To account for the temporal and spatial dependence of $\mathbf{u}$ in calculating the acceleration of the fluid element, we use the chain rule. In particular, if $\mathbf{a} = \{a_x, a_y, a_z\}^T$ is the acceleration of the fluid element, then a_x can be calculated as

$$a_x = \frac{du(x, y, z, t)}{dt} = \frac{\partial u}{\partial t} + \frac{\partial u}{\partial x}\frac{\partial x}{\partial t} + \frac{\partial u}{\partial y}\frac{\partial y}{\partial t} + \frac{\partial u}{\partial z}\frac{\partial z}{\partial t}. \tag{4.28}$$

The partial derivatives $\partial x/\partial t$, $\partial y/\partial t$, and $\partial z/\partial t$ give the instantaneous change in position of the fluid element with respect to time in the x-, y-, and z-directions, respectively. Because the fluid element is moving with velocity vector $\mathbf{u}$, then these expressions are equivalent to the velocity of the fluid,

$$u = \frac{\partial x}{\partial t}, \quad v = \frac{\partial y}{\partial t}, \quad \text{and} \quad w = \frac{\partial z}{\partial t}, \tag{4.29}$$

and Equation 4.28 can be rewritten as

$$a_x = \frac{\partial u}{\partial t} + u\frac{\partial u}{\partial x} + v\frac{\partial u}{\partial y} + w\frac{\partial u}{\partial z}. \tag{4.30}$$

With the external forces and fluid element acceleration in hand, we now apply Newton's second law. The mass of the element is

$$m = \rho \Delta x \Delta y \Delta z. \tag{4.31}$$

Setting the sum of forces in Equations 4.26 and 4.27 equal to the product of the mass in Equation 4.31 and acceleration in Equation 4.30 we find

$$\left(\frac{\partial \sigma_{xx}}{\partial x} + \frac{\partial \sigma_{yx}}{\partial y} + \frac{\partial \sigma_{zx}}{\partial z} \right) \Delta x \Delta y \Delta z + f_x \rho \Delta x \Delta y \Delta z$$

$$= \rho \Delta x \Delta y \Delta z \left(\frac{\partial v}{\partial t} + u\frac{\partial v}{\partial x} + v\frac{\partial v}{\partial y} + w\frac{\partial v}{\partial z} \right), \tag{4.32}$$

which simplifies to

$$\left(\frac{\partial \sigma_{xx}}{\partial x} + \frac{\partial \sigma_{yx}}{\partial y} + \frac{\partial \sigma_{zx}}{\partial z}\right) + \rho f_x = \rho\left(\frac{\partial v}{\partial t} + u\frac{\partial v}{\partial x} + v\frac{\partial v}{\partial y} + w\frac{\partial v}{\partial z}\right) \tag{4.33}$$

when we divide out the volume of the control volume, $\Delta x \Delta y \Delta z$. Repeating these steps for the forces and accelerations in the y- and z-directions,

$$\left(\frac{\partial \sigma_{xy}}{\partial x} + \frac{\partial \sigma_{yy}}{\partial y} + \frac{\partial \sigma_{zy}}{\partial z}\right) + \rho f_y = \rho\left(\frac{\partial v}{\partial t} + u\frac{\partial v}{\partial x} + v\frac{\partial v}{\partial y} + w\frac{\partial v}{\partial z}\right) \tag{4.34}$$

$$\left(\frac{\partial \sigma_{xz}}{\partial x} + \frac{\partial \sigma_{yz}}{\partial y} + \frac{\partial \sigma_{zz}}{\partial z}\right) + \rho f_z = \rho\left(\frac{\partial w}{\partial t} + u\frac{\partial w}{\partial x} + v\frac{\partial w}{\partial y} + w\frac{\partial w}{\partial z}\right). \tag{4.35}$$

Equations 4.33, 4.34, and 4.35 are called Navier's equations, named after Claude-Louis Navier.

Constitutive relations and the continuity equation are necessary to make Navier's equations solvable

Upon inspection of Navier's equations, we see that there are more unknowns (six independent stress components and three velocity components) than equations (three), which means they cannot be solved unless further relations are specified. To make these equations solvable, George Gabriel Stokes proposed a set of constitutive relations that related stress to fluid velocity, viscosity, and pressure. These are a more general form of the constitutive relation we introduced in Section 4.2 for a Newtonian fluid. Specifically,

$$\sigma_{xy} = \sigma_{yx} = \mu\left(\frac{\partial u}{\partial y} + \frac{\partial v}{\partial x}\right) \tag{4.36}$$

$$\sigma_{yz} = \sigma_{zy} = \mu\left(\frac{\partial v}{\partial z} + \frac{\partial w}{\partial y}\right) \tag{4.37}$$

$$\sigma_{xz} = \sigma_{zx} = \mu\left(\frac{\partial u}{\partial z} + \frac{\partial w}{\partial x}\right) \tag{4.38}$$

$$\sigma_{xx} = -P - \frac{2}{3}\mu\left(\frac{\partial u}{\partial x} + \frac{\partial v}{\partial y} + \frac{\partial w}{\partial z}\right) + 2\mu\frac{\partial u}{\partial x} \tag{4.39}$$

$$\sigma_{yy} = -P - \frac{2}{3}\mu\left(\frac{\partial u}{\partial x} + \frac{\partial v}{\partial y} + \frac{\partial w}{\partial z}\right) + 2\mu\frac{\partial v}{\partial y} \tag{4.40}$$

$$\sigma_{zz} = -P - \frac{2}{3}\mu\left(\frac{\partial u}{\partial x} + \frac{\partial v}{\partial y} + \frac{\partial w}{\partial z}\right) + 2\mu\frac{\partial w}{\partial z}. \tag{4.41}$$

The above six equations, together with Navier's equations, give nine equations in total. They also introduce another unknown, pressure, making 10 unknowns in total. We still require one more equation to make the system solvable. Because we have used the equilibrium and constitutive relations, you may not be surprised that what is missing is a statement of kinematics. Specifically, this is known as the *continuity equation* and is a mathematical statement of the conservation of mass. As we discussed previously, for incompressible fluids, the flow rate into a fixed volume must be equal to the flow rate out of it,

$$V_{\text{in}}A_{\text{in}} = V_{\text{out}}A_{\text{out}}. \tag{4.42}$$

If $A_{\text{in}} = A_{\text{out}} = A$, then Equation 4.42 can be rewritten as

$$(\Delta V)A = 0, \tag{4.43}$$

where $\Delta V = V_{\text{out}} - V_{\text{in}}$. A differential form of Equation 4.43 is

$$\frac{\partial u}{\partial x} + \frac{\partial v}{\partial y} + \frac{\partial w}{\partial z} = 0. \tag{4.44}$$

Equation 4.44 is called the differential form of the continuity equation for incompressible fluids.

Navier–Stokes equations: putting it all together

With Navier's equations (three total), the constitutive relations proposed by Stokes (six total), and the continuity equation, we have 10 equations and 10 unknowns (u, v, w, P, and the six independent components of stress). Although this is a solvable system of equations, it can be greatly simplified by substituting in the relations for stress in Equations 4.36–4.41 into the stress derivatives in Equations 4.33–4.35. Taking the term in parentheses in the left-hand side of Equation 4.33, we can substitute in Equations 4.36, 4.38, 4.39, and after some manipulation and invoking Equation 4.44, we arrive at the following expression

$$-\frac{\partial P}{\partial x} + \mu\left(\frac{\partial^2 u}{\partial x^2} + \frac{\partial^2 u}{\partial y^2} + \frac{\partial^2 u}{\partial z^2}\right) + Pf_x = P\left(\frac{\partial u}{\partial t} + u\frac{\partial u}{\partial x} + v\frac{\partial u}{\partial y} + w\frac{\partial u}{\partial z}\right). \tag{4.46}$$

It can be similarly be shown that

$$-\frac{\partial P}{\partial y} + \mu\left(\frac{\partial^2 v}{\partial x^2} + \frac{\partial^2 v}{\partial y^2} + \frac{\partial^2 v}{\partial z^2}\right) + Pf_y = P\left(\frac{\partial v}{\partial t} + u\frac{\partial v}{\partial x} + v\frac{\partial v}{\partial y} + w\frac{\partial v}{\partial z}\right) \tag{4.47}$$

$$-\frac{\partial P}{\partial z} + \mu\left(\frac{\partial^2 w}{\partial x^2} + \frac{\partial^2 w}{\partial y^2} + \frac{\partial^2 w}{\partial z^2}\right) + Pf_z = P\left(\frac{\partial w}{\partial t} + u\frac{\partial w}{\partial x} + v\frac{\partial w}{\partial y} + w\frac{\partial w}{\partial z}\right). \tag{4.48}$$

Equations 4.47 and 4.48 are the Navier–Stokes equations. Together with the continuity equation, they form a set of four equations to solve for the four unknowns u, v, w, and P.

4.4 RHEOLOGICAL ANALYSIS

So far, we have discussed the mechanics of two fundamentally different materials elastic solids (Section 3.2) and viscous materials (Section 4.3). Although we have discussed these two types of material in isolation, many materials cannot be characterized as being, purely elastic or viscous, as they may exhibit both solid- or fluidlike behavior, depending on the circumstances. For instance, like a solid, a dab of toothpaste can hold its shape under its own weight. However, we can still squeeze it out of the tube because, like a fluid, toothpaste has little capacity to

---

Nota Bene

Continuity equation. A simple though somewhat unrigorous analysis can be used to demonstrate the relation between Equations 4.43 and 4.44. Consider a small, imaginary cubical volume immersed in a flow field. The volume is aligned with the coordinate system and with dimensions Δx, Δy, and Δz in the x-, y-, and z-directions. The volume is assumed to be fixed in space, with fluid entering it at some velocity, and leaving it at a different velocity. Let Δu, Δv, and Δw be the changes in fluid velocity in the x-, y-, and z-directions that occur within the volume. From Equation 4.42, we know that

$$\Delta u \Delta y \Delta z + \Delta v \Delta x \Delta z + \Delta w \Delta x \Delta y = 0. \tag{4.45}$$

Dividing Equation 4.45 by the volume of $\Delta x \Delta y \Delta z$, and letting the volume go to zero, we obtain Equation 4.44.

Example 4.2: Fluid flow within parallel plates

In Section 4.2, we derived the flow profile for steady, incompressible, fully–developed laminar flow between two infinite parallel plates, driven by pressure. We now seek to perform the same calculation using the Navier–Stokes equations. Consider plates separated by height h, with one plate at $y = 0$, and another at $y = h$. Calculate $u(y)$ assuming a pressure gradient in the x-direction of $\partial P/\partial x$, with no body forces present.

To calculate u, we begin with Equation 4.46. We can simplify this expression in several ways. First, we can set $\partial u/\partial t = 0$ because the flow is steady. Second, there should be no flow in the y- and z-directions, thus we can set $v = 0$ and $w = 0$. Third, because the flow is fully developed, u does not depend on x, and we can set any x derivatives equal to zero. Finally, because we assume no body forces, then $f_x = 0$, and Equation 4.46 simplifies to

$$\frac{\partial P}{\partial x} = \mu \frac{\partial^2 u}{\partial y^2}.$$

To solve for $u(y)$, we integrate u with respect to y twice and obtain

$$u(y) = \frac{1}{\mu} \frac{\partial P}{\partial x} \frac{y^2}{2} + Ay + B.$$

We can solve for A and B by requiring that $u = 0$ at $y = 0$ and $y = h$, the no-slip condition. Doing so, we arrive at our final solution, which is identical to that achieved via differential analysis

$$u(y) = \frac{1}{2\mu} \frac{\partial P}{\partial x} (y^2 - hy).$$

sustain shear or recover from deformations. Within the body, tissues and organs, as well as the cells residing within them, may be composed of both fluid and solidlike materials. The cytoplasm of a given cell may contain a solidlike cytoskeletal network immersed in an aqueous environment in which numerous proteins are densely dispersed. In this case, one can see why cells would be expected to exhibit both solid and fluidlike mechanical behavior.

Rheology is a scientific discipline that can be broadly characterized as the study of materials that have some capacity to flow, but which cannot be adequately described using classical fluid mechanics. Rheology is considered a distinct branch of continuum mechanics that bridges solid and fluid mechanics. The need for such a discipline can be better understood by revisiting our examples of non-Newtonian fluids; power-law fluids and Bingham plastics. Although both models are able to capture some aspects of nonlinear fluid behaviors, they do not contain any fundamental solidlike behaviors. No matter how you adjust the exponent in a power-law fluid, it will always continuously deform under shear. A Bingham plastic is able to resist shear stress if it is below a critical level, but here the material is completely rigid and does not exhibit elastic behavior. Rheological methods allow us to better understand materials that exhibit solid and fluidlike behavior, such as cells (Figure 4.12). A subset of rheology is the study of viscoelasticity, in which one seeks to decompose mechanical behavior into purely elastic and purely viscous components. We will discuss some basic rheological approaches for investigating viscoelastic substances.

The mechanical behavior of viscoelastic materials can be decomposed into elastic and viscous components

Because a viscoelastic substance is one that exhibits both elastic and viscous mechanical behavior, studies of such substances often rely on oscillatory stimuli. Recall from Section 3.2, we introduced the notion of a linearly elastic material. Specifically, when subjected to stress, such a material would undergo strain in a manner proportional to stress, and would recover completely upon removal of the load. So, consider a material that is exposed to an oscillatory stress of the form

$$\sigma = \sigma_0 \cos(\omega t), \tag{4.49}$$

Figure 4.12 Silly putty is a substance that exhibits both elastic and fluid responses. It can clearly hold its shape, as seen on the left where it is molded into a rectangular solid. However, within 30 min it has flowed under gravity, forming a "puddle" on the table, a result of its fluidlike behavior. Characterizing it as a non-Newtonian fluid is somewhat limiting. If one were to mold this putty into a sphere, one could bounce it like a ball, which requires elastic behavior.

where σ_0 is the magnitude of the stress, and ω is the frequency of oscillatory loading (in radians per unit time). If the material is linearly elastic, it would deform proportionally to the stress as

$$\varepsilon = A\cos(\omega t), \tag{4.50}$$

where A is a constant. Unlike an elastic material, the stress in a purely viscous material is not dependent on strain, but on strain rate. This is similar in concept to a Newtonian fluid, in which shear stress is proportional to shear rate. For the oscillatory stress described in Equation 4.49, a purely viscous material would deform such that the time derivative of strain is proportional to stress,

$$\varepsilon = B\sin(\omega t). \tag{4.51}$$

Notice that the strain profile of a viscous material given by Equation 4.51 can be rewritten as

$$\varepsilon = B\cos(\omega t - \pi/2), \tag{4.52}$$

indicating that the strain in Equation 4.52 is exactly $\pi/2$ radians, or 90 degrees, out of phase with the stress. Here the strain is said to be completely out of phase with the stress. Materials that exhibit some combination of elastic and viscous behavior will exhibit a phase shift that is between 0 and $\pi/2$ radians. In particular, if

$$\varepsilon = \varepsilon_0 \cos(\omega t - \delta) \tag{4.53}$$

is the strain of the material, then the phase shift is δ radians. The parameter δ is called the *phase lag* (Figure 4.13). The phase lag is useful because it gives a single

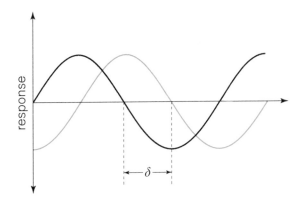

Figure 4.13 If the stimulus applied to a material is a pure sine wave, then the response of a viscoelastic material, in blue, will typically exhibit a phase lag, δ, which can be used as a measure of how solid- or fluidlike the material is.

value that characterizes the degree to which the mechanical behavior is elastic relative to that which is viscous. If the phase lag is close to zero, then the material is behaving primarily like an elastic material. If the phase lag is close to $\pi/2$, then the material is behaving primarily like a viscous material.

Next, consider the case in which we subject a material to oscillatory loading and obtain the strain profile given in Equation 4.53. We can decompose this profile into purely in-phase and out-of-phase components. First, we rewrite Equation 4.53 as a sum of cosine and sine functions using the identity

$$\cos(u + v) = \cos u \cos v - \sin u \sin v. \tag{4.54}$$

Equation 4.54 can then be rewritten as

$$\varepsilon = \varepsilon_0' \cos(\omega t) - \varepsilon_0'' \sin(\omega t), \tag{4.55}$$

where

$$\varepsilon_0' \varepsilon_0 \cos(\delta), \tag{4.56}$$

and

$$\varepsilon_0'' = -\varepsilon_0 \sin(\delta). \tag{4.57}$$

Equations 4.55, 4.56, and 4.57 demonstrate that the mechanical response given by Equation 4.53 can be decomposed into two components, one that is *in phase* (the first term, with $\cos(\omega t)$) with the driving stress, and one that is *out of phase* (the second term, with $\sin(\omega t)$). The relative magnitudes of the in-phase and out-of-phase terms are related to the magnitude of the phase lag.

Complex moduli can be defined for viscoelastic materials

Equation 4.55 indicates that, given a strain profile for an oscillatory loaded material, we can decompose this strain into in-phase and out-of-phase components and identify the degree to which the material is deforming like an elastic versus a viscous material. However, such a decomposition would not necessarily allow us to predict how the material would deform if subjected to a different oscillatory stress profile, such as with a different frequency or magnitude. To make such predictions, we require some information regarding the material properties. In Section 3.2, we described the concept of an elastic modulus, which was a material property that related stress to strain for a linearly elastic material. We introduced a similar relationship earlier in this chapter for fluids, using viscosity. We now seek to define an analogous "modulus" for viscoelastic materials. This task is not as simple as it was for the pure elastic or pure fluid cases. In particular, for a linearly elastic material, the ratio between stress and strain is fixed, and we could simply define the elastic modulus as the ratio of the two quantities. For our viscoelastic material, the ratio between stress and strain changes with time, and we are unable to define a modulus in the same way.

A solution to this problem is to use complex numbers and define quantities called the complex stress, complex strain, and complex modulus. The reason why complex numbers provide a nice solution to our dilemma is that functions involving sines and cosines can be dealt with in an elegant fashion through the use of Euler's formula. In particular, let $\wp$ be a complex function of time,

$$\wp = \cos(\omega t) + i\sin(\omega t). \tag{4.58}$$

In the above expression, the real part of $\wp$ is

$$\text{Re}\{\wp\} = \cos(\omega t), \tag{4.59}$$

the imaginary part is

$$\text{Im}\{\wp\} = \sin(\omega t), \tag{4.60}$$

and i is the imaginary unit satisfying $i^2 = -1$. We can rewrite $\wp$ using Euler's formula, which states that complex functions like Equation 4.58 can be expressed as an exponential imaginary function,

$$e^{ix} = \cos(x) + i\sin(x). \tag{4.61}$$

Nota Bene

Euler's Formula. Leonhard Euler published this formula in the mid-1700s. Euler's formula was described by Richard Feynman as "one of the most remarkable, almost astounding, formulas of all mathematics."

Using Euler's formula, $\wp$ can be written as

$$\wp = \cos(\omega t) + i\sin(\omega t) = e^{i\omega t}. \tag{4.62}$$

Given the relation in Equation 4.62, we can now define a complex stress and complex strain. Let the complex stress be defined as

$$\sigma^* = \sigma_0 \cos(\omega t) + i\sigma_0 \sin(\omega t) = \sigma_0 e^{i\omega t}. \tag{4.63}$$

We have intentionally defined the complex stress such that the real part of σ^* is the stress σ applied to the material as in Equation 4.59,

$$\text{Re}\{\sigma^*\} = \sigma_0 \cos(\omega t). \tag{4.64}$$

Similarly, the complex strain is defined as

$$\varepsilon^* = \varepsilon_0 \cos(\omega t - \delta) + i\varepsilon_0 \sin(\omega t - \delta) = \varepsilon_0 e^{i(\omega t - \delta)}, \tag{4.65}$$

where again, we have intentionally defined the complex strain such that the real part of ε^* is the observed strain in Equation 4.54,

$$\text{Re}\{\varepsilon^*\} = \varepsilon_0 \cos(\omega t - \delta). \tag{4.66}$$

With the complex stress and complex strain defined, we can now define a *complex modulus* as the ratio of these two quantities,

$$E^* = \frac{\sigma^*}{\varepsilon^*}. \tag{4.67}$$

Substituting expressions for the complex stress and strain in Equations 4.63 and 4.65 into Equation 4.67, and following some simplification, we arrive at a compact expression for the complex modulus,

$$E^* = \frac{\sigma_0 e^{i\omega t}}{\varepsilon_0 e^{i(\omega t - \delta)}} = \frac{\sigma_0 e^{i\omega t}}{\varepsilon_0 e^{i\omega t} e^{-i\delta}} = \frac{\sigma_0}{\varepsilon_0} e^{i\delta}. \tag{4.68}$$

Because E^* is a complex number, we can define E' as the real part of E^* and E'' as the imaginary part, in other words

$$E^* = E' + iE'' \tag{4.69}$$

with

$$E' = \frac{\sigma_0}{\varepsilon_0} \cos(\delta) \tag{4.70}$$

and

$$E'' = \frac{\sigma_0}{\varepsilon_0} \sin(\delta). \qquad (4.71)$$

E' is known as the *elastic modulus* or *storage modulus*. The storage modulus E' is associated with the in-phase component of resistance to stress. This becomes apparent by substituting $\delta = 0$ into Equations 4.69, 4.70, and 4.71. Here, the complex modulus is equivalent to the storage modulus, $E^* = E'$. Because in-phase deformation is a characteristic of elastic materials, the storage modulus can be considered a measure of the elastic behavior of the material. E'' is referred to as the *damping* or *loss* modulus, and is associated with the out-of-phase resistance to stress. For instance, for $\delta = \pi/2$, $E^* = iE''$. In this case, the magnitude of the complex modulus is equal to the loss modulus. Because this phase lag is associated with viscous materials, the loss modulus can be considered a measure of the viscous behavior of the material.

Before concluding this section, it is worthwhile to note that in our development, we designated the complex modulus as E^*, suggesting a modulus analogous to Young's modulus. There is a corresponding complex shear modulus G^*, which is actually more common in the literature because dynamic rheological measurements are often made under shear. For the rest of the chapter, we will use the complex shear modulus G^* and associated complex shear stress τ^* and complex shear strain γ^* in our developments.

Nota Bene

Historical roots of complex modulus. The complex modulus was introduced by the German physicist Carl Gauss in the early 1800s—the same Gauss for whom the SI unit of magnetic strength was named.

Power laws can be used to model frequency-dependent changes in storage and loss moduli

In the last section, we saw that for a viscoelastic material subject to oscillatory loading, the storage modulus and loss modulus capture the in-phase versus out-of-phase resistance to stress, respectively. For many viscoelastic materials, these moduli depend on the frequency of loading. Furthermore, the frequency dependence for each modulus may be different, such that the relative amount of in-phase versus out-of-phase deformation changes at different frequencies. Such frequency dependence is common in many biological materials, including cells. A variety of models have been proposed to capture this frequency dependence. One such approach involves modeling the frequency dependence of the storage and loss moduli as power laws. For instance, the following relation has been demonstrated to accurately describe the stiffening of cells subjected to oscillatory loading over several orders of magnitudes

Example 4.3: Calculating the behavior of semi-fluid substances

Let us say you stimulate a cell with a pure sine wave via shear, and the shear strain response is shifted exactly 45 degrees ($\pi/4$ radians). Does it have exactly equal contributions from solid and fluid behavior? If so, does this mean that the storage loss moduli is half what the elastic modulus would be if the cell were a pure solid with the same response amplitude, but no phase shift?

Our input is $\tau = \tau_0 \sin(\omega t)$, and the output is $\gamma = \gamma_0 \sin(\omega t - \pi/4)$. Our δ is therefore $\pi/4$, and we can write the complex shear modulus $G^* = G' + iG''$ with $G' = G_0 \cos(\delta)$

and $G'' = G_0 \sin(\delta)$. Thus, the complex shear modulus has equal contributions from both the storage modulus (G') and the loss modulus (G''), but these moduli are individually more than half the value of G_0, which would be the elastic modulus were the cell to be a pure solid. The reason for this is that G^* has magnitude of G_0, so that the individual components must be $2\sqrt{2}$ times G_0. That is, a substance that is half solid and half-fluid has more than half the elastic modulus of a pure solid responding the same way.

Advanced Material: Complex viscosity

Just as we defined a complex modulus, we can also define a complex viscosity. More specifically, the complex viscosity can be defined as the ratio of the complex shear stress and complex shear strain rate as

$$\mu^* = \frac{\tau^*}{\gamma^*}, \qquad (4.72)$$

where

$$\tau^* = \tau_0 e^{i\omega t} \qquad (4.73)$$

and

$$\gamma^* = \frac{d\gamma^*}{dt} = \frac{d}{dt}\gamma_0 e^{i(\omega t - \delta)} = \gamma_0 i\omega e^{i(\omega t - \delta)}. \qquad (4.74)$$

Substituting Equations 4.73 and 4.74 into 4.72, we obtain

$$\mu^* = \frac{\tau_0 e^{i\omega t}}{\gamma_0 i\omega e^{i(\omega t - \delta)}} = \frac{\tau_0}{\gamma_0 i\omega} e^{i\delta}. \qquad (4.75)$$

Similar to Equation 4.68, the complex shear modulus can be computed as

$$G^* = \frac{\tau_0}{\gamma_0} e^{i\delta}. \qquad (4.76)$$

Comparing the expressions for Equations 4.75 and 4.76, it becomes apparent that Equation 4.75 can be rewritten as

$$\mu^* = \frac{G^*}{i\omega} = \frac{G + iG'}{i\omega} = \frac{iG'}{\omega} + \frac{G''}{\omega}. \qquad (4.77)$$

Letting $\mu^* = \mu' + i\mu''$, then

$$\mu' = \frac{G''}{\omega} \qquad (4.78)$$

and

$$\mu'' = \frac{G'}{\omega}, \qquad (4.79)$$

where μ' is the dynamic viscosity, and μ'' is the "out-of-phase" viscosity that yields the change of the storage modulus with respect to frequency. Upon inspection of Equations 4.78, and 4.79, one might see that the complex viscosity is closely related to the complex shear modulus. In fact, μ' and μ'' are the loss modulus and storage modulus normalized by frequency, respectively. This fact allows one to use these parameters to better understand how the storage and loss moduli change with frequency. For instance, assume μ' and μ'' are relatively constant over some frequency range, and $\mu' > \mu''$. Increasing the loading frequency within this range would result in greater increases in G'' relative to that in G', and the material would be expected to exhibit more viscous behavior and less elastic behavior with increasing frequency.

$$G^*(\omega) = G_0 \left(\frac{\omega}{\Phi}\right)^\alpha (1 + i\xi)\Gamma(1 - \alpha)\cos\left(\frac{\pi\alpha}{2}\right) + i\mu\omega \qquad (4.80)$$

where

$$\xi = \tan\left(\frac{\pi\alpha}{2}\right) \qquad (4.81)$$

and

$$\Gamma(n) = (n - 1)!, \qquad (4.82)$$

G_0 is a parameter that gives the frequency-independent component of the elastic response, Φ is a normalization factor, ξ is a structural damping coefficient, μ is a viscous coefficient, and α is the scaling exponent. The real part of G^* is the storage modulus G', and the imaginary part is the loss modulus G''. The coefficients μ and ξ represent distinct viscous components; the term $i\mu\omega$ is meant to ensure

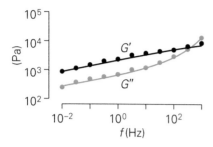

Figure 4.14 Storage and loss moduli as a function of frequency in cells subjected to oscillatory mechanical loading. Because the results are plotted on a log–log scale, the slope of G' gives the scaling exponent α, which is approximately 0.2 in this case. (Adapted from, Fabry et al., [2011] *Phys. Rev. Lett.* 87, 148102.)

Order-of-magnitude analysis. Another analytical approach that can reveal quantitative information in the absence of an exact solution is order-of-magnitude analysis. With this strategy, instead of trying to find precise functional relationships, we estimate how parameters might be related within an order of magnitude or two. As an example, say you are designing a skyscraper and need an estimate for the predicted weight of the building. Assume that the building will be approximately 100 stories high and made of steel. To within one order of magnitude, one could estimate the square footage for each floor as 100 m × 100 m, and the height of a single story as 1 m, which results in a volume of 100,000 m³. One could also estimate the average density of the building as 1 ton/m³, based on the fact that a cubic meter of water weighs 1000 kg, which is approximately 1 ton, and although steel is denser than water, much of the inside of the building is air. An order-of-magnitude estimate for the weight is 100,000 tons. This turns out to be a good estimate, as the Empire State Building has 102 floors and weighs 365,000 tons; the Sears Tower is 108 floors and weighs 223,000 tons.

Similar calculations can be done for estimating the number of receptors in a cell for signaling, the speed of molecular signals within cells, and some properties of cells that are otherwise difficult to measure.

---

that regardless of the scaling exponent α, high-frequency stimuli will result in viscous effects dominating the behavior. The ξ term is much larger than the μ term for most relevant frequency ranges, and in most cases is the primary contributor to the loss modulus. The scaling exponent determines not only how G' and G'' change with frequency but the relative magnitudes of each modulus. If we assume that μ is small, we can see that as α approaches zero, ξ approaches zero as well, and G^* approaches G_0, which gives the elastic component of the response. However, as α approaches one, ξ increases without bound, indicating that the imaginary term will dominate and the material will exhibit strongly viscous effects.

Experiments in a variety of cell types and using various techniques for mechanical loading have demonstrated that cells exhibit a scaling exponent of around 0.2–0.3 (Figure 4.14). This power-law dependence is typical of a class of materials called soft glassy materials, which includes materials such as emulsions, slurries, and pastes. Soft glassy materials are characterized by their possession of some degree of disorder in which the discrete elements of which they are composed are entangled or aggregated via weak interactions. Therefore cells have often been described in the literature as soft glassy materials.

4.5 DIMENSIONAL ANALYSIS

Within fluid mechanics, most flows cannot be determined analytically—that is, the flow profile cannot be derived from first principles alone, and one must turn to experiments to investigate the flow of interest. In such situations, proper experimental design may be difficult if one does not have *a priori* knowledge of relevant parameters that affect a parameter of interest. *Dimensional analysis* is a mathematical technique used to collapse a set of experimental parameters into a reduced set of dimensionless quantities that influence a parameter of interest. In dimensional analysis, the goal is not to obtain an exact formula for the parameter in question. Though dimensional analysis does not provide an exact analytical solution, it provides useful quantitative relationships that often allow one to draw valuable conclusions.

Dimensional analysis requires the determination of base parameters

Consider the case where we wish to perform an experiment to determine the fluid in which drag forces on a swimming bacterium. We are interested in dimensionless quantities that may affect the fluid drag force, as this will aid in the experimental design.

To begin, we must identify all the potential factors affecting our parameter of interest, drag force, and specify their units. To establish this list of *base parameters*, we require some level of intuition for the problem. Because we are interested in the fluid drag force on a swimming bacterium, one might make an educated guess that this force would depend on parameters such as the density of the fluid (units of kg/m³), the viscosity of the fluid (units of kg/ms), the velocity of the bacterium (units of m/s), and some characteristic length scale of the bacterium (units of m). Including drag force (units of kg · m/s²), we have five parameters that we consider to be directly relevant to the modeling of the problem, the base parameters (Figure 4.15). We can express the fact that we expect drag force to be dependent on the selected parameters as

$$f(F, L, \mu, \rho, v) = 0, \tag{4.83}$$

where f represents one or more as yet unknown function(s).

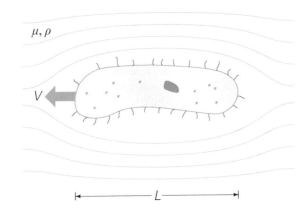

μ, ρ

V

L

Figure 4.15 A bacterium swimming in a fluid at velocity _V_ (moving to the left) in a fluid of density ρ and viscosity μ with length _L_ experiences some drag slowing it down. We wish to characterize the drag force, but because the geometry is odd, it is difficult to obtain an exact analytic solution. However, we can obtain some basic relationships using dimensional analysis.

The Buckingham Pi Theorem gives the number of dimensionless parameters that can be formed from base parameters

With our list of base parameters established, we may now use them to form dimensionless parameters. The number of dimensionless parameters that we can form from our base parameters is given by the _Buckingham Pi Theorem_, which states that the number of dimensionless parameters is equal to the number of base parameters minus the number of independent physical units (dimensions). In our example, we have five base parameters and three independent units (length, speed, and mass), leaving two dimensionless parameters to be found. Bear in mind that an independent unit in this case does not have to be a fundamental unit such as (m), (s), or (kg). They can be derived units such as (m/s), so long as they are independent. Using velocity (m/s) and length (m) would be a valid alternative to using length (m) and time (s). One would not use velocity, length, _and_ time, as they are not independent. Similarly, you would not use area (m²) and length (m) since one is a power of another and therefore not independent.

Dimensionless parameters can be found through solving a system of equations

At this point in the analysis, we have identified our base parameters and dimensions (units). The Buckingham Pi Theorem tells us how many dimensionless parameters we are looking for and the form they will take. The next step is actually to identify the dimensionless parameters. Although a hard and fast procedure is elusive, there is a strategic outline that is effective. First, identify _j_ base parameters (generally independent) that _span_ the dimensions. In other words, these base parameters should involve all of the physical units and will be termed _repeating_ parameters. The procedure is simply to combine the remaining parameters with the repeating parameters one at a time.

In our bacterium example, we have identified five base parameters F, L, ρ, μ, and v and three units, so we expect to find two dimensionless parameters. Because $j = 3$, we need to identify three repeating parameters that should generally be simple and independent. Let us start with length and velocity. They are simple and span two of our units (m and s). For our third repeating parameter we need to add kg. Density or viscosity would both be acceptable, and in fact, both produce the same dimensionless groups. Here we arbitrarily select viscosity, μ.

Now we need to combine each of our remaining parameters with the repeating parameters one at a time. Starting with force, we want to develop a product of the dependent parameters F, L, μ, and v such that the end expression has no units. That is, for the product $F^a L^b \mu^c v^d$, we wish to find the values for the exponents a, b, c, and d such that the product is unitless, or

Nota Bene

Roots of the Buckingham Pi Theorem. The Buckingham Pi Theorem is a formalization of a nondimensionalization method introduced by Rayleigh. Edgar Buckingham proposed the theorem in 1914 and he termed each of the nondimensional parameters Pi (π_1, π_2, ..., π_p), showing that if j is the number of base parameters and k is the number of dimensions (independent units), $p = j - k$. The theorem also shows that the general form for each nondimensional parameter is the product of base parameters with each raised to an unknown integer power. The use of Pi denoted that the nondimensional parameters were _products_ of base parameters.

It should be noted that determining the number of Pi groups is a bit more involved than we are describing here, but the method used here is applicable to enough situations as a very good rule of thumb.

$$F^a L^b \mu^c v^d = 1. \tag{4.84}$$

The "1" on the right side of Equation 4.84 does not refer to its numerical value, but to the fact that the product has no units. To do this, we decompose each of these parameters in Equation 4.84 into their basic units and then pick the exponents so that they all cancel out. If we replace the parameters in Equation 4.84 with their basic units, then

$$\left(\frac{\text{kg} \cdot \text{m}}{\text{s}^2}\right)^a (\text{m})^b \left(\frac{\text{kg}}{\text{m} \cdot \text{s}}\right)^c \left(\frac{\text{m}}{\text{s}}\right)^d = 1. \tag{4.85}$$

For this relation to hold, each unit must have a final exponent of zero. If we consider the exponents for length (m), we find that the following relation must be true:

$$a + b - c + d = 0. \tag{4.86}$$

Similarly, for mass

$$a + c = 0, \tag{4.87}$$

and for time

$$-2a - b - d = 0. \tag{4.88}$$

Equations 4.86, 4.87, and 4.88 form a system of simultaneous algebraic equations. The system can be simplified to yield

$$a = -b$$
$$a = -c \tag{4.89}$$
$$a = -d.$$

This system has three equations and four unknowns, so it is indeterminate. We can arbitrarily set a, the exponent for force, equal to 1 (a general rule of thumb is to choose small whole numbers for one exponent). In this case, the exponents for the rest of the parameters are equal to -1. We have arrived at our first dimensionless term,

$$F^1 \mu^{-1} L^{-1} v^{-1} = \frac{F}{L\mu v}. \tag{4.90}$$

Next, we examine density, ρ, our second remaining parameter. We need to combine it with L, μ, and v, and do a similar analysis, in which

$$\left(\frac{\text{kg}}{\text{m}^3}\right)^e (\text{m})^f \left(\frac{\text{kg}}{\text{m} \cdot \text{s}}\right)^g \left(\frac{\text{m}}{\text{s}}\right)^h = 1. \tag{4.91}$$

Following the same steps as before, we obtain our second dimensionless term,

$$\rho^1 \mu^{-1} L^1 v^1 = \frac{\rho L v}{\mu}. \tag{4.92}$$

The second term is the Reynolds number, which we previously discussed. Equation 4.84 can be recast as a function of dimensionless quantities as

$$f\left(\frac{F}{L\mu v}, \frac{\rho L v}{\mu}\right) = 0, \tag{4.93}$$

which is a simpler functional form because it has fewer terms compared to the original expression.

Similitude is a practical use of dimensional analysis

Now that we have derived our dimensional parameters, we can use them in a variety of ways to facilitate experimental design or interpretation of results. Dimensional analysis is useful from a modeling perspective, because it informs us not only about the number of parameters we need to pay attention to (the dimensionless numbers), but also the relationship between the variables that appear in those parameters. This allows us greater flexibility in creating experimental simulations.

Suppose we wish to measure drag force on our bacterium by building a large-scale model and submerging it in a defined flow. We do this because it is easier to handle larger objects (bacteria being submicrometer in size) and because it is difficult to measure such small forces (both in terms of the magnitude of the forces and in terms of how we can attach instruments to a bacterium). We seek to measure the drag force on a large-scale model and then scale the observed forces to what we would expect at normal scale. This is known as a *similitude experiment*. For the results to scale accurately, a strict requirement is that dimensionless parameters be held constant.

Example 4.4: Deformation of a cell under shear

Assume that a cell is well anchored to its substrate and that fluid shear is deforming the cell. It can be treated as an elastic solid for this problem. Determine the Pi groups for estimating the angle of deformation of the cell, and estimate the shear strain.

Base parameters are fluid properties—density ρ and viscosity μ, the cell height h, the fluid velocity V, the cell area exposed to shear A, the cell shear modulus G, and the shear strain, which is what we want to find.

Density is kg/m^3, viscosity is kg/ms, height is m, area is m^2, velocity is m/s, and the shear modulus is $N/m^2 = kg/ms^2$. As our repeating parameters, we pick height, viscosity, and velocity. The remaining parameters are density, area, shear strain, and shear modulus. Right away, we recognize that the Reynolds number is one parameter. Area is normalized easily by A/h^2. The units of shear strain and radians are length/length or dimensionless.

However, the shear modulus takes some work. Let (m^a) $(m^b/s^b)(kg^c/m^c s^c)(kg^d/m^d s^{2d}) = 1$. The algebraic equations lead to

$$a + b - c - d = 0$$

$$- b - c - 2d = 0$$

$$c + d = 0.$$

Let us pick $c = 1$. Note that we are starting, not with a, but with c, so that it is easier to solve for the rest of the variables. From $c = 1$, we get $d = -1$. Then that leads

to $b = -c - 2d = -1 + 2 = 1$ using the second equation. Finally, using the first equation, $a + 1 - 1 + 1 = 0$ means $a = -1$. So our last Pi group is $V\mu/hG$. Our expression is then

$$f(\gamma, Re, A/h^2, V\mu/hG) = 0.$$

Note that instead, we can select another set of repeating parameters. We can select, for example, shear modulus, height, and density—remember, we need kg, m, and s all represented. If we go through the algebra with these new independent parameters, we get the following Pi groups in our functional form

$$F(\gamma, V\rho/G, A/h^2, \mu/h\rho G).$$

Here, the expression is much messier, and the functional form looks different. Could we have made a mistake? The answer is that despite their different appearances, we can convert from one to the other. The shear strain, γ, and the A/h^2 term remains the same. But what happens when we multiply the second and last Pi groups in the bottom expression? We get $V\mu/hG$, which is just the last Pi group of the first expression. Similarly, if we divide the second Pi group by the last Pi group in the bottom expression, we get $\rho hV/\mu$, which is Re, the second Pi group in the first expression. So, they are equivalent, but one was easier than the other. And it was not obvious in the beginning which parameters to select. Thus, there can be an element of art involved.

Suppose we wish to build a model of the bacterium 1000 times its normal size, expose the model bacterium to fluid flow, and measure the drag force exerted on the model. Recall that our second dimensionless term is $\rho L v/\mu$, which is the Reynolds number. If we simply increase the length of the bacterium by a factor of 1000, the Reynolds number will also increase by a factor of 1000. To scale the experiment correctly and keep the Reynolds number constant, one could use a fluid that has the same density but with 1000 times the viscosity and run the bacterium model at the same velocity as the actual bacterium. We can use our first dimensionless term to see how this would affect our measured force. In particular, the first dimensionless parameter was $F/L\mu v$, indicating that if L and μ are scaled up by 1000, the drag force will be scaled up by a factor of $1000 \times 1000 = 1 \times 10^6$. In this case, the measured drag must be scaled down by the same amount—a million, not a thousand—to obtain the actual full-scale drag acting on the bacterium.

Dimensional parameters can be used to check analytical expressions

Another use of dimensionless quantities is that they can be used to check derived analytical expressions for inconsistencies. Consider the case in which we use simple physical arguments to derive a scaling relationship for the drag force imparted on a fixed object of cross-sectional area L^2 subjected to a flow of velocity V. If we assume viscous forces are small, we know from Bernoulli's equation that for a streamline which passes near the front of the object,

$$P \sim \rho V^2, \tag{4.94}$$

where P is pressure, and we have assumed that the change in height is negligible. The resultant force from this pressure would be expected to scale as the product of pressure and area,

$$F \sim PL^2. \tag{4.95}$$

Combining Equations 4.94 and 4.95, we obtain the scaling relationship

$$F \sim \rho V^2 L^2. \tag{4.96}$$

We now wish to check whether this relation is correct from a functional perspective by expressing the solution in terms of the dimensionless parameters we previously found. To do this, we rewrite Equation 4.96 as

$$F \sim \left(\frac{\rho VL}{\mu}\right)(\mu VL), \tag{4.97}$$

where we have separated the V^2 term and inserted viscosity. Finally, dividing by the first term on the right-hand side we obtain

$$\left(\frac{F}{\mu VL}\right) \sim \left(\frac{\rho VL}{\mu}\right), \tag{4.98}$$

which implies that the relation in Equation 4.96 can be expressed as a function of our dimensionless parameters. Although this does not necessarily mean that Equation 4.96 is correct, it gives confidence that the relation is accurate, since as Equation 4.98 is consistent with what we expect (it turns out that Equation 4.98 is indeed correct, as it gives the same scaling found in the so-called drag equation, which gives the drag force experienced by objects caused by movement through fluid). If the relation could not be expressed as a function of our dimensionless quantities, that would imply there was an error in either the derivation of Equation 4.96 or our dimensionless parameters.

Nota Bene

Scaling. Like dimensional analysis, the goal in scaling analysis is to obtain functional relationships that allow one to see roughly how some dependent parameter varies with other independent parameters. In this case, one may use a somewhat unrigorous analytical approach (as we did in deriving Equation 4.96). The goal is not an exact formula, but rather a relation that describes how one parameter changes with the rest. Here, scaling problems can be considered a combination of dimensional analysis and order-of-magnitude estimation. In scaling, as in dimensional analysis, constants such as π or ½ are commonly dropped.

Example 4.5: Determining the Reynolds number using a scaling relationship from the Navier–Stokes Equations

For a one-dimensional flow, we assume that the flows in y- and z-directions are negligible, and so are changes in those directions. The Navier–Stokes equations in x simplifies to

$$\rho(\partial u \partial t + u \partial u \partial x) = \rho g - \partial P \partial x + \mu \partial^2 u \partial x^2.$$

We are interested in the viscous versus inertial terms, so we ignore the time-varying term (the first term), and the gravitation and pressure terms. This simplifies to

$$\rho u \partial u \partial x = \mu \partial^2 u \partial x^2.$$

Let us use U as a characteristic velocity and L as a characteristic length scale. Then, $\partial u \partial x$ can be estimated as U/L, and, $\partial^2 u \partial x^2$ can be estimated as $(U/L)/L = U/L^2$. The inertial terms (associated with momentum) are on the left, which simplifies to $\rho U^2/L$, and the viscous terms are on the right, which simplifies to $\mu U/L^2$. If we want to form a ratio of inertial to viscous terms, we make the quotient of the left side versus the right side

$$(\rho U^2/L)/(\mu U/L^2) = \rho U L/\mu,$$

which is the Reynolds number. You can make similar ratios using the time-varying term to get the Womersley parameter, and other parameters based on gravity, pressure, etc. This also explains why in pipe flows, the transitional Reynolds number can be above 1; in fully developed, steady, one-dimensional pipe flows, the left-hand side is actually zero, because the term $\partial u \partial x$ is zero and there are technically no inertial terms at all. At the microscopic scale, however, there are imperfections in the pipe surface, which leads to non-zero v and w, as well as changing u. Those perturbations can (with increased velocity) induce turbulence, but usually at Re much greater than 1, more typically around 1000–2000.

Key Concepts

- A fluid differs from a solid in that it takes on the shape of the container in which it is placed.
- Hydrostatics is the study of pressure in a stationary fluid, and exerts pressure equally in all directions.
- Newtonian fluids are those in which the velocity gradient is proportional to shear stress. The proportionality constant is called the dynamic viscosity.
- Laminar flow involves ordered or aligned streamlines. In contrast, turbulent flows involve tortuous streamlines and mixing. The dimensionless quantity Reynolds number (*Re*) is a measure of velocity and governs the transition from laminar to turbulent flow.
- Generally, Reynolds numbers much less than one are dominated by viscosity and are laminar. Reynolds numbers much greater than one are dominated by inertia and are turbulent. The specific Reynolds number at which the transition occurs depends on the specific geometry. At the cellular scale, flow is typically laminar.
- The velocity profile of laminar Newtonian flow for simple geometries can sometimes be solved in closed form.
- If viscosity is small enough to be neglected, the flow is termed inviscid and is governed by Bernoulli's equation.

- The Navier–Stokes equations results from combining equilibrium (conservation of momentum), constitutive (Newtonian behavior), and kinematic (compatibility assumption) relationships. They describe the behavior of a very wide class of problems in fluid mechanics.
- For non-Newtonian fluids there is a nonlinear relationship between shear stress and velocity gradient. Power law fluids and Bingham plastics are two examples.
- Rheology is the study of materials with some capacity to flow. Viscoelastic materials exhibit both solid- and fluidlike behavior. The complex modulus can be used to describe viscoelastic behavior. It may depend nonlinearly on frequency, and power-law relationships can be effective descriptions of complicated viscoelasticity.
- Dimensional analysis, scaling, and estimation are methods that can provide insight into functional relationships when a specific equation is not available. Often these relationships are sufficient to solve important problems.
- Dimensional analysis exploits the requirement for units to be consistent in order to arrive at potential functional relationships and novel dimensionless numbers.

Problems

1. Determine the flow profile in a pipe with circular cross section, with inner radius r_0, using differential analysis, assuming pressure-driven flow with the same flow conditions as we used for the parallel-plate problem. Determine the ratio of the peak velocity to the average velocity. Is this ratio the same or different from that of the parallel-plate solution?

2. Show that for a Newtonian fluid, the complex shear modulus and complex viscosity simplify to pure fluid values under an oscillatory input.

3. Plot the flow profiles of a power-law fluid where $\beta = 1$ and $\alpha = 0.5$ or 2. You may select the plate gap and peak velocity for convenience, but make sure you specify them in your response.

4. Using the Navier–Stokes equations, derive the shear stress acting on the bottom plate of a parallel-plate setup, with gap h, if there is both a pressure gradient ($dp/dx = C$) and an upper plate moving with velocity V_0, with the bottom plate stationary. The Newtonian fluid between the plates has density ρ and viscosity μ.

5. If a cell can be treated as a fluid-filled bag with the fluid having viscosity about 10 times that of water, determine whether the flow within a cell during micropipette aspiration is laminar, turbulent, or transitional. You may assume the pressure gradient is applied so that the cell takes about 1 min to enter the pipette completely, at roughly uniform velocity.

6. Determine a relationship for the radii of blood vessels that branch out from a parent vessel of radius R to two progeny branches of radii R_1 and R_2 (not necessarily equal) such that the shear stress resulting from laminar blood flow on the inside of these vessels remains constant. Ignore fluid effects at the bifurcation itself. You may assume that blood is Newtonian and incompressible.

7. In the text, we discussed the number of base parameters and units. A unit is independent if it cannot be expressed in terms of the other units. To derive the number of independent units, one can create a matrix based on the units at the top of each column and the parameters beside each row. Example:

$$
\begin{array}{c}
\text{time} \\
\text{velocity} \\
\text{acceleration}
\end{array}
\begin{bmatrix}
0 & 1 \\
1 & -1 \\
1 & -2
\end{bmatrix}
$$

Where the first column is meters and the second column is seconds. So time has no meters units and a single time component ($t = s^1$). The velocity has a meters component and a 1/s component ($v = m/s = m^1s^{-1}$), and similarly for acceleration. The rank of the matrix tells you the number of independent units, and if you perform Gaussian elimination, you can derive terms that can serve as the parameters on which you will base the dimensional analysis (these are not canceled out). So in the example, we can eliminate the time component of velocity and acceleration, showing them to be redundant with respect to time. If we eliminate acceleration, we get the rank = 2, and time and velocity are left.

(a) Based on this, go back to the fluid drag problem presented in the text and determine the unit matrix for that scenario, and show that the rank is 2. What other base parameter groups could we use? (Hint: in general, the dependent variable, force in this case, is typically not used in the base parameter groups, to avoid having it show up in multiple places in the function, thereby making it hard to isolate. It is partly for this reason some people prefer leaving the dependent variable outside the function.)

(b) Using a second base parameter group, derive the dimensionless form of the function. Based on this new set, if you decrease the length by a factor of 1000, keep the viscosity the same, keep the velocity the same, increase the density by 1000, and multiply the end force by 1000, can you still scale the experiment correctly? Does that make sense? Why?

8. Suppose you are considering a spring-dashpot system as a model to describe some process. The spring behavior can be described by Hooke's law, $F = kx$, where x is a displacement, F is the force, and k is the spring constant. The Newtonian dashpot behavior is described by $F = \mu v$, where v is the velocity, and μ is the viscous coefficient (not viscosity). You wish to determine the characteristic frequency of the system. Using dimensional analysis, find the dimensionless parameters governing this frequency.

9. Dimensional analysis is frequently used in fluid simulation. Let us say you are given a smooth sphere of radius r, suspended in a fluid of density ρ and viscosity μ with a spring of constant k, under gravity g. You stretch the spring out and release it to let the system oscillate. Use dimensional analysis to determine the relationship of the "half-life" time of damping to the other parameters. This half-life damping time refers to the time it takes the oscillations to reach one-half of the initial amplitude when the system is initialized.

10. How much power do you give off in body heat? That is, your presence in a room will warm the room from your body heat. If I were to replace you with a lamp of some wattage to achieve the same warming rate, what wattage bulb would I need?

11. Each cell in your body is very roughly the same density as water (in reality, a bit higher, but ignore that for this question). If half of a typical body weight is in matrix (bones, cartilage, etc.), how many cells does a typical adult have?

12. How much force could your biceps muscle generate if it were directly connected to a weight? Note that in the body, the bicep is actually levered to your forearm bones with the elbow as a fulcrum.

Annotated References

Fabry B, Maksym GN, Butler JP et al. (2001) Scaling the microrheology of living cells. *Phys. Rev. Lett. 87*, 1481–2. *This article describes how cells may be characterized as soft, glassy materials by analysis of the frequency response of beads attached the cells.*

Kamm R (2001) Molecular, Cellular, and Tissue Biomechanics. Lecture notes from course number 20.310, Massachusetts Institute of Technology. *This course introduced the concept of scaling analysis and estimation methods applied to cell mechanics. It was the inspiration for several of the authors to undertake this text, which has been heavily influenced by it as a result.*

Kollmannsberger P & Fabry B (2011) Linear and nonlinear rheology of living cells. *Annu. Rev. Mater. Res.* 41, 75–97. *A review of the rheological findings as applied to cells. This article contains many references to trace how this sort of modeling developed.*

Pritchard PJ (2011) Introduction to Fluid Mechanics. John Wiley. *This textbook on fluid mechanics covers in much more mathematical detail topics including hydrostatics, differential analysis, mass conservation, dimensional analysis, and the Navier–Stokes equations.*

Stamenovic D, Suki B, Fabry B, et al. (2004) Rheology of airway smooth muscle cells is associated with cytoskeletal contractile stress. *J. Appl. Physiol.* 96, 1600–1605. *This article uses the power-law presented in Equation 4.17 and shows an application of rheological analysis to studying cell response.*

Vogel S (1996) Life in Moving Fluids. Princeton University Press. *This book has a minimally mathematical description of biological fluids, with more focus as the organisms scale. It covers many key concepts about biology and fluids in an accessible way.*

CHAPTER 5

Statistical Mechanics Primer

The structural components of cells can often be considered as collections of many smaller, individual pieces. As we discussed in Chapter 2, polymers are large molecules consisting of individual monomers or groups of individual monomers joined together. It can be beneficial to determine how the properties of the smaller pieces affect the collective behavior of the whole. We might want to determine how much a given polymer curves (an experimentally observable, "macroscopic" property) as a function of some "microscopic" property, such as the number, size, charge, of the monomers. Alternatively, we may want to determine the force required to straighten the polymer, or to extend it to a certain length, as a function of these same microscopic properties. These questions can be addressed using the analytical framework of statistical mechanics. In statistical mechanics, our goal is to relate the behavior of a system's macroscopic behavior (characterized by "macrostates") to what we know about its microscopic properties and behavior (characterized by "microstates").

Statistical mechanics relies on the use of probabilistic distributions

Statistical mechanics is so named because it relies on the use of probabilistic distributions to form these relationships. The use of probability distributions allows for the analysis of systems with large numbers of degrees of freedom by assuming their collective behavior can be described by an appropriate statistical distribution. Consider the case in which we come across an imaginary (and very large) pool table with 1000 balls colliding. We wish to calculate the total kinetic energy of the balls at some moment in time. In classical mechanics, we would track each ball's velocity, calculate each ball's kinetic energy, and sum these values to get the total kinetic energy. In statistical mechanics, our approach is to assume some probability distribution for the velocities (10% of the balls have a velocity of 0–1 m/s, 15% of the balls have a velocity of 1–2 m/s, etc.), and use this distribution to calculate an expected value for the total kinetic energy.

Statistical mechanics can be used to investigate the influence of random molecular forces on mechanical behavior

The ability to formulate relationships between microscopic and macroscopic behavior is not the only beneficial feature of statistical mechanics. As we will see, statistical mechanics is particularly useful for analyzing the behavior of very soft structures, such as biopolymers and membranes under the influence of random molecular forces. The notion of random molecular forces influencing mechanical behavior can perhaps best be conveyed by considering a phenomenon called *Brownian* motion. Suppose we were to observe a very small particle (such as a pollen grain, which was used by Einstein in his seminal studies of this phenomenon; see Section 5.6) suspended in water at room temperature. If we were to observe its movements under a microscope, we would see that the particle is not stationary, but experiences small random fluctuations in its position.

Figure 5.1 Schematic depicting Brownian motion of a particle in fluid and thermal fluctuations of a polymer. In both cases, the behavior or configurations are caused by random forces due to the surrounding molecules.

particle: Brownian motion   polymer: thermal fluctuations

This phenomenon is termed Brownian motion and is caused by collisions between the water molecules and the particle, with small, instantaneous imbalances in the forces causing small fluctuations in position. Because the kinetic energy of the water molecules is associated with temperature, it comes as no surprise that the fluctuations are dependent on temperature. If the temperature is increased, the motion of the particle will also be increased.

These same random molecular forces that give rise to Brownian motion also exert an influence on the mechanical behavior of soft structures within cells such as actin microfilaments (as well as other biopolymers) which have been observed to undergo a phenomenon called *thermal fluctuations*. These fluctuations are manifested as polymer movements in the form of small wiggles or undulations. The tendency of the polymer to wiggle is due to the same molecular forces that cause Brownian motion. Just as increasing the temperature leads to greater Brownian motion–induced displacements, it also leads to a greater degree of thermal fluctuations. If we were to observe an actin polymer suspended in solution, we would see that as temperature is increased, the polymer would tend to exhibit more curvy or wiggly configurations (Figure 5.1). Conversely, if we were to lower the temperature, we would observe configurations that tended to be straighter. We will see that statistical mechanics is able to account for the influence of random molecular forces in this and other phenomena by accounting for the influence of entropy on equilibrium behavior.

We now present some of the basic analytical tools of statistical mechanics. We will be covering fundamental concepts and relations in statistical mechanics, such as internal energy, entropy, free energy, the Boltzmann distribution, and the partition function. We'll also be discussing random walks, a large class of mathematical problems often used in statistical mechanics analyses of membranes and polymers. By covering these topics, our goal is not only to give insight into the mathematical and/or physical origins of some of the polymer models in Chapters 7 and 8, but also to provide you with a basic understanding of statistical mechanics sufficient for more advanced topics in polymer physics in the future. Because statistical mechanics considers the thermodynamic energy of an object, we will begin this chapter with a discussion of a particular form of energy called internal energy.

5.1 INTERNAL ENERGY

Potential energy can be used to make predictions of mechanical behavior

As you may know, there are many forms of energy: kinetic, potential, thermal, electromagnetic, etc. These different forms of energy are unified by one common

---

Nota Bene

Degree of freedom. A degree of freedom refers to the number of independent "ways" something can maneuver. If the object in question can be treated as a small point, then typically the degrees of freedom refer to the number of dimensions available for motion. A bead on an abacus is provided one degree of motion, along the abacus rod. A single helium atom in the air would have three degrees of motion. However, if the object in question has geometry, then the degrees of freedom must also account for rotations around different axes. A book, for example, has six degrees of freedom because, in addition to being able to translate along the three spatial axes; it can also rotate about each axis.

principle: within a closed system, any form of energy can be transformed into another form, but the total energy of the system remains constant. In thermodynamics, we are concerned with several forms of energy, one of the most important being *internal energy*. For now, we will simply consider it as a form of energy that is the sum of multiple types of energy, one of the primary types being potential energy.

In mechanics, potential energy is defined as the capacity to do work. It is a particularly important form of energy in analyzing mechanical systems because of the *principle of minimum total potential energy*. This principle states that when a structure is subject to mechanical loading, it shall deform in such a way as to minimize the total potential energy in reaching equilibrium. The implications of this principle are that it does not matter if a structure is relatively simple (such as a single polymer) or relatively complex (such as an entangled network of one million polymers), potential energy can take the form of a single scalar quantity that represents the mechanical state of the structure. In addition, by finding the configuration that minimizes the potential energy, we can determine the equilibrium configuration of the structure under a given mechanical load.

Let us consider an example using springs to demonstrate how potential energy can be used to determine the equilibrium state of a mechanical system. Consider a Hookean spring. The force required to separate the ends of the spring by a distance x is given by $F = k_1(x - x_1)$, with a spring constant k_1 and equilibrium length x_1. If we have a second Hookean spring, we know that $F = k_2(x - x_2)$. In this case, the potential energies of the two springs are $W_1 = \frac{1}{2}k_1(x - x_1)^2$ and $W_2 = 0.5k_2(x - x_2)^2$. If we link the ends of both springs together so that their lengths are the same, and pull them by their ends so they begin to stretch, the total potential energy is

$$W = W_1 + W_2 = \frac{1}{2}k_1(x - x_1)^2 + \frac{1}{2}k_2(x - x_2)^2, \tag{5.1}$$

because their lengths are the same (Figure 5.2). As the length x is changed, the total potential energy of our spring system also changes. We can see that although this system contains multiple bodies (that is, two springs), the potential energy gives a single (scalar) representation of the mechanical state of this system. This quantity has a useful physical meaning, as it is equal to the work this system is capable of performing.

Now, suppose we allow the system to equilibrate without applying any external forces—we just let the system relax. We know that each spring has its own equilibrium length, however, now that their ends are joined, the two-spring system has its own equilibrium length. As we mentioned earlier, the principle of minimum total potential energy states that we can find the equilibrium state by finding the configuration that minimizes the total potential energy. For our spring system, this means finding the length x at which Equation 5.1 is minimized. We can find this length by taking the derivative of W with respect to x, and setting this equation to zero:

$$\mathrm{d}W/\mathrm{d}x = k_1(x - x_1) + k_2(x - x_2) = 0. \tag{5.2}$$

Solving for x, we get

$$x = (k_1x_1 + k_2x_2)/(k_1 + k_2), \tag{5.3}$$

which is the equilibrium length of our system.

Note that finding the equilibrium length via the principle of minimum total potential energy gives the same answer as that obtained by performing a force balance. We know that in equilibrium, without any external forces, there should be no net

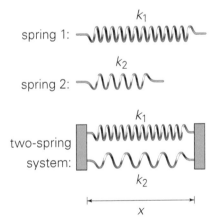

Figure 5.2 Schematic of a two-spring system. The equilibrium length of the system, x, can be found by minimizing the potential energy.

internal forces. Therefore any compression from one spring should balance the extension from the other. Here, the equilibrium length x can be computed as

$$F = k_1(x - x_1) = -k_2(x - x_2),$$ (5.4)

where we use a negative sign on the second spring to indicate that it is mechanically opposing the first spring (the second spring is compressing if the first spring is extending and vice versa). As expected, Equation 5.4 is identical to Equation 5.3. The reason performing a force balance is equivalent to potential energy minimization is that fundamentally, forces are the gradient of potential energy. Forces are balanced (net forces are zero) when there is no gradient of potential energy, which occurs at maxima or minima. Energy maxima are technically equilibrium points but are unstable, so they are omitted from the discussion (the classic example is an inverted pendulum). While performing a force balance and potential energy minimization produce equivalent results, in analyzing complex mechanical systems, the latter approach is often preferred due to mathematical simplicity. Potential energy has the added advantage that it allows us to marry the analytical machinery of continuum mechanics with that of statistical mechanics, as we will see in the next section.

Strain energy is potential energy stored in elastic deformations

When a body is subjected to a mechanical force, the body deforms. These deformations may either be elastic (in other words, they self-reverse upon removal of the force) or plastic (permanent deformations that do not reverse upon removal of the force). When a structure is subjected to elastic deformations, potential energy is stored in the structure; this form of potential energy is called *strain energy*. Note that all strain energy is a form of potential energy, but not vice versa. For example, raising an object a certain height increases its potential energy via opposition of the gravitational forces, but does not increase its strain energy, because the object is not deformed.

In Chapter 3, we learned about the concepts of stress and strain. Conveniently, strain energy can be described in terms of these quantities. To demonstrate this, we first derive the strain energy for a rod axially loaded at the tip. In Chapter 3, we found that the force was proportional to the displacement,

$$F = EA\Delta L/L.$$ (5.5)

The potential energy input into the rod to achieve this displacement is

$$W = \tfrac{1}{2}F\Delta L.$$ (5.6)

In the rod example we considered the material response from the force and displacement at the tip to be distributed throughout the material in the rod, in the form of stress and strain. The potential energy input into the rod by the force at the tip is analogously distributed through the material in the rod. For each small volume in the rod, a small element of potential energy is stored that we denote as dW. Thus, the total potential energy stored in the rod is the product of the volume of the rod and this small element of energy

$$W = LA dW.$$ (5.7)

If we equate the potential energy input at the tip with the internal potential energy stored, we get,

$$dW = \tfrac{1}{2}F/A\,\Delta L/L = \tfrac{1}{2}\sigma\varepsilon.$$ (5.8)

The internal potential energy is strain energy, because it is the energy stored in the form of strain (similar to the way the potential energy of a spring depends on its stretch). The internal potential energy per unit volume is the strain energy density.

Nota Bene

Microscopic strain energy.
Microscopically, by strain energy we mean that when we displace the tip of the rod a bit, all of the molecules or atoms making up the rod are separated or deformed a bit more from their equilibrium distances and shapes, resulting in spring-like potential energy storage throughout the rod.

We showed for an axially loaded rod, the strain energy density can be found as one-half of the product of stress and strain. In general, it can be shown that the strain energy density is one half of the component-wise product of stress and strain

$$dW = \tfrac{1}{2}(\sigma_{11}\varepsilon_{11} + \sigma_{12}\varepsilon_{12} + \sigma_{13}\varepsilon_{13} + \sigma_{21}\varepsilon_{21} + \sigma_{22}\varepsilon_{22}$$

$$+ \sigma_{23}\varepsilon_{23} + \sigma_{31}\varepsilon_{31} + \sigma_{32}\varepsilon_{32} + \sigma_{33}\varepsilon_{33}). \tag{5.9}$$

Equilibrium in continuum mechanics is a problem of strain energy minimization

In the development above, we assumed that the energy input into the rod by the force at the tip was equal to the stored internal strain energy. This assumption is actually a statement of the conservation of energy principal of classical mechanics. Outside of nuclear processes, energy can neither be created nor destroyed. We have also seen that the three fundamental relations that make up a problem in continuum mechanics are kinematics, constitutive, and equilibrium. However, there is an even more fundamental problem statement in continuum mechanics that arises from the principle of minimum total potential energy. Specifically, the problem can be given as finding the internal deformation state that both minimizes the internal strain energy and satisfies the boundary conditions (just as the two-spring problem could be given as finding the length that minimized the potential energy). Indeed, it can be shown that the equilibrium equation follows from assuming the minimization of strain energy and that the two are equivalent.

Changes in mechanical state alter internal energy

With our discussion of potential energy and strain energy in hand, we now introduce the concept of internal energy. In thermodynamics, *internal energy* is the total energy contained within the system and is defined as the capacity for a thermodynamic system to do work plus release heat. The two major components of internal energy are potential energy and kinetic energy.

We have already seen that elastic deformations are associated with changes in potential energy. These changes in potential energy arise out of configuration-dependent changes in potential energy between interacting atoms or molecules. The potential energy due to interactions between two molecules can vary greatly, depending on the distance between them. There may be van der Waal forces that result in long-range attractive forces, if they are dipoles. The potential energy due to these van der Waals forces would decrease as the distance between the molecules decreased. However, if the two molecules came too close together, the potential energy would quickly rise, owing to steric interactions between atoms of neighboring molecules (such as the energetically unfavorable interaction between the electron clouds of atoms). If the atoms of each molecule were charged, there would also be a Coulomb potential that would contribute to the energy. If the atoms within each molecule both had an equal charge of the same sign, the potential energy would increase as they came closer together.

In contrast to potential energy, which changes with alterations in mechanical configuration, changes in kinetic energy arise out of alterations in the velocity of the system's particles. There is an association between kinetic energy and temperature, because the temperature of an object is related to the speed of its fundamental particles. If we were to raise the temperature of an ideal gas, the average velocity of the gas particles would increase.

In our calculations, we will generally be interested in obtaining the equilibrium mechanical configuration for some system of interest. We will assume a reference state with zero internal energy, with changes in configuration to this reference state

Nota Bene

Stress can be defined through the strain energy. One of the most useful properties of strain energy is that it can be used to define stress. In our initial discussion of stress, we defined it in an intuitive fashion as a distributed or normalized force. However, this is a somewhat dissatisfying approach in that it is conceptually far from a rigorous thermodynamic quantity. On the other hand, the strain energy is the increase in potential energy stored locally in a material due to its deformation (or strain) and thus is a thermodynamic quantity with a clear definition. Critically, stress is simply the first derivative of the strain energy density with respect to strain,

$$\boldsymbol{\sigma} = dW/d\boldsymbol{\varepsilon}.$$

Often, this is a very useful definition of stress. Indeed the constitutive behavior of many complex materials is defined not in terms of a relationship between stress and strain, but in terms of their strain energy functions.

resulting in increases or decreases in internal energy. This will allow us to ignore contributions to the internal energy that (1) do not change (or which change relatively little) compared with the reference state, or (2) are largely decoupled from changes in configuration. To illustrate this, in many of our calculations, we will assume that changes in internal energy arise solely out of changes in potential energy, and ignore changes in internal energy arising from alterations in kinetic energy.

5.2 ENTROPY

Entropy is directly defined within statistical mechanics

Our focus now turns to another important thermodynamic quantity, entropy. You may have heard the vague explanation that entropy is related to disorder. Many people have heard the "messy room" analogy: a messy room is more disordered than a clean room, and so has higher entropy. One reason why the concept of entropy can be difficult to grasp is that no direct relation for entropy exists within thermodynamics. Instead of a direct definition, an incremental definition relates the change in entropy ΔS to a change in heat Δq of a system. In particular, for a constant temperature (isothermal) and reversible process (one that can be reversed without changing the system or its surroundings), the change in entropy is defined as

$$\Delta S = \Delta q / T, \tag{5.10}$$

where T is the absolute temperature and Δq is the amount of heat absorbed. We see that this is an indirect, incremental definition of entropy, as it only relates the change in entropy to thermodynamic quantities, rather than directly defining entropy itself.

In contrast, within statistical mechanics, entropy is directly defined. This definition of entropy was developed by Ludwig Boltzmann and was one his most important contributions. In particular, the Boltzmann definition of entropy is

$$S = k_B \ln \Omega. \tag{5.11}$$

Here, k_B is the Boltzmann constant and is equal to 1.38×10^{-23} J/K. Ω is defined as the density of states and is equal to the number of microstates for a given macrostate. Each of these terms—density of states, microstates, and macrostates—will be discussed in detail next.

Microstates, macrostates, and density of states can be exemplified in a three-coin system

Before formally defining microstates, macrostates, and density of states, we present an example to gain some intuition for these quantities. Consider a set of three coins placed inside a container (such as a coffee can): a nickel, a dime, and a quarter. We designate the nickel as coin 1, the dime as coin 2, and the quarter as coin 3. When we shake the container, this flips all three coins simultaneously. Each of the coins can land heads (denoted as h) or tails (denoted as t). Whether each coin lands heads or tail is random (mathematically, this is a random variable).

When we shake the container, we consider each microstate to be a possible outcome of a shaking event specified by whether each coin comes up heads or tails. If we were to shake the can, and the nickel comes up heads, the dime comes up tails, and the quarter comes up heads, this microstate, which we denote as m, can be written as

$$m = \text{hth}$$

Figure 5.3 Ludwig Boltzmann's tombstone in Vienna, Austria. (Courtesy of Thomas D. Schneider).

where the ith letter designates the outcome (that is, h or t) of coin number i. We say that this system's current microscopic state, or microstate, is "hth"

Given that there are three coins and two possible outcomes per coin, there are eight possible microstates. These individual microstates, denoted by m_x, are:

$$m_1 = \text{hhh}$$

$$m_2 = \text{hht}$$

$$m_3 = \text{thh}$$

$$m_4 = \text{hth}$$

$$m_5 = \text{htt}$$

$$m_6 = \text{tht}$$

$$m_7 = \text{tth}$$

$$m_8 = \text{ttt.}$$

We now turn our attention to macrostates. Imagine a case where in which we are not able to observe whether each coin lands heads or tail, but only some property that is dependent on the number of coins that lands heads. For instance, imagine that we place a small elf into the coffee container with the coins and cover the top of the coffee can with a lid (Figure 5.4). We tell the elf inside to count the number of coins that lands heads after each shake of the container, and yell out that number. We shake the container, and after the dime and the nickel (but not the quarter) lands heads, we hear a diminutive voice yell out, "two!" We shake the container again, and after all the coins land heads, we hear "three!"

We consider this number, the total number of heads, to be representative of the current system's macroscopic state, or macrostate. It is considered a macroscopic value because we cannot observe the exact configuration of the microscopic constituents that led to this macrostate (in other words, whether each coin landed heads or tails). Rather, we are only able to observe some macroscopic property. If

Figure 5.4 Example of a macroscopic property. Three coins and an elf are placed inside a coffee can. The can is covered and shaken. The elf yells out the number of coins that landed heads. The number of heads is considered a macroscopic value, because we cannot observe the exact configuration of the microscopic constituents associated with this macrostate (in other words, which coins came up heads or tails).

we define $W(m_x)$ to be the number of heads in microstate m_x, then for each of the eight microstates, the corresponding system macrostate is

$$m_1 = \text{hhh}: W(m_1) = 3$$

$$m_2 = \text{hht}: W(m_2) = 2$$

$$m_3 = \text{thh}: W(m_3) = 2$$

$$m_4 = \text{hth}: W(m_4) = 2$$

$$m_5 = \text{htt}: W(m_5) = 1$$

$$m_6 = \text{tht}: W(m_6) = 1$$

$$m_7 = \text{tth}: W(m_7) = 1$$

$$m_8 = \text{ttt}: W(m_8) = 0.$$

Notice that W can range from 0 to 3, and that the number of microstates associated with each macrostate is different. In particular, the macrostate corresponding to $W = 3$ has only one microstate associated with it (hhh), whereas the macrostate corresponding to $W = 2$ has three microstates associated with it (hht, thh, and hth). The specific microstates associated with each macrostate are:

$$W = 3: \text{hhh}$$

$$W = 2: \text{hht,thh,hth}$$

$$W = 1: \text{htt,tht,tth}$$

$$W = 0 : \text{ttt}.$$

We can now introduce the density of states (sometimes called the *multiplicity of states*), $\Omega(W)$, as the total number of microstates associated with each macrostate W. In our three-coin system, we know that for $W = 3$ (three heads), there is only one such microstate (hhh). In this case, $\Omega(W = 3) = 1$. For $W = 2$, there are three such microstates: hht, thh, and hth. Therefore, $\Omega(W = 2) = 3$. Similarly, $\Omega(W = 1) = 3$ and $\Omega(W = 0) = 1$.

Microstates, macrostates, and density of states provide insight to macroscopic system behavior

Now let us introduce more formal definitions for *microstates, macrostates,* and *density of states*. Remember, our goal is to take knowledge of the microscopic behavior and properties of some system (for example, a structure or body of interest) and gain insights into its macroscopic behavior. We do this by considering the microscopic behavior in a statistical, rather than a deterministic, fashion. In other words, we want to describe the system's average large-scale behavior and not its specific microscopic behavior over time. With this in mind, we provide the following definitions:

- **Microstate.** Short for microscopic state, it is the state of the system with regard to the specific (detailed) configuration of its microscopic components.
- **Macrostate.** Short for macroscopic state, it is the state of the system with regard to some macroscopic property. There may be multiple microstates associated with a given macrostate.
- **Macroscopic property.** A scalar property that describes the thermodynamic state of a multi-bodied system. These properties are considered macroscopic in that they are typically observable thermodynamic variables, such as pressure, temperature, or volume. Another example might include end-to-end length for a polymer.
- **Density of states.** A function of some macrostate characteristic, it describes the number of microstates associated with each macrostate with regard to some macroscopic property of interest.

Ensembles are collections of microstates sharing a common property

Before concluding this section, it is worthwhile to introduce the concept of ensembles. An *ensemble* refers to a collection of microstates that share a common property. In our three-coin system, one ensemble we could define is the collection of all microstates with at least one coin that is heads. This would consist of all possible microstates (hhh, hht, thh, hth, htt, tht, and tth) except for one (ttt). Another ensemble we could define is the collection of all microstates in which the nickel (coin 1) is heads (hhh, hht, hth, and htt). Although these may not seem like terribly useful ensembles, within statistical mechanics there are three important ensembles. The first, known as the *microcanonical ensemble*, consists of all microstates that share the same internal energy. The microcanonical ensemble is used in analyses of systems that are energetically isolated. The second, the *canonical ensemble*, consists of all microstates associated with a constant temperature system. This ensemble is used in analyses of systems that are allowed to exchange energy with a large thermal reservoir. The last, the *grand canonical ensemble*, is used in analyses of systems in which exchange of both energy and mass (or particles) can occur. We will use both the microcanonical ensemble and canonical ensemble, but the grand canonical ensemble is beyond the scope of this text.

Entropy is related to the number of microstates associated with a given macrostate

With our definitions of microstate, macrostate, and density of states in hand, we now revisit the Boltzmann definition of entropy, Equation 5.11. We saw that entropy is equal to the product of the Boltzmann constant and the log of the density of states. Because the density of states is a function of some macroscopic characteristic (in particular, it gives the number of microstates associated with each value of some macroscopic property), we can use this definition to calculate how entropy varies with a macroscopic property, so long as we can enumerate the density of states. Upon inspection of Equation 5.11, it is easy to see that for a given macroscopic property W, $S(W)$ is large when $\Omega(W)$ is large, and conversely, $S(W)$

is small when $\Omega(W)$ is small. Note that the minimum number of microstates associated with a particular macrostate is one. Therefore, the lowest possible value of Ω is 1; in this case, $S = k_B\ln(1) = 0$.

Example 5.1: Entropy calculation

Consider the density of states from our three-coin system:

$$\Omega(W = 0) = 1$$

$$\Omega(W = 1) = 2$$

$$\Omega(W = 2) = 2$$

$$\Omega(W = 3) = 1$$

Calculate the entropy for each value of W. For which value(s) of W is entropy the highest? The lowest?

We know that $S = k_B\ln\Omega$, thus

$$S(W = 0) = k_B\ln(1) = 0$$

$$S(W = 1) = k_B\ln(2) = 0.7k$$

$$S(W = 2) = k_B\ln(2) = 0.7k$$

$$S(W = 3) = k_B\ln(1) = 0$$

In this case, entropy is highest when W is equal to 1 or 2 (it is equally high in both cases), and lowest when W is equal to 0 or 3 (it is equally low in both cases).

5.3 FREE ENERGY

Equilibrium behavior for thermodynamic systems can be obtained via free energy minimization

We now introduce the concept of free energy, and explore ways in which this thermodynamic potential can be used to predict equilibrium behavior of thermodynamic systems. In Section 5.1, we learned about the principle of minimum total potential energy, which states that when a structure is subject to mechanical loading, it shall deform in such a way as to minimize the total potential energy. Within thermodynamics, there is an analogous principle called the *principle of minimum free energy*. The free energy Ψ (also known as the *Helmholtz* free energy or the energy available to do work at a constant temperature and volume) is the sum of internal energy W and the entropy S, where the latter is weighted by the negative absolute temperature T:

$$\Psi = W - TS. \tag{5.12}$$

The principle of minimum free energy states that a closed system (namely, a system which can exchange energy in the form of heat or work, but not matter, with its surroundings) at constant temperature will spontaneously adapt itself to lower its free energy. A thermodynamic equilibrium is reached when free energy is minimized. Just as potential energy minimization can be used to find the equilibrium state of mechanical systems, free energy minimization can be used to find the equilibrium state of thermodynamic systems.

Upon inspection of Equation 5.12, it becomes evident that for a given body, the free energy can be minimized in two distinct ways: through decreasing the internal energy W, or increasing the entropy S. The particular influence of energy versus entropy in the process of free energy minimization is determined by the relative magnitudes of W and S, as well as the temperature T. For example, at zero temperature, the entropy term vanishes, and the free energy is equal to the internal energy. Conversely, at very high temperatures, the TS term will dominate the free energy. When the entropic contribution to the free energy TS is comparable to the energetic contribution W (when $W = TS$.), a transition occurs between energy-dominated and entropy-dominated behavior.

One question that may arise is if we have two principles of energy minimization, one based on potential energy and one based on free energy, how do we know which one is appropriate for a given system? Also, would these two principles give contradictory behaviors? For everyday (nonmicroscopic) structures such as a steel beam in a bridge, we know that the principle of minimum potential energy dictates its deformation under loading. However, we could also consider the beam to be a thermodynamic body subject to the laws of thermodynamics, which state that its deformation should be dictated by the principle of minimum free energy. So which one is correct?

The answer is they are both correct. It is important to point out that the principle of potential energy minimization is not contradicted by the principle of free energy minimization; rather, it is encompassed in it. In particular, energetic effects for everyday structures are so dominant that they mask any entropic influences on their behavior. Mathematically, this can be expressed as $W \gg TS$; the free energy is equivalent to the internal energy. Remember from Section 5.1 that the primary components of internal energy are potential energy and kinetic energy; however, in many cases we can assume that for the microscopic structural components of cells such as polymers and membranes, kinetic energy is generally not dependent on their mechanical configurations. The same is true for everyday structures as well. Therefore, changes in internal energy can be attributed solely to changes in potential energy. In this case, for bodies in which $W \gg TS$, free energy minimization is equivalent to potential energy minimization. For brevity, we will refer to systems in which $W \gg TS$ as mechanical systems; systems in which $W \sim TS$ or $W \ll TS$ will be referred to as thermodynamic systems.

Temperature-dependence of end-to-end length in polymers arises out of competition between energy and entropy

We now explain the concept of the competing influences between energy and entropy as a system spontaneously minimizes its free energy. Let us revisit our example of a polymer subject to thermal fluctuations. If we were to observe a suspended actin polymer over time at room temperature, it would not be completely rigid, rather, it would continuously wiggle. Recall that the tendency of the polymer to wiggle is due to random molecular forces similar to those that cause Brownian motion. In the same way that increasing the temperature increases the small fluctuations in position of a suspended particle during Brownian motion, increasing the temperature increases the tendency of the polymer to take on an increasingly wiggly configuration (Figure 5.5).

To develop some intuition for the competition between energy and entropy in this temperature-dependent phenomenon, we present an example of a model polymer subject to thermal fluctuations. Note that the formal mathematical framework for this introductory example will be given in subsequent sections. We first need to define two lengths (Figure 5.6). The *contour length*, L, is the length of the polymer if it were completely straight. The *end-to-end length*, R, which, as the name implies, is the length of a straight line between one end of the polymer and the other, and depends on the actual configuration of the polymer. If the polymer is straight, $R = L$; in all other cases, $R < L$. Therefore, R is always less than or equal to L but greater than or equal to 0. If we were to observe a thermally fluctuating

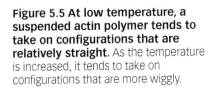

low temperature   high temperature

Figure 5.5 At low temperature, a suspended actin polymer tends to take on configurations that are relatively straight. As the temperature is increased, it tends to take on configurations that are more wiggly.

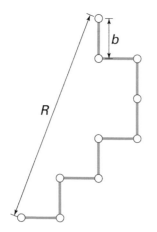

Figure 5.6 Model polymer. The links of size b are connected to each other by freely rotating hinges. The contour length L is the length of the polymer if it were completely straight. In this case, if there are n links, $L = nb$. The second length is the end-to-end length R, which is the length from one end of the polymer to the other, and depends on the polymer configuration.

actin polymer and quantify its end-to-end length as a function of temperature, we would see that on average, its end-to-end length decreases with increasing temperature, indicating that as temperature is increased, the configuration of the polymer tends to go from a straight configuration to a more wiggly one.

Now, consider a model polymer consisting of n rigid links of size b. The links are connected to each other by freely rotating hinges. We constrain the potential orientations of the links to simplify our analysis. In particular, we confine the orientations to be within a two-dimensional plane, and further, the links can be oriented only vertically or horizontally (so at each joint the relative angle between two adjacent segments is either 180°, 90°, or 0°; note that in the case of the former, the polymer is overlapping onto itself, which we allow for simplicity). Note that because there are n links of size b, then $L = nb$.

Remember that entropy is defined as the product of the Boltzmann constant and the log of the density of states Ω, and that Ω is defined as the number of microstates for a given macroscopic quantity. In our example, the microstates are the different possible configurations of the polymer, and the macroscopic quantity of interest is the end-to-end length R. Thus, finding Ω boils down to counting the number of ways in which our polymer can have a particular end-to-end length. Computing the density of states for every possible value of end-to-end length ($\Omega(R)$ for $0 \leq R \leq L$) would then allow us to determine the entropy associated with each value of R, $S(R)$.

While the mathematics for enumerating $\Omega(R)$ will be presented in later sections, a simple thought experiment will allow us to obtain a qualitative sense for how the density of states and entropy change with different values of R. Suppose the polymer is completely straightened out. The polymer will have an end-to-end length equal to the contour length, or $R = L$. Because we are only interested in the dependence of Ω on R, we do not distinguish between rotated states (to illustrate, a polymer straightened in the horizontal direction is considered to be equivalent to a polymer straightened in the vertical direction) or reflected states. In this case, there is only one polymer conformation in which this can occur, and so $\Omega(R = L) = 1$. Now suppose that the polymer is not completely straightened out, or $R < L$. There are multiple polymer conformations that may have the same end-to-end length (**Figure 5.7**) (in fact, mathematically, it can be shown that Ω is a

Figure 5.7 The model polymer in fully extended and less extended configurations. If $R = L$, there is only one possible configuration. In contrast, if $R < L$, there are multiple possible configurations.

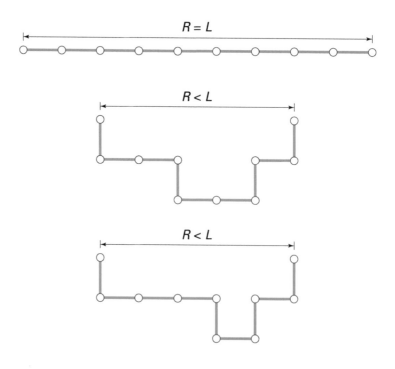

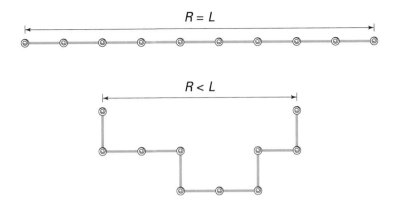

Figure 5.8 Schematic of a polymer in which the links are connected by rotational springs. No energy is stored in the springs (swirls) when the polymer is straight ($R = L$). However, work is required to bend the polymer ($R < L$), and this work is stored in the springs as strain energy.

maximum when $R = 0$), therefore,

$$\Omega(R < L) > \Omega(R = L). \tag{5.13}$$

The entropy is proportional to ln Ω, so the higher the value of Ω, the higher the entropy. Therefore, polymer macrostates with end-to-end lengths of $R < L$ are associated with higher values of entropy than the polymer macrostate $R = L$. Recall that equilibrium is achieved when the free energy is minimized, and that increasing the entropy decreases the free energy. In our example, the smaller the end-to-end length, the higher the entropy, and we therefore say that entropically, a coiled configuration is more favorable than a straight configuration.

Now let us look at the effect of internal energy on the end-to-end length. Consider the same polymer model as before, but this time imagine that instead of the links being connected by freely rotating hinges, we now straighten out the polymer and attach rotational springs between each link such that the springs are at their equilibrium position when adjacent links are straight (180° between the links) (Figure 5.8). The addition of these springs is a theoretical construct that represents an increase in internal energy if the polymer is bent. In this case, bending the polymer requires work, which becomes stored in the springs as strain energy. Thus, the polymer macrostate associated with $R = L$ is associated with zero internal energy ($W = 0$), while all other macrostates are associated with non-zero internal energy ($W > 0$). Because the free energy is decreased when the internal energy is reduced, we say that energetically, a straight configuration is favorable.

Now we know that entropically, a bent or coiled configuration is favorable in our model polymer, while energetically, a straight configuration is favorable. So the question becomes, what configuration will it tend to adopt? The answer is it depends on the temperature. If we look at the expression for free energy, we can see that the influence of energy or entropy on the free energy is determined by the temperature. At lower temperatures, internal energy dominates the free energy. The polymer will favor macrostates that are energetically favorable and will tend to be straighter. At higher temperatures, entropy dominates the free energy; the polymer will favor macrostates that are entropically favorable and will tend to be more wiggly.

In the next several sections, we will quantify this behavior by formulating a relationship that relates the equilibrium configuration of the polymer with temperature and its microscopic properties, such as the number of links and the energy of bending. We will perform the same calculation using two distinct analytical approaches: the microcanonical ensemble and the canonical ensemble.

5.4 MICROCANONICAL ENSEMBLE

In the previous section, we used a model polymer system to gain a qualitative understanding of the competition between energy and entropy in minimizing the system's free energy. The next section of this chapter focuses on developing an analytical framework to quantify this behavior. Indeed, determining macroscopic behavior from microscopic properties is a general goal of statistical mechanics. As we will see, there are several different ways within statistical mechanics to do this. In the next two sections, we present two separate analytical approaches for performing the same calculation: the microcanonical ensemble and the canonical ensemble. In analyzing our model polymer, both approaches ultimately give the same answer; however, we will see that their mathematical complexity can be vastly different.

The hairpinned polymer as a non-interacting two-level system

Before we begin with our calculations using the microcanonical ensemble, we first establish our model system. In this section, we will analyze a system of N particles called a *non-interacting two-level system*. It is called a two-level system because each particle can be in only one of two energy levels. The system is called non-interacting because each of the particles is considered independent. The energy level that each particle takes on is unaffected by the energy levels of the other particles. This type of system can be used to model a variety of biological and physical phenomena. In this example, we use it to idealize the behavior of a slightly modified version of the "hinged" polymer discussed earlier.

Consider a polymer consisting of $N + 1$ monomers and N linkages. The monomers are rigid links, with torsional springs between adjacent links. For the sake of simplicity, assume that the polymer is constrained to move in only one dimension. Instead of sites of 90° bending between adjacent links, we can think of the polymer as being able to undergo a 180° "hairpin" turn between adjacent links. Although this does not seem realistic, there are certain cases in which this can occur (Figure 5.9).

Because there are N linkages, there are a total of N possible hairpin sites. For simplicity, we shall assume that the energetic cost of a hairpin bend is ϵ (assume ϵ is positive). This energetic cost is the work that is necessary to twist the torsional springs 180°, and is stored in the springs as strain energy. If N_h is the number of sites that contain a hairpin, then the internal energy of a polymer with N_h hairpins is

$$W = N_h \epsilon. \tag{5.14}$$

A microcanonical ensemble can be used to determine constant energy microstates

With our model established, we now attempt to find the equilibrium configuration using an analytical tool called the microcanonical ensemble. Recall that the microcanonical ensemble refers to a group of microstates, all of which are associated

Figure 5.9 Schematic and micrograph of hairpin polymers. (A) Depiction of the model polymer undergoing 180° "hairpins." Between each pair of subunits, there can either be a hairpin, or not. The polymer is constrained to one dimension. (B) Sequential images of a hairpin within an actin filament that is straightening with time. (B, from, Dogic Z, Zhang J, Lau AWC et al. (2004) *Phys Rev Letter 92*. With permission from the American Physical Society.)

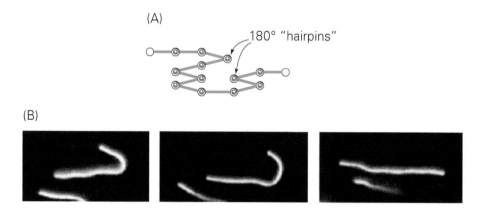

(A)

180° "hairpins"

(B)

with the same internal energy W. To apply the microcanonical ensemble to our model polymer, we consider it to be thermally and mechanically isolated from the surroundings. We consider a constant energy ensemble of microstates, with each microstate within the ensemble having the same energy W. The ensemble can be considered to be the collection of all microstates associated with the macrostate having energy W. Within such an ensemble, each microstate is equally likely to occur. Therefore, if there are Ω microstates at a particular energy W, each microstate within the ensemble occurs with the probability of $p = 1/\Omega$. Our goal is to calculate $S(W)$, or the entropy for this ensemble of microstates having energy W.

Entropy can be calculated via combinatorial enumeration of the density of states

Because entropy depends on the density of states, we must first calculate $\Omega(W)$, or the density of states Ω for each ensemble with energy W. We can find this using combinatorics. If we have N possible hairpin sites, the total number of configurations can be found using the bionomial coefficient. The binomial coefficient is

$$\binom{m}{n} = \frac{m!}{n!(m-n)!},\tag{5.15}$$

which gives the number of ways in which a group of n items can be chosen from a set of m. In our problem, we can use the binomial coefficient to find the total number of microstates in which N_h hairpins occur, given N total sites

$$\Omega(N_h) = \binom{N}{N_h} = \frac{N!}{N_h!(N-N_h)!}.\tag{5.16}$$

Now, we know that the entropy $S = k_B \ln\Omega$, or

$$S(N_h) = k_B\left(\ln N! - \ln N_h! - \ln(N-N_h)!\right).\tag{5.17}$$

But from Stirling's approximation, assuming N is very large,

$$S(N_h) = k_B\left(N\ln N - N_h\ln N_h - (N-N_h)\ln(N-N_h)\right).\tag{5.18}$$

We know that $W = N_h\epsilon$, so we can write the entropy in terms of the energy by substituting the expression $N_h = W/\epsilon$

$$S(W) = k_B\left(N\ln N - \frac{W}{\epsilon}\ln\frac{W}{\epsilon} - \left(N - \frac{W}{\epsilon}\right)\ln\left(N - \frac{W}{\epsilon}\right)\right).\tag{5.19}$$

Entropy is maximal when half the sites contain hairpins

Let us analyze our expression for $S(W)$, which is plotted in Figure 5.10. We see that the expression for entropy is equal to zero at its bounds, that is at $W = 0$ and $W = N\epsilon$. These points correspond to the cases in which none or all of the sites contain a hairpin, respectively. In both cases, the corresponding value for the density of states is one, resulting in zero entropy. The entropy is maximal between these two points, at $W = 0.5N\epsilon$. This value of internal energy corresponds to the case when half the sites contain a hairpin. This indicates the number of possible microstates is maximized when half the sites contain a hairpin.

S(W) can be used to predict equilibrium behavior

Now that we have invoked the microcanonical ensemble to form a relationship for $S(W)$, we can use thermodynamic relations to see how a group of these polymers

Figure 5.10 Entropy *S* as a function of internal energy *W*. For the hairpin model, polymer entropy is maximal when $W = 0.5N\epsilon$, which corresponds to the case when half the sites contain a hairpin.

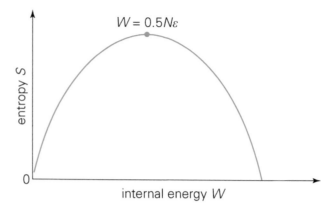

$W = 0.5N\epsilon$

entropy S

internal energy W

Nota Bene

Thermodynamic equations of state imply relations between state variables. The expression $\frac{1}{T} = \frac{\partial S}{\partial W}$ is called an equation of state because it gives a relation between state variables (macroscopic variables that describe the thermodynamic state of a system). Equations of state can be derived from the fundamental thermodynamic relation

$$dW = TdS - PdV + \mu dN,$$

where P is pressure, μ is chemical potential, and N is the number of particles. In particular, dW can equivalently be written as

$$dW = \left(\frac{\partial W}{\partial S}\right)_{V,N} dS + \left(\frac{\partial W}{\partial V}\right)_{S,N} dV$$
$$+ \left(\frac{\partial W}{\partial N}\right)_{S,V} dN,$$

where the subscripts following the derivative terms denote that those terms are kept constant. Comparing the two expressions, we obtain the equations of state

$$T = \left(\frac{\partial W}{\partial S}\right)_{V,N}, \quad P = -\left(\frac{\partial W}{\partial V}\right)_{S,N},$$

$$\text{and} \quad \mu = \left(\frac{\partial W}{\partial N}\right)_{S,V}.$$

would behave at thermal equilibrium. In particular, we know from thermodynamics at thermal equilibrium,

$$\frac{1}{T} = \frac{\partial S(W)}{\partial W}. \tag{5.20}$$

Combining Equation 5.19 and 5.20, we obtain

$$\frac{1}{T} = \frac{\partial S(W)}{\partial W}$$

$$= -\frac{k_B}{\epsilon}\left(\ln\frac{W}{\epsilon} - \frac{1}{N\epsilon}\ln\left(N - \frac{W}{\epsilon}\right)\right)$$

$$= -\frac{k_B}{\epsilon}\ln\left(\frac{\frac{W}{\epsilon}}{N - \frac{W}{\epsilon}}\right) \tag{5.21}$$

$$= -\frac{k_B}{\epsilon}\ln\left(\frac{1}{\frac{\epsilon N}{W} - 1}\right).$$

Solving for W yields

$$W = \frac{N\epsilon}{e^{\frac{\epsilon}{k_B T}} + 1}. \tag{5.22}$$

Because $W = N_h\epsilon$, the fraction of hairpin sites which contain hairpins can be written as

$$\frac{N_h}{N} = \frac{1}{e^{\frac{\epsilon}{k_B T}} + 1}. \tag{5.23}$$

The number of hairpins at equilibrium is dependent on temperature

Equation 5.23 gives the fraction of sites that contain a hairpin when the polymer is at thermal equilibrium (**Figure 5.11**). We see that at very high temperatures, the fraction asymptotically approaches the value of $N_h/N = 0.5$. Based on what we know about the entropy for this polymer, this makes intuitive sense. In particular, we know that at thermal equilibrium, the polymer will seek a state that minimizes its free energy, and that in the high temperature limit, the free energy is dominated by the entropy term. The free energy will be minimized when entropy is maximal and $N_h = N/2$, because entropy is maximal when $W = 0.5N\epsilon$.

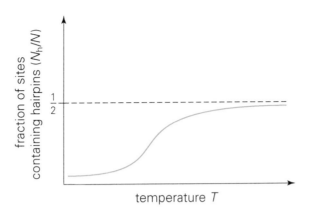

Figure 5.11 Fraction of sites containing a hairpin as a function of temperature. At low temperature, energy dominates the behavior, and very few hairpins occur. At high temperature, entropy dominates, and hairpins occur more frequently. In the limit of infinitely high temperature, half of the sites contain a hairpin.

What happens when you decrease the temperature? In the limit of $T = 0$, we know that

$$\lim_{T \to 0} \frac{N_{\mathrm{h}}}{N} = \lim_{T \to 0} \frac{1}{e^{\frac{\epsilon}{k_{\mathrm{B}} T}} + 1} = 0. \tag{5.24}$$

Because there are no hairpins, $W = 0$. In other words, in the low-temperature limit, the free energy is dominated by the internal energy, and the free energy will be minimized when internal energy is zero. This occurs when no sites contain a hairpin (for instance, when the polymer is "straight"). The transition between these two limits is sigmoidal, as can be seen in Figure 5.11.

Equilibrium obtained via the microcanonical ensemble is identical to that obtained via free energy minimization

In the last section we used the microcanonical ensemble to calculate $S(W)$, and then used thermodynamic relations to predict the behavior of the polymer when it is at thermal equilibrium at some constant temperature. Recall that the principle of free energy minimization states that a closed system at constant temperature will spontaneously adapt itself to lower its free energy, with thermodynamic equilibrium attained when free energy is minimized. This suggests that the equilibrium behavior obtained in Equation 5.23 using the equation of state Equation 5.20 should be identical to that obtained via free energy minimization. To confirm this, we must find the free energy as a function of the number of hairpins that occur. Recall Equation 5.18, in which we found the dependence of entropy on N_{h}. Because we also know the dependence of internal energy on N_{h} (see Equation 5.14), we can write an expression for the free energy as a function of the number of hairpins

$$\Psi(N_{\mathrm{h}}) = N_{\mathrm{h}}\epsilon - k_{\mathrm{B}}T\big(N \ln N - N_{\mathrm{h}} \ln N_{\mathrm{h}} - (N - N_{\mathrm{h}})\ln(N - N_{\mathrm{h}})\big). \tag{5.25}$$

Now that we have an expression for the free energy in terms of the number of hairpins, we can compute the number of hairpins at equilibrium by finding the value of N_{h} that minimizes the free energy. To do this, we take the derivative of the free energy with respect to N_{h} and set this equal to zero,

$$\frac{\partial \Psi}{\partial N_{\mathrm{h}}} = \frac{\partial W}{\partial N_{\mathrm{h}}} - T\frac{\partial S}{\partial N_{\mathrm{h}}} = 0. \tag{5.26}$$

Analyzing the internal energy term in this expression, we find

$$\frac{\partial W}{\partial N_{\mathrm{h}}} = \epsilon. \tag{5.27}$$

For the entropy term,

$$\frac{\partial S}{\partial N_{h}} = -k_{B}\left[\ln N_{h} - \ln(N - N_{h})\right] = -k_{B}\ln\left(\frac{N_{h}}{N - N_{h}}\right). \tag{5.28}$$

Plugging Equations 5.27 and 5.28 into W and solving for N_{h}/N,

$$\frac{N_{h}}{N} = \frac{1}{e^{\frac{\epsilon}{k_{B}T}} + 1}, \tag{5.29}$$

which is equivalent to Equation 5.23.

5.5 CANONICAL ENSEMBLE

In the last section, we derived an expression for the number of hairpins in our model polymer when it is at thermal equilibrium. Specifically, we used the microcanonical ensemble to obtain an expression for $S(W)$ and used this relation in conjunction with thermodynamic relations to derive an expression for the average energy/number of hairpins at thermal equilibrium. In this section, we seek to derive this same expression using an alternative approach for the calculation—the canonical ensemble. In the canonical approach, we consider an ensemble of microstates associated with a system at constant temperature. Although both the microcanoncial and canonical ensembles are theoretical constructs that can be used to perform the same calculations, the latter may be considered a more "natural" method for analyzing many problems, because experiments are typically performed at constant temperature rather than constant energy. In addition, we will see that unlike the microcanonical approach, the canonical approach permits analysis to be performed in a manner that does not necessitate the enumeration of the density of states, which can be mathematically difficult.

Canonical ensemble starting from the microcanonical ensemble

We now derive the relevant relations of the canonical ensemble. In particular, we derive two important relations, *Boltzmann's distribution* and the *partition function*. Later, these relations will be used to analyze our model polymer.

Consider a system of interest contained within a heat bath (**Figure 5.12**). The system is in thermal contact with the heat bath, such that it is maintained at constant temperature (in other words, the bath and the system of interest are in thermal

Figure 5.12 Schematic demonstrating the system of interest, heat bath, and total system. In the canonical ensemble, the system of interest and heat bath are in thermal contact, and we consider combinations of system of interest and bath microstates such that the energy of the total system is fixed.

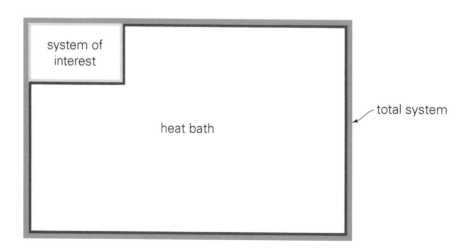

equilibrium). Both the system of interest and heat bath have associated microstates and internal energies. If we were to consider a polymer (the system of interest) placed in a very large chamber filled with an ideal gas (the heat bath), a system microstate would be a particular configuration of the polymer, and a bath microstate would be a particular state of the gas in which each gas particle had a particular position and velocity. Both microstates would be associated with a particular value of internal energy, one for the polymer, and one for the ideal gas.

Let the subscript "s" denote the system of interest and "b" denote the bath. In addition, let $Q_s(m_s)$ be the internal energy of the system microstate m_s, and $Q_b(m_b)$ be the internal energy of the bath microstate m_b. If we idealize the system and bath together as a single entity, we can consider an ensemble of different combinations of system and bath microstates such that the total internal energy (the sum of the internal energy of the system and the bath) is W_{tot}. In this manner, by considering an ensemble of microstates with a fixed value of energy, this approach is similar to the microcanonical ensemble example presented previously in this chapter.

Consider the case in which the system is in microstate m_s. The system's internal energy is $Q_s(m_s)$, and the internal energy of the heat bath is $W_{tot} - Q_s(m_s)$. An important realization (and a key step in the derivation) is that the probability for the system to be in microstate m_s is proportional to the number of heat bath microstates with energy $W_{tot} - Q_s(m_s)$. This can be expressed as

$$p(m_s) \propto \Omega_b \big(W_{tot} - Q_s(m_s) \big), \qquad (5.30)$$

where Ω_b is the bath density of states. We can write the right-hand side in the above expression as a function of entropy. In particular, $S = k_B \ln\Omega$, so the right-hand side of Equation 5.30 can be written as

$$\Omega_b(W_{tot} - Q_s(m_s)) = e^{\frac{S_b(W_{tot} - Q_s(m_s))}{k_B}}. \qquad (5.31)$$

Combining Equations 5.30 and 5.31, we find

$$p(m_s) \propto e^{\frac{S_b(W_{tot} - Q_s(m_s))}{k_B}}. \qquad (5.32)$$

Next, require the temperature of our system to remain constant. We can accomplish this by making the assumption that the heat bath is very large, such that changes in energy and entropy of the system do not affect the energy and entropy of the bath. Mathematically, this implies that for any system microstate m_s, $Q_s/W_{tot} \ll 1$, allowing us to perform a Taylor expansion. In particular, for very small displacements Δx between two points x and x_0, it can be shown that a "suitably smooth" function $f(x)$ can be approximated as

$$f(x) \approx f(x_0) + \Delta x f'(x), \qquad (5.33)$$

when $\Delta x / x_0 \ll 1$.

In our case, we can Taylor-expand S_b (assuming $x_0 = W_{tot}$, $x = W_{tot} - Q_s$, and $\Delta x = -Q_s$) using Equation 5.33 as

$$S_b \big(W_{tot} - Q_s(m_s) \big) \approx S_b \big(W_{tot} \big) - \frac{\partial S_b}{\partial W} Q_s(m_s). \qquad (5.34)$$

But, because at equilibrium,

$$\frac{\partial S_b}{\partial W} = \frac{1}{T_b}, \qquad (5.35)$$

where T_b is the temperature of the heat bath, Equation 5.34 can be rewritten as

$$S_b(W_{tot} - Q_s(m_s)) = S_b(W_{tot}) - \frac{Q_s(m_s)}{T_b}. \tag{5.36}$$

Combining Equations 5.32, 5.36, we finally find

$$p(m_s) \propto e^{\frac{S_b(W_{tot})}{k_B} - \frac{Q_s(m_s)}{k_B T}} = \frac{1}{Z} e^{\frac{-Q_s(m_s)}{k_B T}}, \tag{5.37}$$

where in the right-hand-most relation in Equation 5.37 we have turned the proportionality for the probability into an equality by introducing a normalization factor,

$$Z = \sum_{m_s} e^{\frac{-Q_s(m_s)}{k_B T}}. \tag{5.38}$$

Note that the summation in Z is over all possible discrete system microstates m_s, and that we have lumped the $\exp(S_b(W_{tot})/k_B)$ term within Z, as it does not depend on m_s.

Although we have thus far only considered discrete microstates, we will often consider systems that possess a continuous distribution of microstates. In this case, Z can equivalently be written as

$$Z = \int_{m_s} e^{\frac{-Q_s(m_s)}{k_B T}} dm_s. \tag{5.39}$$

The probability distribution in Equation 5.37 and normalization factor in Equations 5.38 and 5.39 are the key relations of the canonical ensemble. Equation 5.37 is called *Boltzmann's distribution*, and Equations 5.37 and 5.38 are called the canonical *partition function*.

Probability distribution from the canonical ensemble gives Boltzmann's law

The probability distribution in Equation 5.36, Boltzmann's distribution (or *Boltzmann's law*), implies that for the system described in the canonical (constant temperature) ensemble, the probabilities of the individual microstates are not evenly distributed. Instead, the probability goes as the inverse exponential of the energy of that microstate. Interestingly, Boltzmann's law can be applied to an amazingly diverse set of systems. For example, it can be used to predict the distribution of electrons under an electric field, or the diffusion of molecules or proteins experiencing Brownian motion. As long as the system is at thermal equilibrium with and can exchange energy, but not mass, with its surroundings, the microstates will be distributed following a Boltzmann's distribution.

It is worthwhile discussing the differences between discrete and continuous probability distributions, such as those that appear within the sum and integral in Equations 5.38 and 5.39, as they are associated with different quantities. In particular, in the discrete case, the probability distribution in Equation 5.38 gives the probability of microstates with energy Q, while in the continuous case, it gives the probability of a range of microstates within a specified interval. Therefore, unlike the discrete distribution in Equation 5.38, we cannot evaluate the probability of a specific microstate in Equation 5.39. Instead, we can calculate the probability of a range of microstates.

Example 5.2: Boltzmann distribution (discrete)

Consider a large number of cells migrating on the surface of a culture dish. Empirically, we observe that the migration velocity of the cells, in an appropriate fixed time interval, can exhibit a Boltzmann distribution, with the energy Q equal to a constant α times the migration velocity (in Section 10.1, we will show that this may be a reasonable model in some cases, as this form of energy arises due to energetic effects associated with adhesion energy and viscous dissipation). For simplicity, assume that the velocities can only take discrete values of 0, 1, or 2 µm/s. Let the energy be constant $\alpha = 10^{-20}$ W/µm and assume the temperature is 37°C. Determine the probability of observing a cell migrating at 0, 1, or 2 µm/s.

Because this is a discrete problem, we first calculate Z using the summation:

$$Z = \sum_{v=0}^{2} \exp\left(\frac{-\alpha v}{k_B T}\right) = 1 + 0.097 + 0.009 = 1.106$$

Knowing Z, we can now compute that the fraction of cells with zero velocity is simply the total number of cells times the probability that the cell will have zero velocity,

$$p(v = 0) = \frac{1}{Z}\exp\left(\frac{0}{k_B T}\right) = \frac{1}{1.106} = 0.904$$

$$p(v = 1) = \frac{1}{Z}\exp\left(\frac{-\alpha}{k_B T}\right) = \frac{0.097}{1.106} = 0.087$$

$$p(v = 2) = \frac{1}{Z}\exp\left(\frac{-\alpha}{k_B T}\right) = \frac{0.009}{1.106} = 0.008.$$

These results show that a Boltzmann distribution is heavily weighted toward the lower end, and there is a exponential drop-off in probability as you increase the energy level being measured. Note that the sum of the probabilities is not exactly 1, due to rounding.

Example 5.3: Boltzmann distribution (continuous)

Now consider the same plate of cells with migration velocities that follow a Boltzmann distribution. However, in this case, assume that the velocities are not discrete values, but instead range continuously between 0 and 2 µm/s. We cannot compute the probability of observing a cell migrating at a specific velocity, but rather within some specified range. Compute the probability of observing a cell with migration velocity in the range 0–1 µm/s or 1–2 µm/s.

First, we need to compute Z,

$$Z = \int_{v=0}^{2} \exp\left(\frac{-\alpha v}{k_B T}\right) dv = \frac{-k_B T}{\alpha}\left(\exp\left(\frac{-2\alpha}{k_B T}\right) - 1\right) = 0.424.$$

Now, we can compute the following probabilities

$$p(0 \le v \le 1) = \frac{1}{Z}\int_{v=0}^{1} \exp\left(\frac{-\alpha v}{k_B T}\right) dv = \frac{0.386}{0.424} = 0.912$$

$$p(1 \le v \le 2) = \frac{1}{Z}\int_{v=1}^{2} \exp\left(\frac{-\alpha v}{k_B T}\right) dv = \frac{0.037}{0.424} = 0.088.$$

You can see that the probabilities when using the continuous distribution decay similarly with energy as compared with the case using the discrete distribution.

The free energy at equilibrium can be found using the partition function

We now turn our attention from the Boltzmann distribution to the other important relation obtained from the canonical approach, the partition function Z from Equations 5.37 and 5.38. At first glance Z looks to be just a normalization constant for the Boltzmann probability. However, it is an extremely valuable quantity that can be used to calculate equilibrium behavior. The partition function is related to the free energy and can be used to calculate the free energy at equilibrium without having to enumerate the density of states.

Consider the case in which we have a system of interest as defined in the canonical ensemble. We know that the Boltzmann distribution gives the probability $p(m_s)$ of a microstate with microscopic energy $Q(m_s)$. To show that this distribution is related to the free energy, consider the probability of the system having internal energy W at thermal equilibrium. We shall denote this probability $p(W)$. We know that for a macroscopic value of W, the system may have multiple microstates. The probability of the system having a macroscopic internal energy W can be calculated as the total number of microstates with energy $Q = W$, multiplied by the associated Boltzmann probability,

$$p(W) = \frac{1}{Z} e^{\frac{-W}{k_B T}} \Omega(W).$$ (5.40)

But because

$$\Omega(W) = e^{S/k_B} = e^{ST/k_B T},$$ (5.41)

we can see that Equation 5.40 is related to the free energy as

$$p(W) = \frac{1}{Z} e^{\frac{-(W-TS)}{k_B T}} = \frac{1}{Z} e^{-\beta \Psi},$$ (5.42)

where $\beta = 1/k_B T$ and Ψ is the free energy. Equation 5.42 implies that analogous to how the Boltzmann distribution implies that the probability of a microstate goes as the inverse exponential of its microscopic energy the probability of a macrostate (which can be associated with a different number of microstates) goes as the inverse exponential of its free energy. With this relation in hand, we are now able to show that Z is related to the free energy.

We previously showed that Z can be obtained by integrating the Boltzmann distribution over all microstates, as in Equations 5.38 and 5.39. But we also know that for Equation 5.42 to yield the correct probability for $p(W)$,

$$Z = \int e^{-\beta \Psi(W)} dW$$ (5.43)

or in the discrete case, Equation 5.43,

$$Z = \sum_W e^{-\beta \Psi(W)}.$$ (5.44)

Collectively, our analysis shows that Z can be computed in two distinct ways, using Equation 5.39 or Equation 5.43. The main difference between these two expressions is that in Equation 5.43, the exponential term contains the free energy Ψ instead of the microstate energy Q, and that the integral is over all macroscopic energies W instead of microstates m_s.

For very large systems, integrals or sums of the form in Equation 5.43 and Equation 5.44 can be approximated using the saddle point method, as long as the quantity being integrated or summed over is an *extensive* quantity. An extensive quantity is one that scales linearly with the number of particles in the system, whereas an *intensive* quantity is one that does not depend on the size of the system. For instance, mass is an extensive quantity, while temperature is an intensive quantity. Free energy is an extensive quantity, so the above expression can be approximated using saddle point integration. In particular,

$$Z = e^{\frac{-\Psi_{min}}{k_B T}},$$ (5.45)

where Ψ_{min} is the free energy evaluated at which $\Psi(W)$ is a minimum. Rearranging this expression,

$$\Psi_{min} = -k_B \ln(Z).$$ (5.46)

Equation 5.46 has important implications, in that it reveals that the partition function has greater utility than just serving as a normalization constant for the Boltzmann distribution. In particular, we know that for closed systems, the free energy is minimized at thermal equilibrium. Therefore, Equation 5.46 implies that to compute this value of free energy at equilibrium for a given system, one merely needs to compute the partition function using Equation 5.38.

Advanced Material: Saddle point approximation can be used to approximate summations of certain exponential quantities

Statistical mechanics deals with determining the overall behavior of systems consisting of many bodies and/or degrees of freedom. Within treatments in statistical mechanics, it is often useful to make mathematical approximations when the number of degrees of freedom is very large. For example, in statistical mechanics, we often have to find summations of exponential quantities of the form

$$S = \sum_{i=1}^{n} e^{y_i N},$$

where N is a large number and y_i is real. Let y_{max} be the maximum value of y_i. Because $y_i N$ is in the exponential, as N is increased, the sum will very quickly become dominated by the term corresponding to $y_i = y_{max}$. In fact, it can be shown that for a very large N,

$$S \approx e^{N y_{max}}.$$

In other words, instead of performing the summation, one simply needs to calculate the largest term in the summation. It also implies that for integrals of the form

$$S = \int e^{N y(x)} dx \approx e^{N y_{max}}$$

for very large N. Specifically, $y(x)$ is a function that is bounded at $\pm \infty$ independent of N, and y_{max} is its global maximum value. Computing the integral in this manner is referred to as saddle point integration.

Now, define a new function $z(x) = N y(x)$. In other words, z is an extensive quantity.

In this case,

$$S = e^{z_{max}}.$$

Thus, if $z(x)$ is an extensive quantity, we can approximate S using saddle point integration, as long as N is sufficiently large.

The internal energy at equilibrium can be determined using the partition function

In the last section, we saw that the partition function can be used to calculate the free energy at equilibrium. We show here that it can also be used to compute the internal energy. This is demonstrated in the following example. Let's say we want to calculate the internal energy W of the system at equilibrium. W can be calculated by summing over each of the microstate energies $Q(m)$, multiplied by the probability $p(m)$ of that microstate occurring:

$$\langle Q \rangle = \sum_{m} p(m) Q(m) = \frac{1}{Z} \sum_{m} e^{-\beta Q(m)} Q(m), \tag{5.47}$$

where $\beta = 1/(k_B T)$. Note that in Equation 5.47, the brackets denote the expected value based on an average over all microstates. Equation 5.47 can equivalently be written as

$$= -\frac{1}{Z} \sum_{m} \frac{\partial}{\partial \beta} e^{-\beta Q(m)} = -\frac{1}{Z} \frac{\partial}{\partial \beta} \sum_{m} e^{-\beta Q(m)} = -\frac{1}{Z} \frac{\partial Z}{\partial \beta}. \tag{5.48}$$

But we know that the partial derivative of $\ln Z$ is defined as

$$\frac{\partial \ln Z}{\partial \beta} = \frac{1}{Z}\frac{\partial Z}{\partial \beta}. \tag{5.49}$$

Therefore,

$$W = \langle Q \rangle = -\frac{\partial \ln Z}{\partial \beta}. \tag{5.50}$$

This shows that, like the free energy, the internal energy at equilibrium can also be found from the partition function. Again, we are able to find the equilibrium internal energy without having to enumerate the density of states.

We now know that if we calculate the partition function, we can calculate Ψ and W at equilibrium. If we want to know S, it can be calculated easily, because $\Psi = W - TS$. Thus, we can specify the thermodynamic state of the system at equilibrium from its microscopic behavior, as long as we can calculate the partition function Z. In the next section, we revisit our model polymer to demonstrate use of these relations.

Using the canonical approach may be preferable for analyzing thermodynamic systems

Let us reconsider the non-interacting two-level system (the hairpin polymer), but this time, analyze it using the relations derived from the canonical ensemble. We know that the energy Q of a given microstate is

$$Q = \epsilon \sum_{i=1}^{N} n_i, \tag{5.51}$$

where n_i equals one if a hairpin occurs at hairpin site n_i, or zero if a hairpin does not occur. Combining the microstate energy in Equation 5.51 with the Boltzmann distribution, we obtain an expression for the probability of a given microstate in thermal equilibrium,

$$p(n_1, n_2, ..., n_N) = \frac{1}{Z} e^{-\beta \epsilon \sum_{i=1}^{N} n_i}, \tag{5.52}$$

where Z is the discrete partition function given in Equation 5.37

$$Z = \sum_m e^{-\beta \epsilon \sum_{i=1}^{N} n_i}. \tag{5.53}$$

To compute Z, we need to perform a sum over all possible microstates. To do this, we perform N summations over each individual hairpin site,

$$Z = \sum_m e^{-\beta \epsilon \sum_{i=1}^{N} n_i} = \sum_{n_1=0}^{1} \sum_{n_2=0}^{1} ... \sum_{n_N=0}^{1} e^{-\beta \epsilon \sum_{i=1}^{N} n_i}. \tag{5.54}$$

The summation in the exponential in the rightmost relation in Equation 5.52 can be written out explicitly as

$$Z = \sum_{n_1=0}^{1} \sum_{n_2=0}^{1} ... \sum_{n_N=0}^{1} e^{-\beta \epsilon n_1} e^{-\beta \epsilon n_2} ... e^{-\beta \epsilon n_N}. \tag{5.55}$$

Now, because $n_1, n_2, \ldots n_N$ are independent variables, we can separate the sums

$$Z = \sum_{n_1=0}^{1} e^{-\beta \epsilon n_1} \sum_{n_2=0}^{1} e^{-\beta \epsilon n_2} \ldots \sum_{n_N=0}^{1} e^{-\beta \epsilon n_N}. \qquad (5.56)$$

In Equation 5.56, each of the sums are numerically equivalent. Therefore, can rewrite the above expression as

$$Z = \left(\sum_{n=0}^{1} e^{-\beta \epsilon n} \right)^N = z^N, \qquad (5.57)$$

where

$$z = \sum_{n=0}^{1} e^{-\beta \epsilon n} = 1 + e^{-\beta \epsilon} \qquad (5.58)$$

is termed the *single partition function*. With the partition function, we can calculate the free energy at equilibrium as

$$\Psi = -\frac{\ln Z}{\beta} = -\frac{N \ln(1 + e^{-\beta \epsilon})}{\beta} \qquad (5.59)$$

and the equilibrium internal energy as

$$W = \frac{\partial \ln Z}{\partial \beta} = N \epsilon \frac{e^{-\beta \epsilon}}{1 + e^{-\beta \epsilon}} = \frac{N \epsilon}{e^{\beta \epsilon} + 1}. \qquad (5.60)$$

Note that Equation 5.60 is identical to Equation 5.22, which we derived through the calculation of $S(W)$ in the microcanonical ensemble (in conjunction with thermodynamic relations). Although we performed the same calculation in two different ways, we can see that mathematics were quite different, with the use of the canonical approach avoiding the need to use combinatorics to compute the density of states. For this reason, the canonical ensemble may often be preferable for analyzing thermodynamic systems.

> **Nota Bene**
>
> **Partition functions for systems of non-interacting particles can be calculated from single partition functions.** In Equation 5.57 we saw the partition function was calculated as
>
> $$Z = z^N,$$
>
> where Z is the single partition function. This will be the case for any system of N identical particles that do not interact. For instance, in the case of the hairpinned polymer, the N hairpin sites are identical (they all can undergo the same range of motion) and they do not interact (whether one site contains a hairpin is not influenced by whether a hairpin occurs elsewhere.)

5.6 RANDOM WALKS

We now leave our discussion of the equilibrium behavior of our model polymer and discuss another classical topic of statistical mechanics: the random walk. Random walks are a large class of mathematical problems in which at discrete points in time a "walker" moves in space a discrete distance. They have application to a wide range of problems in fields as diverse as physics, chemistry, computer science, biology, and even economics. In this section we will introduce them in their most basic form with some simple analysis. We will also show how they relate to another fundamental statistical process: diffusion. Note that we will revisit the concept of random walks in Section 7.4 in our discussion of polymer mechanics.

A simple random walk can be demonstrated using soccer

Consider a soccer player standing at midfield on a soccer field, facing one of the goals. We define midfield to be at location $r = 0$. The player takes out a coin, and flips it. If it comes up heads, the player steps forward a distance b toward one goal; if it comes up tails, the player steps backward toward the opposite goal. This "flip and step" process is repeated a certain number of times. We will now calculate the probability of the player being at a certain location after a given number of steps. In particular, after n random, b-sized steps, what is the probability that the player will end up at location r? Note that if the player has taken more backward steps than forward steps, r is negative, and vice versa.

If the player takes n steps, and for each step they can only move backwards or forwards, then there are 2^n different paths that can be taken (although the player may end up in the same position with many different paths). The number M of those paths in which only n_+ of the steps are forward can be calculated using the binomial coefficient from combinatorics. To see why the binomial coefficient can be used, imagine that the "objects" are the coin flips, and that the "picks," or the selection of an object, are how many heads (or positive steps) occur. Then, if there are n total flips and n_+ of them come out heads, M must be

$$M = \binom{n}{n_+}. \tag{5.61}$$

Our goal is to calculate the probability of the player being at point r after n flips and steps. We have the following relationships, $n = n_+ + n_-$ and $r = b(n_+ - n_-)$, where n_- is the number of backward steps. Combining these yields

$$n_+ = \frac{n + \dfrac{r}{b}}{2}. \tag{5.62}$$

And,

$$M(n, r) = \binom{n}{\dfrac{n + \dfrac{r}{b}}{2}} = \frac{n!}{\dfrac{n + \dfrac{r}{b}}{2}! \, \dfrac{n - \dfrac{r}{b}}{2}!}. \tag{5.63}$$

The probability that the player will end up at position r after n steps ($p(n, r)$) is simply the number of ways or "paths" the player could use to get to point r divided by the total number of paths the player might take. M is the total number of ways to get to r and, as we noted above, there are 2^n total paths available to the player. Therefore, the probability is

$$p(n, r) = \frac{M(n, r)}{2^n} = \frac{1}{2^n} \frac{n!}{\dfrac{n + \dfrac{r}{b}}{2}! \, \dfrac{n - \dfrac{r}{b}}{2}!}. \tag{5.64}$$

Equation 5.64 is the exact expression for the probability that the player will be at a position r on the field after n steps of size b. To get a feel for what the random walk distribution looks like for different step numbers, we calculate $p(n, r)$ for various values of n in **Figure 5.13**. Note that the appearance of the distribution begins to resemble a normal distribution as n gets larger. This is perhaps not unexpected, as the position r results from the sum of several independent coin-flipping events, with the probability of heads or tails the same at each step (in other words a 50/50 chance). The central limit theorem states that for a sufficiently large number of steps, r will be approximated by a normal distribution, with this approximation becoming better as the number of steps are increased. This is an important result, as in many cases we are interested in the behavior of random walks in the limit as n becomes very large. We can rewrite Equation 5.64 in the limit of large n in the form of a normal or Gaussian distribution,

$$p(n, r) = \frac{1}{\sqrt{2\pi n b^2}} e^{-r^2/2nb^2}. \tag{5.65}$$

Equation 5.65 gives the probability distribution for a one-dimensional random walk for n steps of length b in the limit of large n; the full derivation for Equation

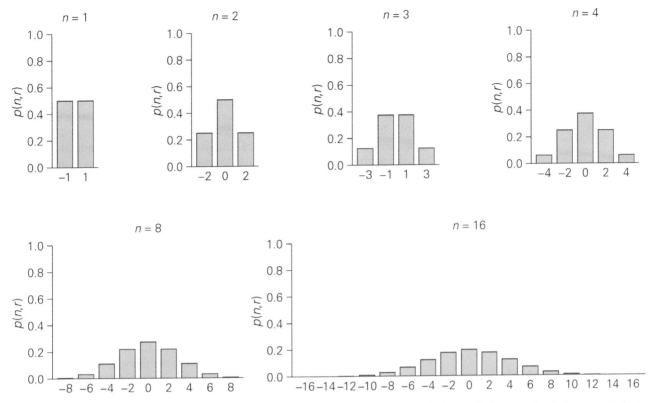

Figure 5.13 Parametric plots of Equation 5.65. As n is increased, the distribution can be increasingly approximated as a normal or Gaussian distribution. For simplicity we have assumed $b=1$.

5.65 is in Section 7.4. It can be shown that for a random walk of n steps of size b, the root mean-square position is

$$\sqrt{\langle r^2 \rangle} = b\sqrt{n}. \tag{5.66}$$

The derivation of Equation 5.66 is left as an exercise for you. Combining Equation 5.65 and 5.66, we obtain an alternative expression for the random walk probability distribution,

$$p(n,r) = \frac{1}{\sqrt{2\pi \langle r^2 \rangle}} e^{-r^2 / \langle r^2 \rangle}. \tag{5.67}$$

The diffusion equation can be derived from the random walk

At the molecular level, the process of diffusion is governed by the random motion of particles. It is perhaps not surprising that the behavior of a random walk scaled up to the continuum level can be described by the diffusion equation. This relationship was first demonstrated by Albert Einstein in a 1905 paper he published while working in the Swiss patent office. He demonstrated that Brownian motion, the random motion of small particles suspended in a fluid, results in a rate of diffusion related to the velocity of the individual particles. His theoretical results fit the observed data so well, that they provided convincing support for the atomic theory of matter before there was direct evidence of the existence of atoms or molecules.

The one-dimensional diffusion equation is known as Fick's second law (this law is covered in more detail in Chapter 9.3), and is given as

$$\frac{dC}{dt} = D \frac{d^2C}{dx^2}, \tag{5.68}$$

Nota Bene

Recurrence in random walks is dependent on the dimension of the walk. Although it is beyond our scope here, it can be shown that the random walk example we have used here is *recurrent*. This means that eventually, the player will visit every point on the playing field, even if the field is infinitely wide. In the context of games of chance, this property of recurrence is sometimes called the *Gambler's Ruin*. The idea is that a gambler's funds at any point in time can be considered analogous to the soccer player's position on the field. The flip and step of the soccer player is the same as each spin of the wheel, roll of the dice, or hand of cards. No matter how large the gambler's initial bankroll (so long as it is finite), he or she will eventually go bust, assuming the bets are always of the same amount, even if the odds were fair (instead of favoring the house, as it usually does). Interestingly, and perhaps surprisingly, the property of recurrence of random walk processes can be shown to be limited to one- and two-dimensional situations only. In three dimensions (or higher), the walker will not visit all available coordinates. Random walks of these types are known as *transient*.

where C is the concentration and D is the diffusion coefficient. We now show that this equation can be derived from a random walk process.

First, let's recast our statement of the random walk problem slightly. Consider a particle that is constrained to move in only one dimension. We define its initial position to be at $r = 0$. After some time t, we are interested in the probability that the particle will end up at some new location r. In recasting our random walk, we replace the number of steps n with time t, which we will initially treat as a discrete variable. In this case, we are able to assess where the particle is at discrete time points. In regards to the particle position, we will treat r as a continuous variable. Thus, the particle can move arbitrary distances. To allow for steps of any distance, not just steps to the right and left of a specified increment (for instance, we do not use a fixed step size b), we need to introduce another probability function, $p(x)$. This function gives the probability distribution for a particle making a step (positive or negative) of distance x in any time step. We can now state the new problem in terms of the probability distribution of finding the walker at a certain position, r, at time $t + 1$ in terms of the distribution at time t. Specifically,

$$p(t + 1, r) = \int_{-\infty}^{\infty} p(x) p(t, r - x) \mathrm{d}x. \tag{5.69}$$

This type of relationship is known as a *recursion* relationship. It defines p at one point in time in terms of its distribution at the prior time. Next we can perform a Taylor series expansion for the $p(n, r - x)$ term in the integral about r to obtain

$$p(t + 1, r) = \int_{-\infty}^{\infty} p(x) \left((pt, r) - x \frac{\mathrm{d}p(t, r)}{\mathrm{d}r} + \frac{1}{2} x^2 \frac{\mathrm{d}^2 p(t, r)}{\mathrm{d}r^2} + O^3(x) \right) \mathrm{d}x$$

$$= \int_{-\infty}^{\infty} p(x) p(t, r) \mathrm{d}x - \int_{-\infty}^{\infty} p(x) x \frac{\mathrm{d}p(t, r)}{\mathrm{d}r} \mathrm{d}x + \frac{1}{2} \int_{-\infty}^{\infty} p(x) x^2 \frac{\mathrm{d}^2 p(t, r)}{\mathrm{d}r^2} \mathrm{d}x + O^3(x),$$

$$\tag{5.70}$$

where $O^3(x)$ represents terms of third order or higher in x. Because $p(t, r)$ and its derivatives do not depend on x, these terms can be removed from the integrals. In this case, Equation 5.70 can be rewritten as

$$p(t + 1, r) = p(t, r) \int_{-\infty}^{\infty} p(x) \mathrm{d}x - \frac{\mathrm{d}p(t, r)}{\mathrm{d}r} \int_{-\infty}^{\infty} p(x) x \, \mathrm{d}x + \frac{1}{2} \frac{\mathrm{d}^2 p(t, r)}{\mathrm{d}r^2} \int_{-\infty}^{\infty} p(x) x^2 \mathrm{d}x + O^3(x). \tag{5.71}$$

Next, we seek to simplify each of the terms on the right-hand side of Equation 5.71. For the first term, we know that probability distributions possess the property that integrals over their entire range must be unity. Therefore, for the first term,

$$p(t, r) \int_{-\infty}^{\infty} p(x) \mathrm{d}x = p(t, r). \tag{5.72}$$

The second term in Equation 5.71 can be simplified if we restrict the probability distribution, $p(x)$, to be isotropic. In other words, there is no bias in the probability of a particle to make a step in one direction or another, $p(x) = p(-x)$. This condition also implies that there is no net motion of the particles, owing, for example, to convection should there be flow in the underlying medium. This restriction implies that the integral in the second term is zero, thus

$$\frac{\mathrm{d}p(t, r)}{\mathrm{d}r} \int_{-\infty}^{\infty} p(x) x \mathrm{d}x = 0. \tag{5.73}$$

Finally, we assume that the higher-order terms from the Taylor series expansion can be neglected, which can be shown to be valid for some small region around r as long as p is a continuous function. The net result is that

$$p(t+1,r) - p(t,r) = \frac{1}{2}\frac{d^2 p(t,r)}{dr^2}\int_{-\infty}^{\infty} p(x)x^2 dx. \tag{5.74}$$

Now, consider the right-hand side in Equation 5.74. It is the integral over all possible positions of the particle of the position squared times the probability of being in that position. This is the definition of the *average* squared position, a term that we encountered earlier in this chapter. In Section 7.4, we show that for a random walk of n steps with fixed step size b, $\langle r^2 \rangle = nb^2$ (see Equation 5.66). It can be similarly shown that for a random walk with variable step size but with an average squared displacement of Δr^2 per step,

$$\langle r^2 \rangle = n\Delta r^2, \tag{5.75}$$

where n is the number steps. Combining Equations 5.74 and 5.75, we get

$$p(t+1,r) - p(t,r) = \frac{t\Delta r^2}{2}\frac{d^2 p(t,r)}{dr^2}. \tag{5.76}$$

Now, in the limit of a large number of particles, all behaving according to this equation, the concentration of particles at a given location is proportional to the probability of finding a single particle at that position. Therefore, we can replace $P(n,r)$ by the concentration C. Also, if we divide the equation by Δt, the left-hand side becomes the time derivative of concentration in the continuous limit and we have

$$\frac{dC}{dt} = \frac{n\Delta r^2}{2\Delta t}\frac{d^2 C}{dr^2}. \tag{5.77}$$

This is, in fact, the one-dimensional diffusion equation with

$$D = \frac{n\Delta r^2}{2\Delta t}. \tag{5.78}$$

We will build on our treatment of random walks in subsequent chapters. We will use the random walk to better understand the behavior of polymers in Section 7.4. We will also learn how confining diffusion into a two-dimensional membrane can greatly enhance the kinetics of diffusion-limited biochemical reaction in Section 9.1.

Key Concepts

- Within statistical mechanics, entropy is directly defined as $S = k_B \ln\Omega$. Ω is the density of states gives the number of microstates associated with a given value of a macrostate.
- Thermal equilibrium of thermodynamic bodies is governed by two competing phenomena: internal energy, and entropy. At low temperatures, internal energy–driven phenomena dominate a body's behavior, whereas the influence of entropy increases with increasing temperature. The competing nature of energy and entropy in equilibrium behavior is captured by a thermodynamic potential called the free energy, defined as $\Psi = W - TS$.
- The principle of minimum total potential energy states that when a structure is subject to mechanical loading, it shall deform in such a way as to minimize

the total potential energy. The principle of minimum free energy states that a closed system at constant temperature will spontaneously adapt itself to lower its free energy. Just as potential energy minimization can be used to find the equilibrium state of mechanical systems, free energy minimization can be used to find the equilibrium state of thermodynamic systems.

- Different analytical approaches can be used to determine behavior of thermodynamic systems at equilibrium, such as the microcanonical ensemble and the canonical ensemble. In the microcanonical ensemble, we analyze microstates at constant energy. Thermal equilibrium can be obtained from relations derived from the microcanonical ensemble.

- In the canonical ensemble, we analyze microstates in a system held at constant temperature. The probability of a given microstate is given by the Boltzmann distribution. The partition function is a normalization factor for the Boltzmann distribution, and is related to the minimum value of free energy. We can specify the thermodynamic state of the system at equilibrium by calculating the partition function without enumerating the density of states.

- Random walks are a class of mathematical problems in which at discrete points in time a "walker" moves in space a specified distance. Through its use we can relate certain macroscopic behavior of thermodynamic bodies to microscopic properties.

- When the behavior of random walks is scaled up to the continuum level, their behavior is described by the diffusion equation.

Problems

1. We know that the mechanical behaviors of everyday rod-like objects, such as steel beams and jump ropes, can be predicted through the principle of minimum total potential energy, but not soft structures such as biopolymers, whose behavior is significantly influenced by entropy. We know that a transition between energy- and entropy-dominated behavior occurs when $W = TS$. Given the statistical mechanics definition of entropy, we can estimate a characteristic energy at which this transition occurs as k_B (why would this be the case?). Calculate this value of energy assuming a temperature of 37°C.

   Now, consider a length of rope and a strand of DNA. Calculate the strain energy stored in a jump rope bent 180 degrees with a constant radius of curvature of 1 m, assuming it can be modeled as an elastic cylindrical beam with a Young's modulus of 100 MPa, and a radius of 0.5 cm. Do the same for a strand of DNA with the same curvature. Again, assume that the strand of DNA can be modeled as an elastic cylindrical beam, assuming a Young's modulus of 1.9 GPa and a radius of 1 nm. How do these values compare to the characteristic transition energy?

2. Two dice are rolled, each of which can generate a number between 1 and 6. Consider this a statistical system. How many microstates are there for this system? How many macrostates are there? What is the density of states, Ω, for each macrostate? What is the probability of each macrostate if the dice are rolled many times?

   Now suppose that the "fair" dice are substituted with a pair of "loaded" dice. In the loaded dice there is a small weight placed under the 1 spot. This makes the probability of getting a 6 slightly higher than 1/6, and the probability of getting a 1 slightly lower than 1/6.

   Assume that the amount of this bias is a small number, ϵ. Now, what is the probability of each macrostate?

3. Consider a strand of DNA which only contains two different nucleotides, A or T. Using the binomial coefficient, calculate the number of different possible sequences if the strand is N nucleotides long, and only N_1 of the nucleotides are T.

4. Consider 100 people in a building. Every time a person moves up one floor, it takes them $1k_BT$ of energy. Assume that given enough time the people distribute themselves according to Boltzmann's distribution. How many people are on each floor on average? How tall does the building need to be to accommodate this average distribution of people? How tall would it need to be if there were 10 people? And if there were 1000 people?

5. Spring is in the air and a mosquito is in your apartment. You are watching it fly and land on a set of stairs. Assuming that it needs $1k_BT$ of energy to fly up from one step to the next, what would a Boltzmann's distribution predict for the probability of finding it on any given step, if there are 10 steps in total?

6. Consider a molecule that has a "home" position and a linear restoring force that is proportional to the distance that it is away from this position. In other words, the molecule acts like it is attached to the tip of a spring. The restoring force is $F = -kx$, where x is the distance from the home position and k is the spring constant. The potential energy is a function of position and can be expressed as

$$W = \frac{1}{2}kx^2.$$

We can treat each molecular position of the molecule, x, as a microstate of the system. In this case, assuming thermal equilibrium, the probability of finding the molecule in any given position is given by the Boltzmann distribution. What is the average energy? Hint: a table of integral identities may be useful.

Note: your finding is known as the *principle of equipartition of energy* and holds for the energy associated with any parameter that has an associated energy that varies with the square of the parameter. For example, it also works for kinetic energy, which varies with the square of speed.

7. In this problem, we show that for an isolated system, equilibrium is attained by maximizing the entropy (this is the second law of thermodynamics and is equivalent to stating that equilibrium is attained by minimizing the free energy, assuming constant energy). Imagine a box that is isolated from the external environment, meaning that the total energy of the contents within the box is constant. The box contains a partition that divides it into two subspaces. The partition is constructed such that it can slide freely along the length of the box, similar to a plunger in a syringe; however, it does not allow energy to transfer between the two subspaces. We now fill both sides of the partition with some amount of gas. The partition slides around until it comes to rest at its equilibrium position. How would you expect the pressures to be related at equilibrium? Now formalize this intuition, by showing that entropy is maximized when the pressures are equal. Hint: we know that the total entropy of the system is

$$S_{\text{tot}} = S_1(N_1, V_1, E_1) + S_2(N_2, V_2, E_2)$$

You will need the thermodynamic definition of pressure to complete the derivation,

$$\frac{p}{T} = \left(\frac{\partial S}{\partial V} \right)_{E,N}.$$

8. Imagine the same box as in Problem 7, except the partition is constructed differently. Specifically, it does not slide; however, energy (for instance, heat) is allowed to transfer freely between the two compartments. How would you expect the temperatures of the two gases to be related at equilibrium? Formalize this result by showing that entropy is maximized when the temperatures of the two gases are equal.

9. Consider a mechanosensitive ion channel that can be in one of two states: open or closed. Define the energy of the channel using a configuration parameter σ that can be equal to 0 (closed) or 1 (open). The energy is

$$E = \sigma \varepsilon_{\text{open}} + (1 - \sigma)\varepsilon_{\text{closed}} - \tau \Delta A$$

where $\varepsilon_{\text{open}}$ and $\varepsilon_{\text{closed}}$ are the energies of the channel in the open and closed configurations, τ is the tension applied to the channel, and ΔA is the change in area of the channel in going from closed to open. Write expressions for the partition function, the probability of the channel being in an open state, and average energy.

What is the probability that the channel is open, assuming that $\tau = 1, 2, 3, 4,$ and 5 pN/nm, $\Delta \tau = -5k_BT$, and $\Delta A = 10$ nm^2?

10. Consider a two-dimensional model polymer composed of $n = 4$ rigid segments, with neighboring segments connected via a rotating joint. Assume that the angle between adjacent links θ can only take discrete values: 0° (hairpin), 90°, or 180° (straight between links). Calculate the value of the density of states for two different end-to-end lengths $R = 2$ and $R = 4$. Calculate the entropy for these same values of R. Which value of R is entropically more favorable?

11. Consider the "90° polymer" depicted in Figure 5.6. Write a MATLAB module that analyzes a polymer made up of N segments of unit length and which determines the number of configurations possible for each end-to-end length between 0 and N.

Annotated References

Berg HC (1993) Random Walks in Biology. Princeton University Press. *This book, first released in 1983 and expanded ten years later, is an excellent source for examples of random walks in biology, including diffusion.*

Chandler D (1987) Introduction to Modern Statistical Mechanics. Oxford University Press. *A concise introduction to elementary statistical mechanics.*

Dogic Z, Zhang J, Lau AWC et al. (2004) Elongation and fluctuations of semiflexible polymers in a nematic solvent. *Phys. Rev. Lett.* 92, 125503. *Reports the formation of hairpin defects in actin filaments (similar to that in our model polymer) suspended in anisotropic solutions of aligned rod-like macromolecules.*

McQuarrie DA (2000) Statistical Mechanics. University Science Books. *A widely used text in introductory statistical mechanics courses.*

Pande VS (2006) Graduate Statistical Mechanics. (lecture notes from course number Chem 275, Stanford University, Stanford, CA.) *The general structure of this chapter is based on a one-semester course on graduate statistical mechanical developed by Dr Pande at Stanford University. The treatments of specific topics in this chapter, including enumeration of states via the three-coin system, and analysis of the two-level noninteracting system via the microcanonical and canonical ensemble, are based on his course notes, which are currently in development as a formal text.*

Phillips R, Kondev J & Theriot J (2009) Physical Biology of the Cell. Garland Science. *This is an excellent text that gives many examples of insights that can be made into biological phenomena through statistical mechanics. (Note: Problem 6 is adapted from the text of Section 5.5.2 in* Physical Biology of the Cell; *Problem 7 is adapted from Section 5.5.2; Problem 8 is adapted from Section 7.1.2.)*

CHAPTER 6

Cell Mechanics in the Laboratory

In scientific investigation, the shortcomings of experimental techniques are often the limiting factor in our ability to gain a better understanding of the principles underlying phenomena of interest. It comes as no surprise, then, that the development of novel experimental approaches can have profound effects on advancing our understanding of cell mechanics and mechanobiology. Engineering principles have long been used to better understand the role of mechanics in regulating biological processes; however, our understanding of the mechanics of cells has generally lagged behind that of tissues. This is due in part to the technical challenges associated with investigating mechanical behavior at the small length scales associated with cells. In this chapter, we present an overview of some common experimental approaches for mechanically manipulating cells and investigating whole-cell mechanical behavior. In particular, we describe experimental techniques for cell probing, such as magnetic micromanipulation, atomic force microscopy, and optical trapping. We also describe approaches for quantifying traction forces in cells and analyzing the mechanics of cells under physiological mechanical stimuli such as matrix deformation and fluid shear stress. Finally, we conclude this chapter with a brief discussion of experimental design. Our goal is to provide a basic understanding of how each technique works and approaches for analyzing data obtained thereby (and, for the interested reader, resources for learning about a particular technique in more detail). Although we give relatively brief descriptions of each technique, the contribution of these techniques to the field of cell mechanics should not be underemphasized. Indeed, the use of these approaches led to most of the insights in cell mechanics and mechanobiology presented in this text.

6.1 PROBING THE MECHANICAL BEHAVIOR OF CELLS THROUGH CELLULAR MICROMANIPULATION

We begin this chapter with techniques for cellular micromanipulation. To better understand why such techniques could be useful for investigating the mechanics of cells, consider a situation in which we are given a sample of some unknown material, and asked to measure its stiffness. How could we do this? One approach is to subject the sample to mechanical loading of known magnitude and measure the deformation. For instance, if we sought to characterize the stiffness of a spring, we could hang a weight on it and measure its elongation. Or, if we had a homogenous, isotropic, linear elastic material, we could determine its Young's modulus by measuring its strain under a known stress. But what if we wished to characterize the mechanical stiffness of an individual cell? Micromanipulation approaches allow for application of known mechanical forces on the small scale of the cell.

Known forces can be applied to cells through the use of cell-bound beads and an electromagnet

Nota Bene

Magnetic beads. The name
"magnetic beads" is commonly used
to refer to beads used in magnetic
micromanipulation experiments. The
beads themselves are not usually
magnetic; they are typically iron-
bearing beads.

In magnetic micromanipulation, localized forces are applied to cells by adhering ferrous beads, typically on the order of micrometers in diameter, to the cell surface, and then applying electromagnetic forces to the beads. In some cases, the beads are introduced into the cell interior by endocytosis. An electromagnet can be constructed by winding wire around a metal core. Shaping the tip of the core to a chisel (typically, a few hundred micrometers across) results in a strong magnetic field that is generally laterally homogeneous with a strength that depends only on distance from the tip.

A representative system for magnetic micromanipulation is shown in Figure 6.1. Cells that have been seeded in a dish and that have had ferrous beads bound to their surface are placed on a temperature-controlled stage of an inverted microscope. An electromagnet is held above the dish with the tip immersed in the media. A cell with a bead attached is located under the microscope, and the tip of the magnet is positioned near it. Current is passed through the electromagnet, generating a magnetic field that causes the bead to be pulled toward the magnet tip. The force applied is determined by the distance from the bead to the magnet tip. Bead movements are determined microscopically (Figure 6.2) with particle-tracking algorithms (see Section 6.2). To study the behavior of specific molecules, so-called *functionalized* (surface modification is explained at the end of Section 6.2.) beads can be used that have been coated with the protein of interest, which can engage specific receptors of interest on the cell surface.

The dependence of force on distance from the magnet tip can be calibrated through Stokes' law

In magnetic micromanipulation experiments, the force applied to the bead is a function of the distance to the tip. This relationship is determined in a calibration procedure. One approach is to suspend the beads in a high-viscosity fluid. When the magnet is activated, the beads will reach a *terminal velocity*, the constant velocity at which the drag forces on the beads balance the magnetic forces pulling on them. The terminal velocity is related to the force applied on them through Stokes' law,

$$F = 3\pi\mu DV,$$

(6.1)

Figure 6.1 A magnetic micromanipulator mounted on a microscope stage. The electromagnet is wound with copper wires and placed in a micromanipulator (in back, on the right side of the image). The tip is immersed in the dish containing cells; the dish is placed on a temperature-controlled unit. (From, Huang H, Kamm RD, Lee RT (2004) *Am. J. Physiol. Cell Physiol.* 287).

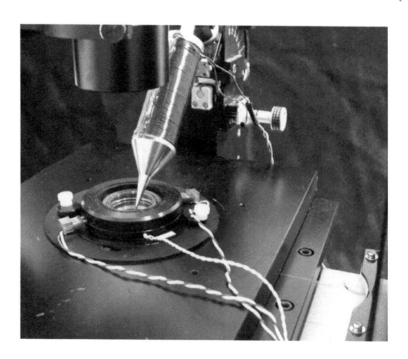

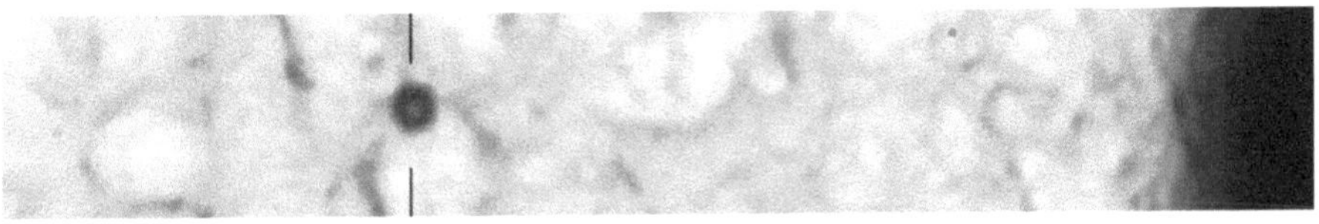

Figure 6.2 The bead is the dark round object left of center, with the lines drawn in to make it easier to visualize bead motion. The shadow on the right is the electromagnet tip. Cells are visible in the background, but typically during experiments the light is increased to white out the cells and obtain clearer tracking of the bead.

where F is the force, μ is the dynamic viscosity, D is the bead diameter, and V is the terminal velocity. By performing this procedure on beads at various distances from the electromagnet tip, one can estimate the bead force as a function of distance. When using this approach, it is important that flow remain laminar, as Stokes' formula is only applicable at very low Reynolds numbers (see Section 4.2).

Magnetic twisting and multiple-pole micromanipulators can apply stresses to many cells simultaneously

Although magnetic micromanipulation is useful for investigating the mechanical behavior of single cells, it is often desirable to apply mechanical loads to many cells simultaneously. Magnetic twisting is an alternative method that uses magnetic beads to apply forces to many simultaneous beads. An electromagnet is used to magnetize beads temporarily in one direction with a powerful pulse. This is quickly followed by a pulse in a perpendicular direction, causing a twisting movement to be applied to the beads to align them with the new field (Figure 6.3). As the beads twist, their individual magnetic fields add together to produce a net magnetic field that is rotated from the original orientation. A magnetometer, which is simply a wire coil attached to a sensitive amplifier, is used to measure the magnitude of the rotated magnetic field. It is possible to obtain the relationship between torque and rotation, in an average sense, for hundreds of beads simultaneously.

Optical traps generate forces on particles through transfer of light momentum

Light is another way to impart forces to cell-bound beads. Perhaps you recall that photons are massless, but nonetheless carry momentum. Light is able to impart momentum to matter with which it interacts. This fundamental property of light is exploited with high-power lasers to build *optical traps* (also termed *optical tweezers*).

To understand how an optical trap works, consider an transparent particle at the center of a focused laser beam (Figure 6.4). The index of refraction of the particle is higher than the surrounding environment, such that as light passes through the particle, the beams change direction or bend. The particle is redirecting the light beam and therefore changing the light's momentum. By conservation of momentum, there must be a force acting on the particle. The bending of the beams leads to a force gradient that pushes the bead toward the focal point and into the center

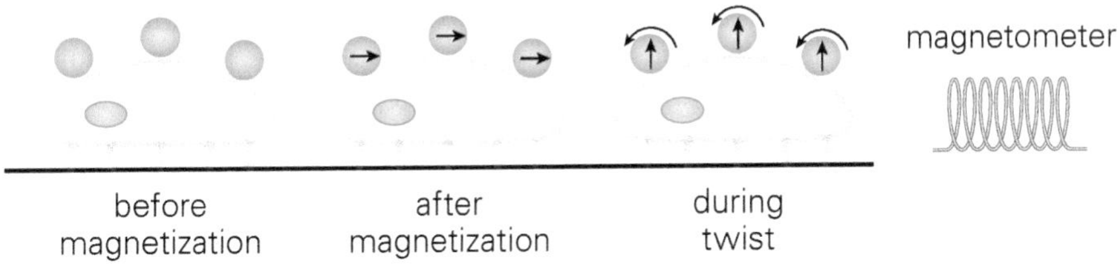

before magnetization   after magnetization   during twist   magnetometer

Nota Bene

Constant force is applied to the magnetic beads. Because force is dependent on the distance from the bead to the magnet tip, the force applied to a cell-bound bead actually increases as the bead is pulled in the direction in the magnet. In most cases, the bead displacement is small enough that the change in force is can be neglected.

Nota Bene

Magnetic bead types. In magnetic micromanipulation experiments, having the bead lose its magnetization is desirable if repeated measurements are being taken. That is, you want to apply some force to the bead, release it, wait some time, and then apply the same force. Here you want to use *paramagnetic* beads, which lose most or all of their magnetization once the magnetic field breaks. A paramagnetic material is also used to construct the electromagnet so that you can quickly control the applied magnetic field. Materials that retain a significant portion of the magnetization are referred to as *ferromagnetic*. In general, paramagnetic materials respond more weakly to magnetic fields, but their ability to de-magnetize is better than ferromagnetic materials.

Figure 6.3 Schematic of magnetic twisting cytometry. After a strong magnetization in the horizontal direction, a weaker magnetic pulse is applied in the vertical direction, resulting in an upward torque on the beads. The extent of bead motion is measured by a magnetometer and is reflective of local cell properties.

Figure 6.4 Schematic of an optical trap. The bead is trapped in a laser beam, with the equilibrium point in the center of the beam just downstream of the focal region. If the bead is displaced, then there is a net restoring force on the bead. Attaching the bead to a cell or other structure allows one to test its mechanical properties. (Adapted from, Huang H, Kamm RD, Lee RT (2004) *Am. J. Physiol. Cell Physiol.* 287).

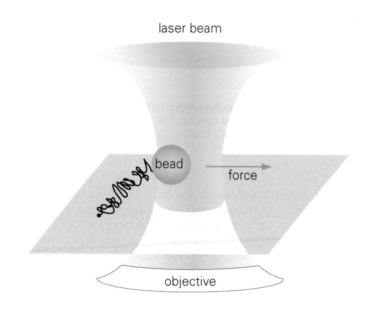

A few key figures of the optical trap. In the late 1960s, Arthur Ashkin from Bell Labs pioneered the use of light to manipulate the position of tiny particles of matter. Eventually this developed into modern-day optical trap systems. It is also the basis of the work of Steven Chu, who used it to cool atoms by selectively trapping slow-moving ones, allowing faster-moving ones to escape. Dr. Chu received a Novel Prize for Physics in 1997 and became United States Secretary of Energy in the Obama administration.

of the beam, where the intensity is highest. There is also a scattering force that results in a force along the direction of the laser beam. The result of these combined forces is that the particle becomes "trapped" within the focused laser beam in a location just downstream of the focal point. If the bead is displaced relative to the beam, the forces will act in a restorative manner to move the bead back into the trapping location, creating a stable equilibrium point.

Ray tracing elucidates the origin of restoring forces in optical tweezers

To quantify the origins of the lateral restoring forces, we can use a technique called *ray tracing*. It is useful for determining such forces, as long as the particle is large compared with the wavelength of light. Consider a spherical bead trapped within a laser beam, as in **Figure 6.5**. When the beam is unfocused it resembles a cylinder. We further assume it to have a constant intensity in the axial direction, but a radially decreasing intensity profile in the lateral direction, with the highest intensity in the center.

If we follow ray 1 in Figure 6.5, it enters the left side of the bead and is deflected rightward. In this rightward deflection, there is transfer of momentum from the beam to the bead that imparts a force on the bead that is primarily leftward (F_1). Ray 2 is located opposite ray 1 and equivalently imparts a primarily rightward force (F_2). Because ray 2 is equal in intensity to ray 1, F_2 is equal in magnitude to F_1, resulting in no net lateral force, if the bead is centered. However, if we displace the bead to the left, the intensity of ray 2 will be less than that of ray 1 owing to the radial gradient of light intensity. In this case, the magnitude of F_1 is greater than that of F_2, resulting in a net rightward restorative force that pushes the bead back to the beam center. Because the forces are the result of the intensity gradient, they are referred to as *gradient forces*.

Although the lateral gradient force example demonstrates why the bead becomes trapped in the lateral direction, what about the axial direction? To understand this phenomenon, consider a focused laser beam as in **Figure 6.6**. The beam forms a cone that converges at the focal point. If we follow a ray entering the left side of a spherical bead located upstream of the focal point, it is deflected rightward upon entering the bead, and again upon exiting the bead. There is a change of momentum as it enters and exits the bead, because the light entering the bead is not purely vertical, the shift of the rays toward the centerline means that the exiting

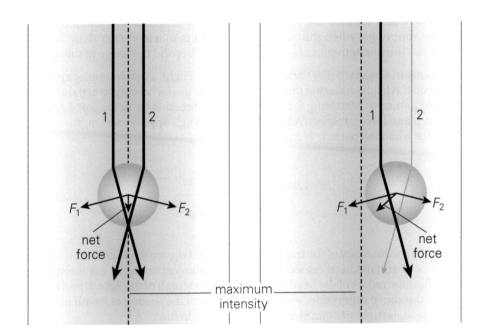

Figure 6.5 A bead situated within a laser beam with highest intensities in the beam center will experience forces toward the beam center. Additionally, the bead will experience forces downstream.

light has a larger vertical momentum component than the entering light. By summing the difference in momentum vectors, one can see that there is a net vertical change in momentum, which is transferred to the bead as an axial force in the downstream direction. A similar analysis reveals a net vertical change in momentum in the upstream direction if the bead were downstream of the focal point, resulting in an upstream restorative force. When the lateral and axial restoring forces are combined with the downstream scattering force, the result is that the bead becomes trapped just downstream of the focal point.

What are the magnitudes of forces in an optical trap?

We now have gained some intuition about the orientation of the restoring forces in optical tweezers that facilitate particle trapping. But what determines the magnitude of forces imparted? In general, they are determined by the physical properties of the bead, the beam, and the medium in which the bead is immersed. Two different approaches can be used to calculate these forces, depending on the size of the particle.

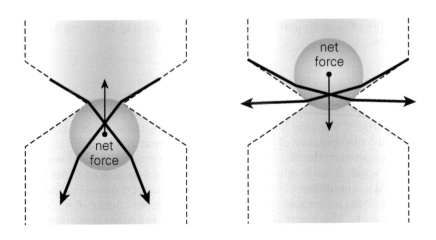

Figure 6.6 A focused light beam interacting with a bead. A bead that moves down in the trap will deflect light beams toward the bottom, resulting in an upward net force. Conversely, a bead moving up will deflect the beams up and experience a downward restoring force.

The generalized Lorenz–Mie theory is a ray optics–based model, and a relatively accurate approach when the wavelength of light is much smaller than the bead diameter. The mathematics are complicated and beyond the scope of this text.

In the case of a bead smaller than the wavelength of light, a much simpler approach based on Rayleigh scattering theory can be used. The forces can be separated into two components: scattering force (in the laser direction) and the force gradient (the in-plane force), respectively. Using the Rayleigh approximation, the scattering force when the particle is trapped is

$$F_z = \frac{128\pi^5}{3} \frac{n_m I}{c} \frac{R_b^6}{\lambda^4} \frac{\left(\dfrac{n_b}{n_m} - 1\right)^2}{\left(\dfrac{n_b}{n_m} + 2\right)^2}, \tag{6.2}$$

where I is the intensity of the trapping light, n_b and n_m are the indices of refraction of the bead and medium respectively, a is the radius of the bead, c is the speed of light, R_b is the radius of the bead, and λ is the wavelength of the light. Upon inspection of the above expression, we can see that given the scaling of F with a, a small increase in particle size can yield a large increase in force.

The second component is the gradient force,

$$F_1 = 2\pi \frac{n_m I}{c} \frac{R_b^3 \left(\left(\dfrac{n_b}{n_m}\right)^2 - 1\right)}{\left(\left(\dfrac{n_b}{n_m}\right)^2 + 2\right)}. \tag{6.3}$$

The gradient force is dependent on the gradient of the light intensity and not the absolute intensity. For a typical experiment, the forces are relatively small (on the order of tens of piconewtons, although some modern configurations can achieve higher forces).

Recall that the generalized Lorenz–Mie theory and Raleigh scattering can be used when the wavelength of the beam is much smaller and larger than the beam diameter, respectively. Often the sizes of particles of interest may be of the same order of magnitude as the wavelength. For example, green light has a wavelength of 510 nm; most cellular targets of interest for trapping range from a few micrometers to 250 nm. Particles much smaller than this are very hard to trap, because they are difficult to image using standard optical techniques and generate much smaller forces. Therefore for many particle sizes and beam wavelengths, the relationship between force and particle/beam properties is not very well understood. In this case, forces generated by the trap can be determined empirically. One approach is to capture a bead of known size, and move the trap at increasing velocity until the capture is lost. Quantifying the fastest velocity before the bead is lost, allows the maximum force generated by the trap to be calculated using Stokes' law.

How does optical trapping compare with magnetic micromanipulation?

Although optical trapping is similar to magnetic micromanipulation in that both allow for the manipulation of particles within or bound to cells, there are important differences between the two that may make one approach more appropriate for a particular application. In optical trapping, the trapped particle need not be ferrous, and spatial manipulation of particles can generally be performed in a more precise manner. Certain customizations allow the independent control of multiple beads simultaneously. Also, some subcellular organelles and other structures like chromosomes, vesicles, bacteria, and even viruses have the correct optical properties that they can be manipulated directly without the need for beads. However, for some biomechanical studies, optical trapping may not be the best method, as the forces are generally low. Additionally, force-based manipulation with optical trapping requires feedback technology, because forces are generated based on the relative position of the bead and the laser beam. Another important difference is that an optical trap, as the name suggests, is a "trap." That is, there is a relatively small equilibrium point and a large gradient of restoration forces around that point. The forces generated magnetically affect a far greater spatial region and are more homogeneous. This allows many beads to be manipulated together, although the manipulations are not independent. Finally, it is not clear whether magnetic fields have effects on biological tissues, but it is well known that focused lasers can heat biological material, and with sufficient power can destroy tissue locally. In fact, this effect has been exploited to generate what are called *laser scissors* or *laser scalpels*, allowing one to disrupt parts of a cell selectively.

Atomic force microscopy involves the direct probing of objects with a small cantilever

Atomic force microscopy (AFM) is an approach for characterizing cells that allows for the direct probing of the cells, using a small cantilever. There is no need to attach a bead or other foreign substance to the cell surface. AFM works by tracking the displacements of a cantilevered beam brought into contact with the cell or other object of interest. Modern designs are able to measure both the shape of a cell and its mechanical properties. If one were to imagine a finger as an AFM cantilever, it is easy to see how a range of physical attributes of an unknown object could be inferred, such as mechanical stiffness (by pushing down on the object and determining how much resistance the object offers), surface geometry (by running one's finger over the surface), and spatial heterogeneity in mechanical stiffness (by probing or poking in several places).

The central component of an atomic force microscope is a cantilever that has a small tip at the free end. The cantilever and tip are typically constructed of silicon or a silicon compound. A typical tip has a radius of curvature on the order of nanometers, which is quite sharp, even at the scale of a cell. In probing a specimen, the cantilever is lowered so that the tip contacts the cell surface. As the tip pushes into the specimen, the cantilever deflects. We saw in Section 3.2 (see Equation 3.31) that for the case of a linear elastic beam of length L clamped at one end, EI is the flexural rigidity, the displacement w of the cantilever at any point x along the beam is a function of the force at the tip, F,

$$ w = \frac{F}{EI}\left(\frac{x^3}{6} - L\frac{x^2}{2} \right). \tag{6.4} $$

We can see from Equation 6.4 that the deflection at the tip ($x = L$) is proportional to the tip force. The displacement at the tip can be related to the force through an effective spring constant. Once this constant is determined, the force being applied at the cantilever tip can be readily calculated from the tip deflection, allowing for simultaneous quantification of deflection and force. In this way, force-displacement curves can be generated by positioning the tip over the surface of the specimen and measuring the deflection of the cantilever as the tip is lowered into the specimen (this mode of AFM use is sometimes referred to as force readout mode). Such curves are conceptually similar to the force-deformation curves we discussed in Section 3.2 (see Figure 3.1) and can be used to infer mechanical properties.

Cantilever deflection is detected using a reflected laser beam

An AFM cantilever is about the same size as a cell (on the order of tens of micrometers), and its deflections are smaller (tens of nanometers). These tiny deflections can be greatly amplified and measured using a geometric trick. A laser is focused on a reflective area on the back of the cantilever and the lateral motion of the beam is measured a large distance away (**Figure 6.7**). A detector is placed in the path of the reflected beam, such that a laser spot is generated on the detector. As the cantilever bends, the laser spot on the detector changes its location. In general, the longer the laser beam path, the more the laser spot will move for a given deflection of the cantilever, and the smaller the deflections are that can be detected. Generally, force sensitivities using AFM are on the order of piconewtons or less.

Scanning and tapping modes can be used to obtain cellular topography

In addition to characterizing mechanical properties and behavior, AFM can be used for specimen visualization, in particular for determining surface topography. There are several modes whereby this can be achieved. *Scanning* mode is used to generate a topographical profile of the surface being examined. Here, the tip is lowered until it just touches the surface of the specimen, then the tip is scanned, or dragged, across the specimen. The deflection of the cantilever at each point is used to generate a height profile. To prevent specimen damage, in many cases a feedback mechanism is incorporated to lower/raise the tip to maintain a constant force. If the material is homogeneous, then the deflection is generally a good representation of the surface geometry. For specimens with heterogeneous mechanical properties, the topography may be reflective of the surface topography as well as the local stiffness. If we were to consider a specimen that had a flat surface but spatially heterogeneous stiffness, a stiffer region may appear "higher" than a more compliant region.

A second AFM mode is *tapping* mode, whereby the tip is actively vibrated up and down (typically by a piezoelectric actuator coupled to the AFM tip holder) close to its resonant frequency, which is usually on the order of 10–100 kHz. The oscillating tip is lowered to the specimen surface until it begins to experience van der Waals forces, resulting in a reduction of the magnitude of the vibration. Similar to scanning mode, during tapping mode operation, feedback is typically used to maintain the cantilever oscillation amplitude constant. A major advantage of tapping mode over scanning mode is that when the tip contacts the surface, it has sufficient oscillation amplitude to overcome "sticking" or adhesion forces between the tip and the sample. It also reduces the chances that a tip will become fouled or coated in cellular debris because, ideally, the tip never comes into physical contact with the cell. One important consideration with AFM is the nature of the probes, which are sometimes unable to detect overhangs or very steep walls.

A Hertz model can be used to estimate mechanical properties

Once we have obtained an AFM force–displacement profile, the next question is how can mechanical properties of the sample be inferred? One mathematical approach is to use the *Hertz contact* model, which describes contact between a

Figure 6.7 Schematic of AFM operation. The tip at the end of the cantilever is brought down on a specimen. As the specimen indents, the cantilever flexes and deflects the laser to a different location as determined by the detector. It is possible to calculate the indentation force using the information about cantilever flexure from the laser spot deflection. (Adapted from, Huang H, Kamm RD & Lee RT (2004) *Am. J. Physiol. Cell Physiol.* 287.)

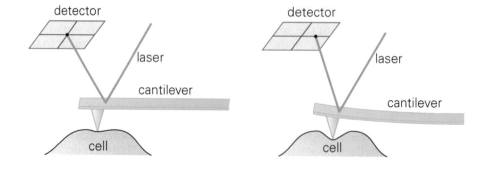

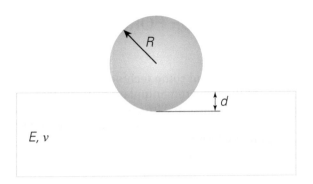

Figure 6.8 Schematic of the indenter. It is depicted here as a sphere of radius R, but as long as the contact region is circular, the rest of the probe can have an arbitrary shape. The relation between d, R, and the material properties of the surface is given in Equation 6.5, with the indenter having infinite rigidity. The material being indented, shown here as a dented rectangle, is treated as having infinite depth and width, which is called a half-space.

rigid sphere and a homogeneous, isotropic, linearly elastic material *half-space* (Figure 6.8). The Hertz solution is

$$F = \frac{4}{3}\left(\frac{E}{\left[1 - v^2\right]}\right)\sqrt{Rd^3},\tag{6.5}$$

where F is the applied force, d is the depth of indentation of the sphere into the material, E is the elastic modulus of the specimen, v is Poisson's ratio of the material, and R is the radius of the sphere/indenter. This mathematical relationship assumes that the indenter is very stiff compared with the sample.

One can approximate a pyramidal or other "sharp" AFM tip as a sphere in the Hertz model if the scale of indentation is small enough that the tip's curvature is evident. Additionally, in some experiments, the cantilever can be customized by attaching a bead to the tip. This way, beads of different diameters can be used to obtain stiffness at different length scales, and sometimes with different forms of stress (e.g., shear vs normal stress).

Nota Bene

The Hertz model can be used to calculate elastic properties for specific locations. In many cases, the assumptions underlying the Hertz model may not be adequately met within cells. For instance, the structural components of cells may not be linearly elastic or isotropic. In addition, due to the spatial heterogeneity of cells, a single elastic modulus is rarely sufficient to capture the spatial variation in material properties. Some studies have focused on using the Hertz model to calculate elastic properties for specific locations and/or structures, such as stress fibers.

Example 6.1: Estimating the elastic modulus of a cell using the Hertz model

Using the Hertz model, estimate the elastic modulus of a cell using $v = 0.5$ (incompressible) and $v = 0.3$ (typical of many biological tissues). The radius of the probe tip is 50 nm.

Using Figure 6.9, we estimate the force for a 100 nm indentation (from 175 nm to 75 nm along the tip approach line) to be 1000 pN = 1 nN. Using the formula $E = 100,000(1 - v^v)$ (actually a bit more, but realistically you can only obtain about two significant figures) yields 75 kPa for incompressible materials and 90 kPa for biological tissues. You can see that the choice of Poisson's ratio does not make a huge difference in our estimate, which is admittedly on the high side (cells typically in the 10–30 kPa range by AFM) if you are not targeting stress fibers.

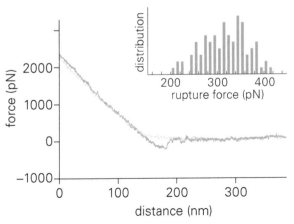

Figure 6.9 Sample plot of an AFM force-displacement curve. The lighter curve shows the approach of the tip above the sample (an endothelial cell), moving from right to left. Contact is made at around 150 nm. Further displacement to the left shows increasing force to deform the cell, such that at 1 nN, the tip has advanced about 75 nm into the cell. The darker line is the retraction of the tip. At about 175 nm, the adhesion of the tip to the cell is broken and there is a sudden "zeroing" of the force. The force magnitude was collected for a number of cells, and the histogram of that adhesive-rupturing force is shown in the inset. (From, Liu J, Weller GE and Zern B (2010) *Proc. Natl. Acad. Sci. USA,* 107).

6.2 MEASUREMENT OF FORCES PRODUCED BY CELLS

In previous sections, we learned some techniques for applying and analyzing mechanical forces through micromanipulation. Such forces are considered exogenous, in that they are generated from outside the cell. In contrast, endogenous forces are those generated from within the cell. Endogenous forces can play critical roles in regulating physiological processes such as cell spreading or locomotion. In these situations, the cell generates traction forces, or endogenous contractions, that are transferred to the substrate through cell adhesion sites. Traction forces can be clearly demonstrated by plating cells on a flexible silicon sheet. The traction forces cause small wrinkles to develop near the sites of adhesion (Figure 6.10). Although traction forces are relatively easily to visualize, determining the magnitude and direction of the forces applied is more difficult. The focus of this section is to detail two methods for quantifying cellular traction forces: traction force microscopy and the use of micropillar arrays.

Traction force microscopy measures the forces exerted by a cell on its underlying surface

In traction force microscopy (TFM), the approach is to characterize the deformation of a substrate induced by traction forces, and to use the deformation field to calculate the traction forces exerted at each point. Small beads placed within the substrate act as displacement markers and facilitate the tracking of deformations and calculation of the displacement field. Typically, bead positions are determined with the cell attached to the substrate, and again after detachment of the cell using trypsin, a serine protease that degrades cell adhesions. An *inverse* problem is then solved to extract forces generated by the cell. We briefly describe these two steps in the next sections.

Cross-correlation can be used for particle tracking

An important requirement for TFM is the ability to track the motions of beads. In this section we will briefly discuss the practicalities of tracking before fully considering TFM. Cross-correlation is a statistical measure that allows the position of particles to be determined with great precision. Consider a set of two images of a cell (with attached fluorescent beads) obtained before and after mechanical

---

Nota Bene

The substrate is often a thin film of a flexible polymer such as polyacrylamide. Polyacrylamide is chosen not only for its flexibility and flexibility that can be adjusted over a wide range, but also because it behaves like an isotropic linear elastic material. This greatly simplifies the analysis. Other gels used in cell culture, such as collagen gels, might be more representative of physiological environments, but are not generally isotropic and linearly elastic. They additionally offer spaces for cells to crawl within the gel itself, which can confound traction mapping.

Figure 6.10 Image of a cell plated on a thin deformable membrane. The cells exert traction forces and locally wrinkle the membrane in a demonstration that in order for cells to spread out, they exert some level of surface forces. (From, Harris AK, Wild P & Stopak D (1980) *Science* 208.)

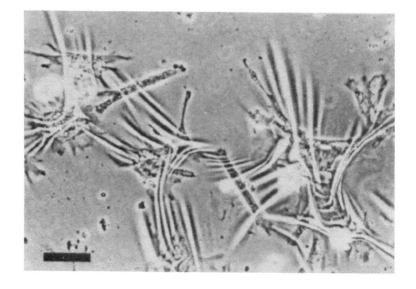

loading. The images contain unique local patterns of intensity that reflect the shape and brightness of the fluorescent objects within that locale. These local intensity patterns can be indentified with image correlation, and changes in position and/or shape used infer the deformations that gave rise to these changes.

In an ideal scenario, a bead will be a perfect sphere, which appears as a perfect circle on our image. We can then locate the center of the sphere, and thus the bead, exactly. If we repeat the same measurement after the bead moves, we can obtain the exact displacement vector of the bead. In the laboratory, however, beads are not perfect spheres and the images will be limited by resolution, illumination, refraction, and other artifacts. *Cross-correlation* allows us to determine locations, and therefore displacements, with high precision by using a mathematical algorithm to fit two profiles against each other.

To illustrate this approach, consider a cell seeded with k fluorescent beads (Figure 6.11). We obtain fluorescence images of each bead before, during, and after loading. Each image can be represented by an array of numerical intensities that we assume, for simplicity, are stored in a square array known as the image intensity function, $I(i, j)$ with $2n + 1$ pixels in each direction such that

$$-n \leq i, j \leq n. \tag{6.6}$$

The intensity pattern for a single bead would be an approximately circular region of high intensity (the bead) surrounded by a dark region (the background) (Figure 6.12). The intensity in the circular region is highest in the center and drops off toward the edges of the bead. As the bead moves, the location of the circular distribution will shift within the array, but because the bead is rigid, the general intensity pattern will not change. In essence, the bead image undergoes a rigid-body translation. Our goal is to determine the magnitude of the translation. One approach might be to track the movement of the brightest pixel. However, the intensity peak can be rather broad, which makes this approach imprecise and sensitive to noise. Imagine a large collection of pixels that are all near the peak with similar high-intensity values but with some fluctuations. In this situation, the

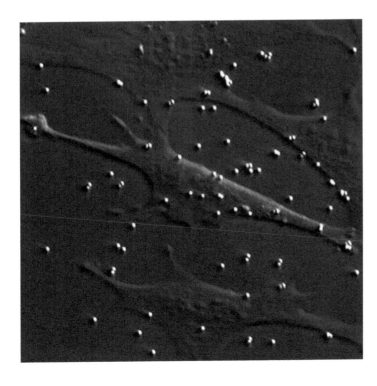

Figure 6.11 Fluorescent beads attached to cells. In this image a phase-contrast image of cells has been superimposed on an epifluorescence image of beads that have been filled with a fluorochrome. (From, Kwon R, et al. (2007) *J. Biomech.* 40)

Figure 6.12 The digital representation of a circular bead can be used for quantifying its displacement. (A) A fluorescent bead appears as a circle of non-uniform intensity. Because the sphere is thicker near the center, there is more fluorescence emanating from that location, so the bead tends to be brighter in the center and dimmer near the edges. (B) When digitized, the circle is approximated by pixels that reflect the intensity. Note that the same intensity pattern is displayed in the digitized version: lighter near the center and darker near the edges. (C) The bead itself can be extracted as a pattern of intensities called a template. (D) When the bead moves, the pattern representing the bead undergoes a rigid-body translation in the image. (E) With the bead at its new location within the image, the location of the pixels representing the bead has shifted, but the pattern (or template) remains the same. In actuality, the intensities will fluctuate over time, owing to instrumentation and other factors, but the rough pattern (brighter near the center) will remain consistent.

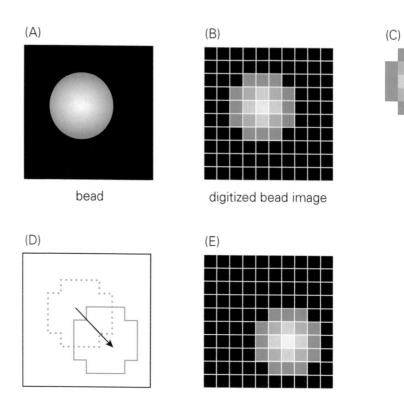

(A)

(B)

(C)

bead

digitized bead image

(D)

(E)

---

brightest pixel can jump from place to place dramatically. To address this problem, a statistical image correlation technique can be applied that is actually so precise it can determine the location of the bead with sub-pixel resolution.

To determine the translation in the intensity pattern between images, we first define a *template* image $K(i, j)$. The template image is the ideal intensity pattern of a bead, which for simplicity we locate at $(0, 0)$. This template could be obtained from the theoretical distribution for an ideal bead, or an actual bead could be imaged before deformation. In either case, we assume that $K(i, j)$ is $2m + 1$ pixels in each dimension, with $m < n$. With K defined, a statistical cross-correlation can be calculated as a measure of the degree of similarity between K and I on a pixel-by-pixel basis. A typical statistical correlation measure is

$$C(i,j) = \sum_{x=-m}^{m} \sum_{y=-m}^{m} I(i + x, \, j + y)K(x, y), \tag{6.7}$$

which is a matrix convolution of I and K. The resulting matrix, $C(i, j)$ is known as the cross-correlation field. It gives the degree of similarity between the intensity pattern surrounding the pixel (i, j) and K. The i and j that maximize $C(i, j)$ give the location at which the intensity pattern of I most closely matches that of K. This location is assumed to be the "correct" location of the bead in I. If the bead is originally at $(0, 0)$, then the distance of the bead displacement can then be calculated from i and j. The location of this peak in C can be computed with an accuracy better than the size of a pixel by calculating the centroid of the cross-correlation field,

$$i_c = \frac{\sum\limits_{i,j} iC(i,j)}{\sum\limits_{i,j} C(i,j)}, \quad j_c = \frac{\sum\limits_{i,j} jC(i,j)}{\sum\limits_{i,j} C(i,j)}. \tag{6.8}$$

Determining the forces that produced a displacement is an inverse problem

Using cross-correlation, we can determine displacement of beads. The problem of determining the set of forces that produced a known displacement field is known as an *inverse problem*. In addition to observed data (displacements in this case), an inverse problem requires a governing equation known as the *forward model*. We need a theory that gives the expected displacements for a given surface load. Luckily, an analytical solution exists, known as the *Boussinesq* solution, which gives the displacement at the surface of an infinite homogeneous linear elastic half-space owing to a point-load at the surface. One issue that arises with the half-space assumption is that the substrates used in TFM are fairly thin (on the order of 100 μm). Thus, the half-space assumption may seem somewhat suspect. However, in general the method produces acceptably accurate results as long as care is taken that the displacements are small relative to the substrate thickness.

Consider a set of n discrete displacement vectors $\mathbf{u}_{i=1,2,...,n}$ determined at discrete points $\mathbf{x}_{i=1,2,...,n}$. The displacement vector $\mathbf{u}_i$ at point $\mathbf{x}_i$ is assumed to arise as a result of the combined influence of a set of j discrete point force vectors $\mathbf{f}_j$ located at another set of points $\mathbf{x}_j$. The Boussinesq solution $\boldsymbol{G}(\boldsymbol{r})$ for a single force can be

Nota Bene

Inverse problems. The field of inverse problems originated with the work of physicist Viktor Ambartsumian, done while he was still a student in the 1920s. His initial paper went unnoticed for 20 years until the mathematics community developed it into the general field it is today. Conceptually, in each inverse problem, there is a set of underlying governing equations that convert a set of model parameters or properties (input) into observable data (output). The inverse problem is to deduce the input from the output. Some typical examples are to determine the distribution of density within a planet from measurements of its gravitational field (using Newton's law of gravity) or to determine the epicenter of an earthquake from seismic waves (using the wave equation).

(A)

bead displacement vector

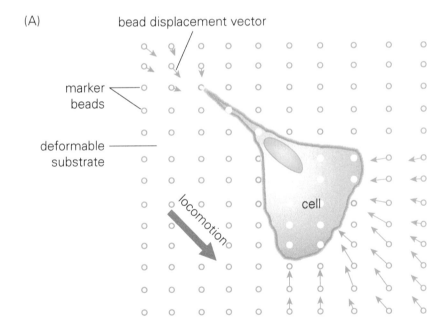

marker beads

deformable substrate

locomotion

cell

(B)

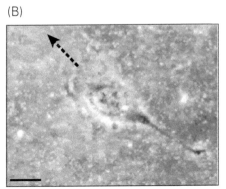

(C)

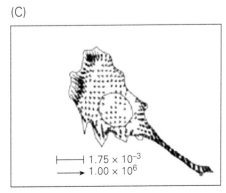

⊢———⊣ 1.75×10^{-3}
———→ 1.00×10^{6}

Figure 6.13 Traction force microscopy takes advantage of the existence of the surface tractions exerted by the cell. Plating a cell on a deformable substrate with embedded beads (A, B) will result in the cell deforming the substrate. Releasing the cell (using trypsin or waiting until the cell has moved away) allows the original location of the beads to be determined. A displacement field and the traction force field can then be calculated (C). (A, adapted from, Roy P, Raifur Z & Pomorski P, (2002) *Nature Cell Biol.* 4. With permission from MacMillan Publishers Ltd., on behalf of Cancer Research UK.; B, from Munevar S et al. *Biophys.* (2001) Biophysical Society.)

used to relate the displacement $\mathbf{u}_i$ to $\mathbf{f}_j$ as

$$\boldsymbol{u}_i = \sum_{j=1}^{m} G(\mathbf{r}_{ij})\boldsymbol{f}_j, \tag{6.9}$$

where $\mathbf{r}_{ij} = \mathbf{x}_i - \mathbf{x}_j$ is a distance vector, and

$$G(\boldsymbol{r}) = \frac{1+\upsilon}{\pi E r^3}\begin{pmatrix} (1-\upsilon)r^2 + \upsilon r_x^2 & \upsilon r_x r_y \\ \upsilon r_x r_y & (1-\upsilon)r^2 + \upsilon r_y^2 \end{pmatrix}. \tag{6.10}$$

In the above expression, r_x and r_y are the x and y component of $\mathbf{r}$ and r is $|\mathbf{r}|$. E and υ are Young's modulus and Poisson's ratio of the elastic substrate, respectively. Computing the above expression for all n displacements gives a simultaneous equation between m force vectors and n displacement vectors. The goal then becomes to find the m force vectors that give rise to the n displacement vectors (Figure 6.13). Although methods for solving the problem are outside the scope of the text, there are nice reviews in the literature.

> **Nota Bene**
>
> **Green's function.** The function $\boldsymbol{G(r)}$ is also known as a Green's function. This approach allows a general solution to be built up by adding many simpler solutions. It is similar to a matrix or integral *convolution*. The simple solution is the Boussinesq solution for a simple force.

Advanced Material: Boussinesq solution

The French mathematician Joseph Boussinesq derived the solution of a force applied to the surface of a half-space in 1885 (Figure 6.14). He started with the solution of a force applied to an infinite continuum. Then he calculated the force along the $z = 0$ plane. He used the principle of superposition to apply an equal and opposite force to this to cancel it out, again using the solution for a point force in an infinite continuum. When an integral is constructed in this way to solve an inhomogeneous differential equation subjected to complex boundary conditions, the integrand is termed a *Green's* function.

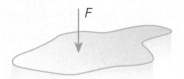

Figure 6.14 Depiction of the Boussinesq problem of a force applied to a half-space. The force is applied to the top surface and the half-space extends infinitely within the plane and in the depth.

Example 6.2: Traction force microscopy

A special case that does not require extensive numerical modeling is a single particle that displaces owing to a single point force. Specifically, suppose you have a force F and displacement u separated by a distance d (Figure 6.15) and that the force and displacement act in the same direction. Our experimental displacement is then $\boldsymbol{u} = [u, 0]$. Our force vector is $\boldsymbol{F} = [F, 0]$. We can calculate the Green's function,

$$G = \frac{(1+\upsilon)}{\pi E d^3}\begin{pmatrix} (1-\upsilon)d^2 + \upsilon d^2 & 0 \\ 0 & (1-\upsilon)d^2 \end{pmatrix}.$$

The off-diagonal terms are zero because we place the origin at the location of applied force, so there is no y-component to this problem. Note that the $(1, 1)$ term can be simplified to just d^2. Further simplification yields

$$G = \begin{pmatrix} \dfrac{(1+\upsilon)}{\pi E d} & 0 \\ 0 & \dfrac{(1-\upsilon^2)}{\pi E d} \end{pmatrix}.$$

Because $u = GF$,

$$u = \frac{(1+\upsilon)}{\pi E d}F.$$

which allows us to solve for F based on our knowledge of u and d. In the more complex non-co-linear condition, there would be y-contributions, and we would need to invert G rather than having a single linear relationship.

However, in many traction-force situations, there is more than a single force to deal with. Generally, multiple forces are active—all of unknown magnitudes, being applied at unknown locations, being exerted in unknown directions. Consider a bead exposed to two different unknown forces. If the bead moves, it could be because of either force or both forces acting in different amounts. To constrain the problem, multiple beads are required. However, the problem then becomes mathematically cumbersome because each bead gets an equation $u = GF$ for each force being applied. Furthermore, d is generally unknown, because the force locations are not proscribed.

To address this there are sometimes logical places to assume forces are being applied, such as points of cellular adhesion that can be determined by immunocytochemistry. A more general approach is to create a regularly spaced grid for the locations of force application. If the grid has high enough density, a fairly complete picture of the force field can be obtained. Mathematically, for each grid location a displacement u and a force F is known. From this one can construct G. This approach involves large arrays of displacements, positions, and components of G, for which optimization routines are applied to extract the traction forces. This technique is particularly sensitive to small errors in displacement because the force–displacement relationship drops rapidly with distance away from the location of the force. Another method is to use Fourier transforms to convert the problem into frequency space. This allows the inverse problem to be solved directly, because

convolutions in frequency space become multiplications. Fourier transforms should be used with caution, as boundary conditions in frequency space are not clear-cut. Typically, there are no traction forces outside the cell, but many solutions using Fourier transforms will generate forces near the border of, but typically outside of, the cell.

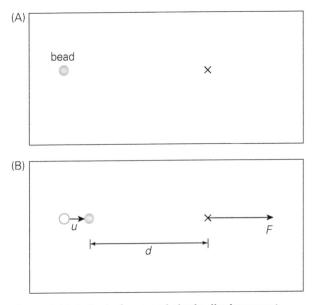

Figure 6.15 A single force and single displacement example. The force and displacement are co-linear.

Microfabricated micropillar arrays can be used to measure traction forces directly

Traction forces can be quantified by solving an inverse problem. An alternative method uses the process of *microfabrication* to create an array of thousands of *micropillars* that essentially act as microscopic force sensors.

Microfabrication is the process of fabricating small structures with dimensions on the orders of micrometers. Particular processing steps—many of which were developed in the semiconductor industry for the manufacturing of integrated circuits—are performed in sequence, resulting in a micromachined structure with desired designs, surface patterns, or topography. Typically, the approach involves three processing steps: thin-film deposition, photolithography (which transfers a defined pattern onto material film using light), and etching (the removal of material in patterns defined through photolithography).

Using these techniques, dense arrays of micropillars (on the order of 1–10 μm in diameter and 10–100 μm in length) can be manufactured, which allows each pillar to function as a force transducer by acting as a cantilever beam. By treating the top surfaces of the micropillars with extracellular matrix proteins, cells seeded onto the arrays adhere to the tops of the pillars and exert traction forces, causing the pillars to bend (Figure 6.16). The deflection of the pillar tip can be related to the force exerted on that pillar as in Equation 3.31

$$w = \frac{F}{EI}\left(\frac{x^3}{6} - L\frac{x^2}{2}\right). \tag{6.11}$$

Figure 6.16 An electron micrograph of a cell sitting on a bed of microneedles. The traction forces exerted by the cell result in the pillars being bent toward the center of the cell. Because the forces exerted on the pillars can be modeled by beam equations, there is no need for the computational complexity associated with classic traction-force microscopy. (From, Tan JL, Tien J & Pirone DM (2003) *Proc. Natl. Acad. Sci. USA* 100.)

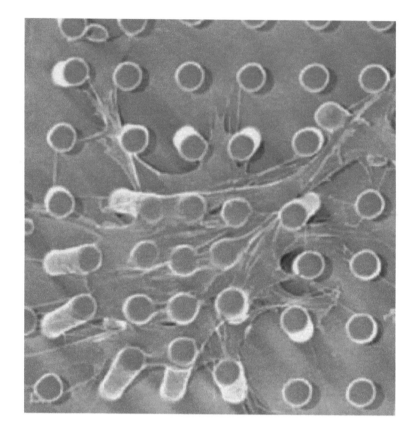

Because we only measure the displacement at the tip δ at $x = L$, the above expression can be simplified to

$$F = \left(\frac{3EI}{L^3}\right)\delta. \tag{6.12}$$

Example 6.3: Deflection of the micropillars in response to cell traction forces

In Figure 6.16, you can see the deflection of the micropillars in response to cell traction forces. Given that the scale bar is 10 μm, the length of the pillars is 11 μm and the elastic modulus E of the pillar material is 2.5 MPa, estimate the magnitude of the forces exerted by the periphery of the cell.

The displacement of the pillar tips is roughly 5 μm. The diameter is roughly 3 μm. So

$$F = \left(\frac{3EI}{L^3}\right)\delta.$$

The moment of inertia of a circular cross section is $I = \pi R^4/4 \approx 4 \times 10^{-24} \text{m}^4$. Thus,

$$F = \frac{3\,(2.5 \times 10^6 \text{PA})(4 \times 10^{-24}\text{m}^4)(5 \times 10^{-6}\text{m})}{1.33 \times 10^{-15}\text{m}^3}$$

$$= 1.12 \times 10^{-7}\text{N} \approx 100\,n\text{N}.$$

This is slightly on the high side but not ridiculously so (the original article by Tan et al. reports values from 0 to 80 nN).

Surface modification can help determine how a cell interacts with its surroundings

Many of the techniques we have just described involve attaching or touching a synthetic structure to the cell, either beads, AFM tips, micropillars, or other structures.

However, cells typically do not encounter glass, plastic, or silicon *in vivo*. How the cells interact with these artificial surfaces is an important consideration. The process of *surface modification* can be a powerful way to modify the artificial material not only to make the material more physiological or biocompatible, but also to gain insight into the role of various properties or even specific proteins and molecules. Techniques are available to alter roughness, charge, surface energy, hydrophobicity, and other physical properties. A particularly useful approach is to *functionalize* a surface, which is the process of coating it with functional groups or whole proteins. These can be extracellular matrix proteins (such as fibronectin or collagen) but they can also be protein fragments, peptides, or antibodies against epitopes of a particular functional interest. Because the cell's interaction with the extracellular matrix can be highly dynamic and time-dependent, selecting the right incubation time is critical. For small beads, a time interval 15 minutes and 1 hour is often used. If the incubation time is too brief, the beads will not have sufficient time to bind; too long, and the cells can endocytose (engulf) the beads, causing loss in receptor specificity.

This process of functionalization can be combined with the photolithographic microfabrication techniques described above. *Microcontact printing* or *micropatterning* involves using a soft polymer such as PDMS (polydimethylsiloxane) to "print" a protein onto a surface with a microscopic pattern. A silicon master is first created by coating a chip with a photoresist that can be removed optically on a microscopic scale. The resulting pattern is then acid-etched into the silicon. Multiple PDMS *stamps* can then be created from the master. The stamp is then dipped into a protein or other solution and pressed onto glass or another surface. Microfabrication techniques are used in "lab-on-a-chip" technologies using miniature channels, pumps, valves, etc. The small scale of these devices provides enormous advantages in terms of small sample size, high throughput, accelerated assay time, and even portability.

6.3 APPLYING FORCES TO CELLS

So far we have examined experimental systems to measure cell mechanical properties and the forces that cells produce. Equally important are systems aimed at investigating how cells sense and respond to mechanical signals. Although we are ultimately interested in the response of a tissue *in vivo*, for cells residing within a tissue, a simple mechanical stimulus such as deformation can lead to many cellular-level physical signals (such as fluid flow, electric fields, and more). One attractive strategy is the reductionist approach that is common in biology (described in Section 2.4). With this strategy the cellular response to simple mechanical stimuli is examined *in vitro*. This allows the signal of interest to be applied in a controlled fashion, and, importantly, many molecular approaches for modulating gene or protein functions can be applied much more easily in cell culture.

At the basic level, there are a few simple mechanical stimuli that have received the most attention, including fluid shear and stretch. We will examine each in turn and consider the key issues and methods involved in their application.

Flow chambers are used for studying cellular responses to fluid shear stress

As we will discuss in Section 11.1, several cell types have been shown to respond to fluid shear stress. Cells that line fluid-transporting vessels, like endothelial and certain epithelial cells, are perhaps expected to respond to shear, but other cell types, such as bone and cartilage cells, are also sensitive to shear. As a result, there is substantial diversity in shear flow application systems, owing to the different physiological conditions. Another reason for a diversity of flow devices is that

Nota Bene

A functional group. A *functional group* is the region or *moiety* of a protein that is responsible for its particular biochemical characteristic or activity.

establishing well-controlled flow can be a challenge in itself; there is generally no way to directly measure the shear stress being applied on the cell monolayer. Although custom devices might circumvent some of these issues, those devices must be designed carefully. A basic understanding of fluid mechanics can greatly ease the development of design specifications based on fluid-flow analysis, such as that presented in Section 4.3. Before fluid mechanics specific to shear devices are addressed, we examine one common consideration in using fluid shear devices.

The transition between laminar and turbulent flow is governed by the Reynolds number

To design the fluid flow device, one normally uses laminar flow to expose the cells to a controlled shear stress. Nevertheless, there are numerous studies involving turbulent flow, such as investigations of how turbulent flow contributes to the development of atherosclerosis. Although a detailed analysis of turbulent fluid mechanics is beyond the scope of this text, one consideration is to ensure that no unintended turbulence occurs. To ensure the flow is laminar, one generally designs the device so that the *Reynolds number* (*Re*) is very low (recall the discussion of the Reynolds number in Section 4.2).

When the Reynolds number is very low, the viscous forces dominate. That means the fluid tends to move uniformly without much mixing. These flows are termed laminar owing to the laminar structure of the streamlines. High Reynolds number flows tend to be inertial: mixing, uneven fluid profiles, and unsteady flows. These flows are turbulent. The actual transition Reynolds number between laminar flow and turbulence is different for different geometries, but for pressure-driven pipe flows (known as *Poiseuille* flow), a Reynolds number of approximately 2000 is a good estimate of the transition point. A Reynolds number of, say, 200 is generally laminar despite having a relatively higher inertial component. If the walls of the pipe are very smooth, it is possible to increase the Reynolds number to even higher values without inducing turbulence, partly because the inertia of pipe flows is one-dimensional. However, if there is roughness on the pipe walls that could "trip" the fluid and induce mixing, a fluid element running close to the wall would move in non-axial directions. This, in turn, causes the adjacent elements to shift, and soon the entire flow is tumbling. This is deliberately done in the dimpling of golf balls to induce turbulence during the flight of the golf ball (Figure 6.17).

Parallel plate flow devices can be designed for low Reynolds number shear flow

One strategy for applying fluid shear to cells is the parallel plate flow chamber. In such a chamber, fluid is pumped through a chamber of rectangular cross section. The height is very small compared with the width, so the fluid profile can be assumed to occur between two infinite parallel plates. The cells are plated on the bottom surface.

In designing such a shear stress apparatus, multiple factors need to be considered. The fixed factors are relatively easily dealt with. The fluid is typically a cell-compatible medium that can be assumed to have nearly the same density and viscosity as water. The temperature is controlled by some custom or commercially available heaters/coolers, or the device is used within a cell culture incubator. The cells are plated on a surface treated with adhesion molecules or otherwise modified to be compatible with cell adhesion. To obtain a low Reynolds number, the chamber dimensions must be carefully selected. To control the shear stress, one usually controls the velocity. However, this also affects the Reynolds number. Therefore, there is an interaction between the size of the chamber and the velocity at which the transition to turbulence will occur.

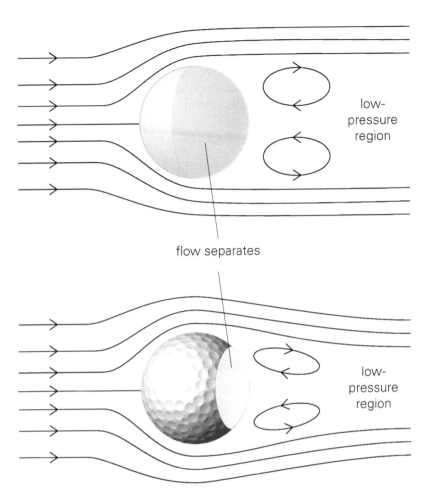

low-pressure region

flow separates

low-pressure region

Figure 6.17 Dimples in golf balls induce turbulence and create a smaller boundary layer that takes longer to separate. The smaller wake helps reduce pressure drag on the ball. Early golfers playing with smooth golf balls found that badly dented balls tended to fly farther. For many cell mechanical applications, though, laminar flows are required and therefore smoother surfaces are desired.

Let us look at an approximation that can be used for a quick estimate of the flow device dimensions. Consider a system using cell culture medium as the fluid ($\rho = 1000$ kg/m^3, $\mu = 1 \times 10^{-3}$ kg/ms). The Reynolds number is then $Vh/(1 \times 10^{-6}$ m^2/s), where V is some characteristic velocity and h is the height of the flow chamber. For $Re < 1$, we then need $Vh < 1 \times 10^{-6}$ m^2/s. Next, consider the shear stress, $\tau = \mu \, du/dy$. We can approximate this as $\tau \sim \mu V/h$, so long as the width is large compared with h. A typical shear stress for physiological modeling of endothelial cells is about 1 Pa. So, $V/h = 1000$ s^{-1}. Assume we express V in m/s and h in m to keep units simple, then $Vh < 10^{-6}$ and $V = 1000 \, h$. This means, $h^2 < 10^{-9}$ m, or $h < 3 \times 10^{-5}$ m = 30 µm, which is just about acceptable, seeing as cells are typically at least a few micrometers (10^{-6} m) large. Although a chamber this short can be difficult to manufacture, generally, the chamber is designed with the shortest height possible. Calculations based on actual flow profiles and relaxing the Reynolds number requirement (to 10–100) can improve this range somewhat.

Fully developed flow occurs past the entrance length

Another consideration of shear flow devices is the *entrance length*. Laminar flow profiles require space to develop, and the entrance length is the size of this region. It changes depending on the velocity and dimensions of the chamber. For a circular pipe, the entrance length is about 0.06 times the Reynolds number multiplied by the diameter, $0.06 \times Re \times d$. When locating cells to examine, it is best to avoid the entrance length to avoid unexpected or unpredictable shear stresses. The cross-wise flow profile (perpendicular to both the height and the flow direction) will also develop over space; this is generally not desired, as the

Nota Bene

Dynamic versus kinematic viscosity. Viscosity is sometimes a confusing term, as the concept of viscous resistance is associated with both dashpots (which we will cover in Section 6.4) and fluids, but these terms have different units and definitions. In this text, we refer to dashpots as having a viscous friction coefficient whereas fluids have viscosity.

In fluid mechanics, there is a further distinction between the dynamic viscosity (which is viscosity that is defined in Section 4.2, represented by µ, and has units of kg/ms) and the kinematic viscosity, which is the dynamic viscosity normalized by the density of the fluid ($v = \mu/\rho$, with units of m^2/s). The reason for using the kinematic viscosity is that it simplifies some calculations (such as Reynolds number), can be written with three terms instead of four, and represents a useful ratio that eliminates density (mass, specifically) from comparisons. That is, denser fluids might appear more viscous because of inertial effects. For this chapter, viscosity will refer to dynamic viscosity.

flow is best modeled as two-dimensional. A rule of thumb requirement is that the width of the chamber should be at least ten times the height, and the height should be as small as possible. The circular pipe formula for entrance length can be used for roughly estimating what the entrance length is for a parallel plate flow chamber, but if validation is required, empirical measurements of the flow profile using beads in the fluid can be performed.

Cone-and-plate flow can be used to study responses to shear

Another shear flow device uses a cone-and-plate viscometer. A cone is positioned above a stationary plate and rotated. A constant shear stress on the bottom plate is established, depending on the gap between the plate and cone bottom and the rotational speed of the cone. The condition of constant shear is obtained by selecting a very shallow cone angle so that the gap between the cone and the plate increases from the apex to the periphery. This configuration facilitates the controlled application of shear stress because it is directly proportional to the rate of rotation of the cone. Note that the Reynolds number must still be kept low, because this configuration can generate turbulence or unsteady transitions if the rotation is too high. Additionally, systems involving moving components can require higher engineering precision, because since small machining errors can have big impacts on the flow. For example, the spacing between parallel plates can be controlled with gaskets sandwiched between the plates. For the cone-and-plate viscometer, ensuring the cone's axis is perpendicular to the plate can be challenging.

Example 6.4: Shear stress in a cone-and-plate viscometer

Show that for a cone-and-plate viscometer (Figure 6.18), the shear stress is the same far from the tip of the cone, if the angle is shallow enough.

Let ω be the rotational speed in radians/second, r the radius away from the tip of the cone (where it contacts the plate), and α the angle between the cone and the plate, with the fluid in the gap and the cells plated on the bottom plate. The distance between the bottom plate and the lower surface of the cone is h, where

$$h = r \tan \alpha.$$

For a very small α, this can be approximated as

$$h = r\alpha.$$

If the rotation of the cone is steady, the flow is laminar, and assuming α is small, we can treat the flow as parallel plate flow; that is, for each r, the flow profile is linear from the cone to the bottom plate. Thus, the shear stress is

$$\tau = \mu(du/dz) = \mu(\omega r)/h = (\mu \omega r)/(r\alpha).$$

And,

$$\tau = (\mu\omega)/\alpha.$$

Note that this is spatially constant, so for a given viscosity, a fixed ω and a fixed α, the shear stress on the bottom plate is the same everywhere. Cone-and-plate viscometers are useful because you can attain good control over the shear stress just by varying the cone rotation rate (ω) and it uses relatively little medium because it is not recirculated. However, careful machining and alignment are crucial, because small errors can result in varying shear stresses along the bottom place or operational instabilities (vibrations, etc.).

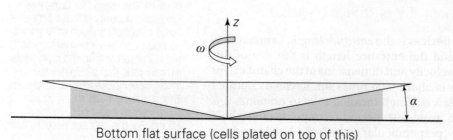

Figure 6.18 Schematic of a cone-and-plate viscometer. As long as the cone-angle is shallow, the shear on the bottom surface is homogeneous.

Bottom flat surface (cells plated on top of this)

(A) rectangular flow chamber

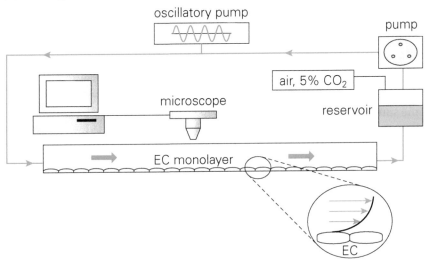

(B) step flow chamber

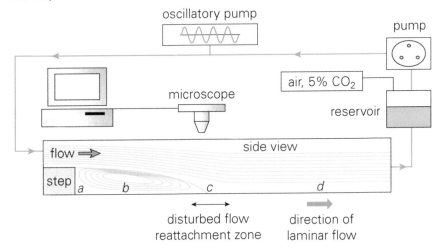

Figure 6.19 Two examples of parallel plate flow chambers. (A) Example of a shear flow setup that conditions the media and allows for imaging during shear application, and (B) the actual flow chamber that has a separation/recirculation region and a laminar flow region to assess effects of disturbed flow on the cell monolayer. (Adapted from, Chien S (2008) *Ann. Biomed. Eng.* 36.)

Although we have focused primarily on steady laminar flow, it is worthwhile to note that oscillatory flows (flows that either halt or reverse), turbulent flows, separated flows (flows that develop recirculation regions), and combinations thereof (Figure 6.19) have also been extensively used to investigate cellular responses to temporally varying shear stress. These studies have become particularly important in the cardiovascular field, in which complex flow patterns have been related to pathological changes.

Diverse device designs can be used to study responses to fluid flow

Before we conclude our discussion of devices for studying responses to fluid flow, it is worthwhile mentioning that although pressure-driven parallel plate and cone-and-plate devices are commonly used designs for investigating flow-mediated mechanotransduction, there is great diversity in shear flow application systems. For example, tubes have been used with success, although it becomes more difficult to image the cells during the application of shear. This geometry may be more physiologically relevant for modeling blood vessels. Another approach is to replace a pressure-driven flow with a driven-plate configuration, known as *Couette* flow. A treadmill used in lieu of the upper plate of a parallel plate chamber will generate a linear flow profile, which mitigates problems with entrance lengths and creates a simpler relationship between the moving wall velocity and shear stress.

Flexible substrates are used for subjecting cells to strain

Although many cells experience fluid shear (or at least, mechanical shearing of some sort), other cells primarily experience stretch. Blood vessel smooth muscle cells, cardiac cells, skin cells, bladder/stomach/intestine, and lung cells are examples of physiological systems in which stretch is vital. Applying controlled stretch to a cell directly is problematic, because the cell membrane is difficult or impossible to grasp. In addition, cells generally experience stretch due to deformation in the surrounding tissue in which they are anchored. The most common way of stretching cells is to plate them on a flexible substrate, which is then stretched. This method has the advantage of allowing many cells to be stretched simultaneously. Substrate stretch is not always a physiologically representative stretch, as it is confined to two dimensions and primarily experienced through the basal surface of the cell. This is less of a concern for cells normally found in monolayers, such as endothelial cells, some epithelial cells, or osteoblasts. However, many key physiological responses can be activated by this technique. Therefore, substrate stretch can be useful for studying these responses, even if the stimulus is not quite physiological.

Confined uniaxial stretching can lead to multiaxial cellular deformations

In stretching cells, some consideration must be given to how the stretch is applied. The simplest form of stretch is unconstrained uniaxial, whereby the substrate is held on two opposing sides and then pulled apart. This lengthens the substrate and cells in the direction being pulled, but will shorten them in the perpendicular direction. Additionally, if the ends being pulled are clamped tightly, then the shortening will be nonuniform, being most pronounced at the center (Figure 6.20).

This partly constrained system can result in strain heterogeneities. A constrained uniaxial design overcomes this by constraining the two stationary sides. This eliminates strain in the perpendicular direction. It is a suitable model for some physiological systems, such as blood vessels, and can demonstrate cellular reorientation responses to stretch quite nicely. However, such designs are trickier to build, because the sides must be kept fixed in one direction but must move in another. Further, constrained uniaxial stretch can result in some degree of heterogeneity in cells that exhibit strong axial orientations. That is, an elongated cell may respond differently if strained along its long axis than along its minor axes.

Cylindrically symmetric deformations generate uniform biaxial stretch

A final, and perhaps the most straightforward, way to expose cells to substrate stretch is with a uniform-biaxial stretch (Figure 6.21). In this strain field, every

Figure 6.20 A flexible membrane that is stretched in one direction will result in a "bowtie" formation that is non-uniform. The shape is exaggerated for clarification.

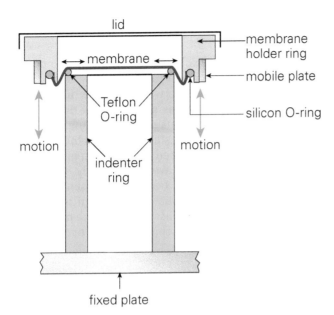

Figure 6.21 An example design for a membrane stretch device. The cells are plated on the upper surface of the membrane, with a lid and rings used to keep the cells immersed in media. The indenter ring can be pushed up (or the membrane pushed down) to stretch the membrane along with cells attached to the membrane surface. (From, Sotoudeh M, Jalali S & Usami S (1998) *Ann. Biomed. Eng.* 26.)

pair of points is expanded by the same percentage strain regardless of orientation. This strain field has no preferred direction, and each cell will experience the same strain in all directions. There are several designs for this sort of stretching, including using air to push up a dome, but a common configuration uses a cylindrical piston.

6.4 ANALYSIS OF DEFORMATION

In the previous sections, we learned about how cells can be exposed to forces to characterize their mechanical properties and responses. One question left open is how can we use such experimental data to describe the mechanical behavior observed? Suppose we obtained displacement profiles for two different cell populations, such as different cell types or cells in the presence/absence of a cytoskeletal inhibitor. How could we convert the displacement profiles into something with which we can compare the mechanical properties of these two populations? One approach is to compare the raw data alone. We might directly compare bead displacement. However, this approach is a simplistic comparison. A broader, potentially more useful, approach is to develop models for a cell whereby parameters that are experimentally obtained can be used to make predictions and comparisons across different experiments.

Viscoelastic behavior in micromanipulation experiments can be parameterized through spring–dashpot models

One of the approaches for mechanical modeling of displacement and/or force profiles generated using micromanipulation is to construct models consisting of spring and dashpot elements. Springs obey Hooke's law, which can be written as

$$F = kx, \tag{6.13}$$

with the spring constant k a measure of the stiffness of the element. Dashpots are viscous damper elements that obey Newton's fluid law

$$F = \eta v, \tag{6.14}$$

where v is the velocity and η is a parameter known as the *viscous friction coefficient*.

In general, a spring or dashpot in isolation is insufficient to model physiological cellular mechanical behavior. Consider the case where we perform a magnetic micromanipulation experiment where we obtain the displacement of a cell-bound bead under a constant force as a function of time. There is no force applied initally; then we turn on the magnet, generating a near-instantaneous step-increase in force that remains constant as long as the magnet is turned on. After a few seconds, the magnet is turned off, resulting in a near-instantaneous step-decrease in force. If the cell behaved as a purely elastic material, like a spring, the displacement curve of the bead would resemble the profile of the forcing function, with an instantaneous displacement when the magnet was turned on and which would remain constant. When the magnet was turned off, an instantaneous jump to zero displacement would occur, as seen in Figure 6.22.

On the other hand, if the cell behaved as a purely viscous material, like a dashpot, the bead's response to a step up or down in force would appear as a line with non-zero slope (Figure 6.23). Unlike a spring, the displacement of the dashpot increases linearly with time with slope η. When the magnet is turned off, the dashpot does not return back to its original length. Instead, it exhibits *creep*, or *permanent* deformation.

In reality, cells do not exhibit purely elastic or viscous behavior. Rather, they are viscoelastic, exhibiting time-dependent mechanical behavior that has attributes of both elastic and viscous materials. Under a constant force, the displacement does not occur instantaneously or slowly increase at a constant rate, but instead typically exhibits asymptotic behavior. Similar asymptotic behavior is exhibited when the force is removed, and in many cases cells exhibit permanent deformation upon removal of the force. As a result, using a spring or dashpot alone is insufficient to capture this mechanical behavior.

Combinations of springs and dashpots can be used to model viscoelastic behavior

One approach to producing more realistic cellular behavior is to combine several springs and dashpots together. Given the fact that cells exhibit behavior that is both elastic- and viscous-like, models constructed of combinations of springs and dashpots in series or parallel can result in mechanical behavior that resembles experimental observations remarkably well. Three examples of such combinations are a spring and dashpot in series (called a *Maxwell body*), a spring and

Figure 6.22 The response of a spring element to a step force being applied at 1 second and turned off at 2 seconds. The displacement reaches a constant as soon as the force is applied. Note that there is no oscillation because there is no inertial term.

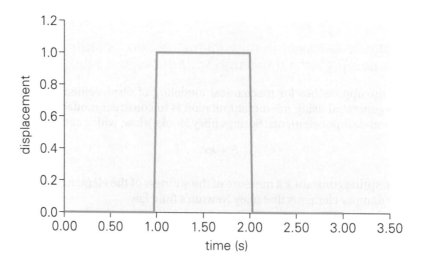

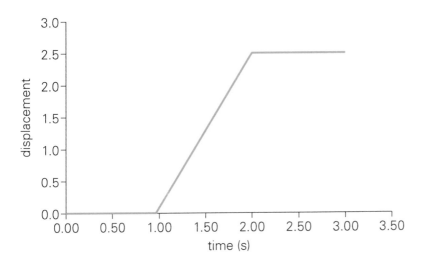

Figure 6.23 At one second, the force is applied, and the response element starts "stretching out." Because the force is proportional to the velocity, a constant force application results in a constant velocity response. At 2 seconds, the force is released, and the element no longer displaces, but does not recoil because here is no elastic restoring force.

dashpot in parallel (a *Kelvin–Voigt body*), and a combination of a Kelvin–Voigt body in series with a dashpot (Figure 6.24). As we will see, the analytical solution for the displacement of these bodies with time under a constant force can be computed in a relatively straightforward manner.

Consider the combination of a Kelvin–Voigt body in series with a dashpot. A free-body analysis shows that the force F applied to the system must equal the force on the right-hand dashpot, and it must equal the sum of the forces in the left-hand spring and dashpot. If we expand these forces with the governing equations (Equations 6.13 and 6.14), we obtain

$$F = \eta_1 v_1 = kx_2 + \eta_2 v_2. \tag{6.15}$$

The total displacement is the sum of the displacement of the single dashpot and the Kelvin–Voigt body, $x(t) = x_1 + x_2$. Thus, to obtain $x(t)$, we can compute $x_1(t)$ and $x_2(t)$ individually, and then add them together to get the displacement of the entire body. For the single dashpot, we know that $x_1 = (F/\eta_1)t$. For the Kelvin–Voigt body the total force across the body is equal to the sum of the force within the spring and dashpot

$$F = kx_2 + \eta \frac{dx_2}{dt}. \tag{6.16}$$

This is a first-order differential equation, which has the solution

$$x_2(t) = \frac{F}{x}\left(1 - e^{\left(\frac{-kt}{\eta_2}\right)}\right). \tag{6.17}$$

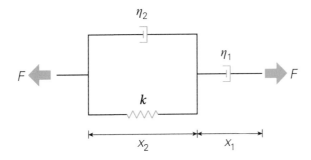

Figure 6.24 A model of cell material properties using spring and dashpot combinations. In this example, a spring and dashpot are in parallel. Together, they are in series with a second dashpot that allows for plastic deformation.

Figure 6.25 Response of the cell model to a step force applied at 1 second and turned off at 2 seconds. Note that after the application of the force, the response slowly approaches a straight line. After the force is released the recoil is not to zero producing residual deformation or strain. This response profile to a step increase and decrease in force is a reasonable approximation of real-cell response to magnetic micromanipulation.

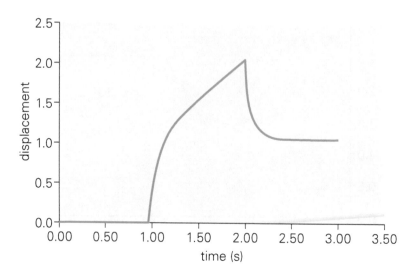

This can be verified by direct substitution, with initial condition $x(0) = 0$. Summing the displacements of the single dashpot and the Kelvin–Voigt body,

$$x(t) = \left(\frac{F}{\eta_1}\right)t + \frac{F}{k}\left(1 - e^{\left(\frac{-kt}{\eta_2}\right)}\right). \tag{6.18}$$

Nota Bene

Lumped parameter models. The models of the type we have been discussing are sometimes called *rheological* models, because they originated from the analysis of viscous fluids (see Section 4.4). They are also sometimes referred to as *lumped parameter* models, because they collect spatially distributed behaviors and lump them into simple discrete elements. For the lumped parameter model in Figure 6.25, all the elastic behavior was captured in a single equivalent spring stiffness parameter (k).

The behavior of this model is depicted in **Figure 6.25**. It is very useful, because it is the simplest model that captures both viscoelastic behavior as well as *plastic* behavior (permanent deformation). To understand this phenomenon, consider what happens as t goes to infinity. The first term (the displacement of the single dashpot) increases without bound at a constant rate of F/η_1. The second term (the displacement of the Kelvin–Voigt body) gives a displacement at t = infinity of F/k. For finite time, the Kelvin–Voigt body never reaches F/k, but approaches it asymptotically. Finally, it is worthwhile noting that the exponential term, $-kt/\eta_2$, is sometimes written as $-t/\tau$, where $\tau = \eta_2/k$ is called the time constant. The time constant is a measure of how long it takes the exponentially decaying quantity in the parentheses to decrease by $1/e$, or about 63%, of the initial amplitude.

Example 6.5: Maxwell body: spring, and dashpot in series

Examine a Maxwell body: a spring and dashpot in series (**Figure 6.26**).

Figure 6.26 Schematic of a Maxwell body. A Maxwell body is a spring and dashpot in parallel.

Let the spring have constant k and the dashpot have constant η.

A. If we were to oscillate the tip of the Maxwell body at very high or very low frequencies, would the response be dominated by the spring or the dashpot?

At very high frequencies, the dashpot velocity will generally be large. Because $F = \eta v$, as the velocity goes to infinity, the force on the dashpot goes to infinity as well.

Therefore, the dashpot will behave like a rigid structure and the response of the Maxwell body will be similar to that of a spring alone at high frequencies.

At very low frequencies, the dashpot velocity will be small, and so the force sustained by the dashpot will be negligible. As a result, the dashpot acts as an open structure, and the spring acts like a rigid structure. Here, Maxwell body will be similar to that of a dashpot.

B. Quantify your response to part A.

Let the stimulus be $F(t) = F_0\sin(\omega t)$, where ω is the frequency of oscillation and F_0 is the amplitude. The response will be $x(t) = A\sin(\omega t + \delta)$, where δ is the phase lag. We know the force through the spring and dashpot are the same (both equal to $F(t)$). We also know the displacement of the entire system is the sum of the displacements of the spring and dashpot. Thus, we can solve for each individually.

For the spring, this is a simple substitution.

Let $x_s(t) = A_s \sin(\omega t + \delta)$

$$F(t) = F_0 \sin(\omega t) = kx_s(t) = kA_s \sin(\omega t + \delta).$$

So we know $\delta = 0$ for the spring and that $A_s = F_0/k$. Therefore,

$$x_s(t) = (F_0/k)\sin(\omega t).$$

For the dashpot, we need a differential equation.

Let $x_d(t) = A_d \sin(\omega t + \delta)$, then

$$F(t) = F_0 \sin(\omega t) = \eta(dx_d/dt) = \eta \omega A_d \cos(\omega t + \delta).$$

In this case, $\delta = -\pi/2$ and $F_0 = \eta \omega A_d$

and thus,

$$x_d(t) = (F_0/\eta\omega)\sin(\omega t - \pi/2).$$

The total displacement is the sum of the spring and dashpot components,

$$x(t) = (F_0/k)\sin(\omega t) + (F_0/\eta\omega)\sin(\omega t - \pi/2).$$

As ω goes to infinity, the dashpot (right-hand) term, which varies as $1/\omega$, goes to zero, and we are left with only the spring term. Similarly, at very low frequencies, the dashpot term overshadows the spring term and we are left only with the dashpot term. At low frequencies, the spring does not significantly influence the behavior of the system.

Microscopy techniques can be adapted to visualize cells subject to mechanical loading

In Section 6.1, we first considered the challenge of determining cellular mechanical behavior, but we limited our discussion to force application at discrete locations, such as beads or AFM probes. As we described in Section 6.3, it is also possible to expose cells to mechanical stimuli more representative of their native environment. Consider, cartilage explants freshly dissected from collected tissue and subjected to the dynamic compression that they would experience *in vivo*. Here, the complex biochemical and mechanical environment of the tissue gives rise to cellular deformations that are the additive result of several different physical and/or mechanical phenomena. These might include loads imparted to the extracellular matrix that are transferred to the cells at the sites of adhesion, fluid shear, hydrostatic pressure, altered osmotic pressure, etc. However, the deformational analysis is more complex, because our readout is not simply the displacement at one location. One approach is to observe cells microscopically during loading. We will discuss here how typical time-lapse image sequences are obtained and how their deformation is analyzed.

To generate image sequences of cells subjected to mechanical loading, structures in living cells can be imaged by epifluorescent or confocal microscopy by using fluorescent stains, or transfecting cells with DNA constructs that encode for fluorescent proteins. Alternatively, certain structures can be visualized with techniques that generate image contrast in transparent specimens, such as phase contrast. By fabricating specialized loading devices that fit on the stage of microscopes, these microscopy techniques can be adapted in a variety of ways to visualize cellular or even subcellular deformations. To illustrate, consider the case of cells subjected to fluid shear stress. Such chambers have been constructed to allow for visualization of the cells from the bottom, or in some cases, from the sides as well. Such multi-plane approaches use mirrors and multiple optic paths to perform simultaneous imaging in two orthogonal planes (Figure 6.27) resulting in quasi-three-dimensional imaging without a confocal microscope.

Figure 6.27 The side-view microscope allows imaging of a cell from the top and in profile simultaneously. This is achieved by having two separate lightpaths and sets of optics. The mirror allows for a more compact design with all optical components in a conventional configuration. A rectangular flow channel is required for side-imaging (From, Cao J, Donell B, Deaver DR, et al. (1998) *Microvasc. Res.* 55.)

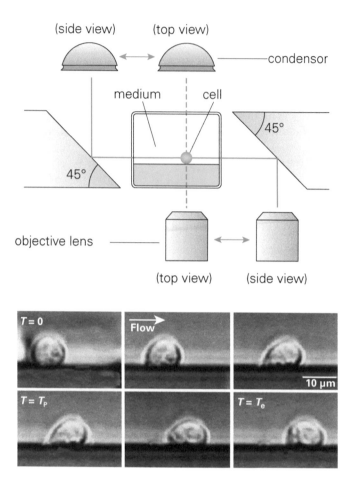

Cellular deformations can be inferred from image sequences through image correlation–based approaches

Once image sequences of mechanically–loaded cells are obtained, they can be analyzed to infer the cellular deformations that occurred through the use of image correlation. As we described previously, this can be accomplished either with texture correlation or by tracking the motion of attached beads. In either case, the result is a set of displacements at discrete spatial locations. Often, we would like to know the whole displacement field. This can be obtained through a process of interpolation. Consider set of k discrete displacement vectors $\mathbf{u}_{1,2,\ldots,k}$ obtained at positions $\mathbf{x}_{1,2,\ldots,k}$. To construct a displacement field, the discrete displacements are interpolated to all the interior locations between the $\mathbf{x}$ positions. A common approach is to interpolate two-dimensional discrete displacements linearly to construct triads, or groups of three displacement vectors that neighbor each other in space. One algorithm for automatically generating triads is *Delaunay triangulation*, which maximizes the minimum angle of all the angles of the triangles in the triangulation (and therefore avoids long aspect ratio triangles), as in **Figure 6.28**. The details of triangulation approaches are beyond the scope of this text but can be found in many image-processing treatments.

Once the triads have been specified, the displacement field can be constructed by linearly interpolating the displacements between each triad. Linear interpretation over the triangular domains results in a plane determined by these three locations. The equation of plane for u and v, the displacements in the x- and y-directions, respectively, are

$$u(x, y) = u_a x + u_b y + u_c \tag{6.19}$$

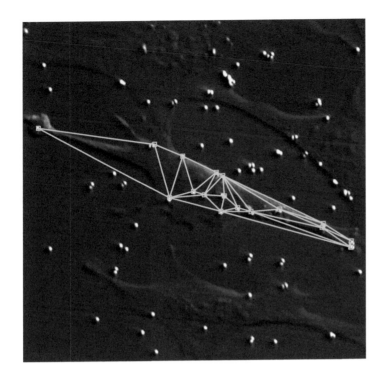

Figure 6.28 Delaunay triangulation of fluorescent beads. Here we have applied a Delaunay triangulation procedure to the cell with attached beads from Figure 6.11. The number of beads was kept small to better illustrate the approach. In practice, more beads and triangles with lower aspect ratios would be required for accurate results. (From, Kwon R, et al. (2007) *J. Biomech.* 40, 3162–3168.)

and

$$v(x, y) = v_a x + v_b y + v_c,$$ (6.20)

where u_a, u_b, u_c, v_a, v_b, and v_c are unknown constants. To solve for u_a, u_b, and u_c, we form the set of simultaneous equations

$$u_a x_1 + u_b y_1 + u_c = u_1$$

$$u_a x_2 + u_b y_2 + u_c = u_2$$ (6.21)

$$u_a x_3 + u_b y_3 + u_c = u_3,$$

which results in three equations and three unknowns. The constants v_a, v_b, and v_c can be solved for using a similar approach. We segment the image into triangles, at which the vertices are locations for which we have displacement data. Then, to estimate the displacement at an arbitrary point, we select the three vertices making up the triangle that encloses the point.

Intracellular strains can be computed from displacement fields

Given the displacement fields $u(x, y)$ and $v(x, y)$, it becomes possible to calculate intracellular strains. In Chapter 3, for small deformations, we derived the infinitesimal strains,

$$\varepsilon_{xx} = \frac{\mathrm{d}u}{\mathrm{d}x}$$

$$\varepsilon_{yy} = \frac{\mathrm{d}v}{\mathrm{d}y}$$ (6.22)

$$\varepsilon_{xy} = \frac{1}{2}\left(\frac{\mathrm{d}u}{\mathrm{d}y} + \frac{\mathrm{d}v}{\mathrm{d}x}\right).$$

For large deformations, the Green–Lagrange strains can be computed as

$$E_{xx} = \frac{du}{dx} + \frac{1}{2}\left(\left(\frac{du}{dx}\right)^2 + \left(\frac{dv}{dx}\right)^2\right)$$

$$E_{yy} = \frac{dv}{dy} + \frac{1}{2}\left(\left(\frac{du}{dy}\right)^2 + \left(\frac{dv}{dy}\right)^2\right) \tag{6.23}$$

$$E_{xy} = \frac{1}{2}\left(\frac{du}{dy} + \frac{dv}{dx}\right) + \frac{1}{2}\left(\frac{du}{dx}\frac{du}{dy} + \frac{dv}{dx}\frac{dv}{dy}\right).$$

An alternative approach can be used to compute the Green–Lagrange strain that does not require the computation of the displacement derivatives. Recall from Section 3.3 that for any differential line segment d**X**, its deformed length after a deformation is given by

$$\mathbf{dx}^2 = \mathbf{dXCdX}, \tag{6.24}$$

where $\boldsymbol{C} = \boldsymbol{F}^{\mathrm{T}}\boldsymbol{F}$ is the symmetric Cauchy–Green deformation tensor. Assuming a two-dimensional linear deformation, there are three unknowns to solve for: C_{11}, C_{22}, and C_{12}. Recalling our triads of displacement vectors, the vertices of each triad located at $\mathbf{x}_1$, $\mathbf{x}_2$, and $\mathbf{x}_3$ define three line segments in the undeformed state. Similarly, the points $X_1 = \mathbf{x}_1 + \mathbf{u}_1$, $X_2 = \mathbf{x}_2 + \mathbf{u}_2$, and $X_3 = \mathbf{x}_3 + \mathbf{u}_3$ define the same three line segments transformed under the deformation described by $\boldsymbol{F}$. Now we can compute $\boldsymbol{F}$ directly. Then we can extract information regarding the principal directions and stretches. Additionally, once $\boldsymbol{C}$ is computed, the Green–Lagrange strain can computed as $\mathbf{E} = 1/2(\boldsymbol{C} - \mathbf{I})$, where $\mathbf{I}$ is the identity tensor.

Example 6.6: Deformation gradient analysis

Let us revisit Example 3.6, but instead of determining the principal strains, we will calculate the principal stretches and principal stretch directions. The analysis is very similar, but the final result is different.

Recall that

$$\boldsymbol{F} = \begin{bmatrix} 2.5 & 0.5 \\ 0.5 & 2.5 \end{bmatrix}.$$

We note that $\boldsymbol{F}$ is symmetric, so that $\boldsymbol{F}^{\mathrm{T}}\boldsymbol{F} = \boldsymbol{F}^2$, but we know $\boldsymbol{F}^{\mathrm{T}}\boldsymbol{F} = \boldsymbol{U}^2$ and therefore $\boldsymbol{U} = \boldsymbol{F}$. Our rotation tensor is the identity matrix and there is no rotation.

To find the principal stretches, we need the eigenvalues of $\boldsymbol{U}$,

$$\boldsymbol{U}\mathbf{v} = \lambda\mathbf{v}$$

is the characteristic equation relating our stretch tensor $\boldsymbol{U}$, the principal stretches, λ, and the principal stretch directions $\mathbf{v}$. Noticing that we can rearrange the equation to read:

$$\boldsymbol{U}\mathbf{v} - \mathbf{I}\mathbf{v}\lambda = 0,$$

where I is the identity matrix. We can then write

$$(\boldsymbol{U} - \mathbf{I}\lambda)\mathbf{v} = 0.$$

This must be true for nontrivial $\boldsymbol{v}$ (our eigenvectors) and the quantity in parenthesis must be non-invertable. That means the determinant of $(\boldsymbol{U} - \mathbf{I}\lambda) = 0$. To set this up,

$$\begin{vmatrix} 2.5 - \lambda & 0.5 \\ 0.5 & 2.5 - \lambda \end{vmatrix} = 0.$$

Some algebra leads to

$$6.25 - 5\lambda + \lambda^2 - 0.25 = 0 = \lambda^2 - 5\lambda + 6.$$

The roots of this equation are $\lambda = 2, 3$. Physically that means if a vector is in a principal stretch direction, the deformation will stretch it out by a factor of 3; if it is in the other principal stretch direction, the deformation will stretch it out by a factor of 2. What are the directions? We simply substitute back into the characteristic equation

$$\boldsymbol{U}\mathbf{v} = \lambda\mathbf{v}$$

for each λ. We find that the eigenvectors are $(1,1)$ (corresponding to $\lambda = 3$) and $(1, -1)$ (corresponding to $\lambda = 2$). Note that when you are solving for the eigenvectors, you will not generally be sufficiently constrained to get the exact values; so you can guess a value for one component.

Typically, the vectors would be expressed in unit form, but we do not do so here to avoid the use of square roots. Our result is also consistent with our findings in Example 3.6 in that the principal stretch directions are the same as the principal strain directions.

6.5 BLINDING AND CONTROLS

Biomedical engineering experimentation can be a delicate field because it straddles more traditional engineering, in which results are often more analytical or quantitative, and biology, where results can be more empirical and experimental. Because of such contrasts in approaches, there are two concepts common to experimental biology that engineers should understand to perform and interpret analysis correctly: *blinding* and *controls*.

A control is a duplicate of the experimental condition for which an intervention or stimulus of interest is left out. If we wished to assess whether the addition of compound A will cause cells to multiply faster, what we could do is to get a dish of cells, add A to the media, and then measure the cell division rate. We could then compare the measured cell division rate to known cell division rates and see if it is increased in the presence of A. The problem is that the cells we have been using may already be dividing faster than average, or perhaps the media we are using to feed the cells causes them to divide faster, or the incubator where the cells are being kept possibly creates an environment whereby the cells will divide faster. The best way to account for all of these effects is to maintain a second dish of cells, to which we do not add A, using the same cell type, media, and incubator. Then, we can compare the division rates between the two dishes and assess whether A, in fact, influences cell division. Although it is accepted that there will be some differences (the dishes will not be kept in exactly the same location, will not be handled exactly the same way, etc.), as many conditions as possible should be kept identical between the two dishes. The ideal experiment is one in which the only difference between experimental conditions is the factor being tested or examined. The clinical analog of the control is the placebo, designed to counter the psychological influence of providing a potential cure to a patient. Patients often feel better when they believe that they have received a medication (this is called the placebo effect).

Blinding refers to a process that reduces potential bias by the investigator. The "Clever Hans" horse is a popular story that illustrates this effect. Essentially, the horse appeared to be able to perform simple arithmetic. A random person would pose a question $(2 + 3)$, and Clever Hans would start stamping his foot. At the correct number of stamps (5), he would stop. However, it was later determined that the horse would simply watch his owner, and when his owner relaxed (at the right number), he would stop stamping his foot. Similar bias can exist in experiments, whereby an investigator who is expecting a certain result may interpret experimental outcomes in favor of preexisting expectations. The best way to avoid this is to blind the investigator to the conditions, so that measurements are taken without this bias. If we were to give three people vitamins, and three people a placebo (as a control), and our hypothesis was that vitamins will prevent headaches, we would need to ensure that the people receiving the substances are not told whether they are receiving the vitamin or placebo. This would be a blinded experiment, because the participants would not know which substance they received and could not use that information to influence the results. For instance, knowing they received the vitamin, they might tend to say they have fewer headaches.

However, our presence might also influence the participants, because we know who received the placebo and who received the vitamins. Therefore, we may act—perhaps unconsciously—more pleasantly toward those who received vitamins. To avoid this error, we would have an assistant interview the participants about their headaches; this assistant could not know who received which substance. The patients and their treatments would be encoded so that nobody involved in the experiment knows who received what. This approach is a *double-blind* design, and is the most rigorous method for uncovering true influences.

Sometimes blinding does not work or is impractical. If we were to perform an operative procedure to relieve pain, the people receiving the operation would clearly know they had it, compared with patients who were left untreated. To account for this, sometimes people use what is called a *sham*. This is a special control whereby a procedure is mimicked, but a key therapeutic step is omitted. So in this situation, the control group would have an operation, but without whatever manipulation was thought to be analgesic. A sham surgery raises important ethical concerns in human studies, but is a common strategy for animal investigations.

The use of controls and blinding applies to many of the techniques that are covered in this chapter. For instance, if you perform an experiment using magnetic twisting by attaching magnetic beads to cells and then applying a torque, you cannot simply obtain a readout (stiffness, upregulation of some gene, activation of some channel, etc.) and decide whether the cells are responding or not. You need to compare the readout to cells that do not have a similarly applied torque. However, the addition of the beads may have caused some changes in cell baseline. The better control is to have a sham in which you add beads but do not torque the beads. Some investigators will go so far as to complete the sham by adding beads to cells, placing the cells in the magnetic twisting device, and turning on the equipment, but will not actually apply the torque. Alternatively, the experiment can be performed with nonmagnetic beads. In this way, vibrations from the equipment, stresses and temperature changes from handling and placing the dish on the apparatus, the magnetic field itself, etc., are all eliminated as potential confounding influences.

Key Concepts

- A variety of techniques allow us to measure cell forces and manipulate cells.
- There are various modes of magnetic micromanipulation (pulling vs twisting).
- Optical traps work by deflection of light and offer contrast to magnetic methods by manipulating bead displacement rather than force. Magnetic forces can be applied to a large number of beads simultaneously.
- In atomic force microscopy cells can be probed by application of a cantilevered tip against the cell. The deflection of the cantilever can be related to the force being applied.
- Microfabrication is useful for assaying small numbers of cells. Micropillars can be used to measure cell–substrate forces based on the deflection of the pillars, which are modeled as beams.
- Traction-force microscopy is another way to measure cell–substrate forces.
- Beads embedded in a flexible substrate allow substrate deformations to be determined. An inverse problem is solved to obtain the applied forces.
- Cells have both solid and viscous behavior. Models that incorporate both behaviors have been developed to describe cellular deformation.
- There are diverse methods for applying fluid shear stress to cells. Consideration must be given not only to maintaining biological compatibility, but to control of turbulence, entrance length, and flow development.
- Stretched samples in which strain levels are large can be analyzed by deformation gradient mathematics. Data are generally acquired at discrete points, which are then interpolated to yield displacement at desired locations.
- Experiments are most useful when only one factor, or one set of factors, is being changed at a time. Use of controls, blinding, and shams can help ensure the readout is most representative of the response to the factor(s) being examined.

Problems

1. Show that a series of n springs in parallel, with spring constants $k_1, k_2, \ldots, k_n$, where k_i does not necessarily equal k_j for $i \neq j$ with a single dashpot of viscous coefficient η can be reduced to a single spring and dashpot in parallel.

2. Below is the response of a bead on a cell experiencing a step–displacement at time $t = t_0$. The vertical axis, F, represents the force on the bead. Devise a spring–dashpot model that is consistent with this response plot, and draw the predicted response for a step release in force (i.e., force is some value F_0 at $t < t_0$ but force = 0 at $t \geq t_0$.

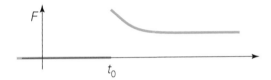

3. Consider a spring and dashpot in parallel subject to an instantaneous step displacement. What is the force through the spring? And through the dashpot? Why would such a model be a poor choice for modeling the behavior of a bead subject to a step displacement?

4. Devise two spring-dashpot models that are initially at rest, experience a step-increase to a constant force F_0 at time t_0, then a step-decrease to zero force at time $t_1 > t_0$, such that (1) the stress–strain (or force-displacement) response curves are identical for loading and unloading and (2) the stress–strain (or force-displacement) response curves are not identical for loading and unloading. The latter case is called *hysteresis*.

5. Determine the energy required to extend a spring of spring constant k from equilibrium ($x = 0$) to a new position $x = x_1$. Next, determine the ratio of energy required to extend a spring of spring constant k in parallel with a dashpot of viscous coefficient η to that of a spring alone. The answer should depend on the rate of loading, which may be assumed to be constant. What happens at low rates? And at high rates?

6. You have just completed imaging a fluorescent bead as part of a traction force microscopy experiment. Your array of intensity values is

2	1	0	2	3	1	1	2	1	0
1	2	1	2	2	3	3	2	1	1
2	1	2	1	6	4	4	2	0	2
2	0	4	5	5	6	5	4	2	0
1	0	3	5	7	8	3	5	2	1
2	1	7	8	9	8	6	4	6	2
3	3	4	6	8	7	6	2	2	1
2	5	3	5	5	6	2	3	1	1
1	3	2	2	3	2	1	4	2	1
1	2	1	1	1	0	2	0	3	0

where the value in the ith row and jth column is the intensity for pixel (i,j).

You have determined that the template image (**K**) should be

1	2	3	2	1
2	7	8	7	2
3	8	9	8	3
2	7	8	7	2
1	2	3	2	1

Using Equation 6.7, compute $C(3,3)$ and $C(5,5)$.

7. Below is a figure of a stretch experiment performed on a piece of tissue. You determine the following coordinates for the markers on the tissue

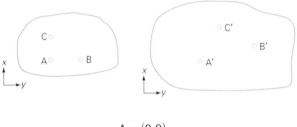

$$A = (0,0)$$

$$B = (1,0)$$

$$C = (0,1)$$

$$A' = (1,1)$$

$$B' = (3,2)$$

$$C' = (2,5)$$

Determine **F**, the deformation gradient tensor.

8. You are working with a deformation gradient tensor given as $\boldsymbol{F} = \begin{bmatrix} 2 & -0.5 \\ 1 & 4 \end{bmatrix}$. A colleague created a special program to extract the rotation tensor **R** (from the decomposition $\boldsymbol{F} = \boldsymbol{RU}$). He input your **F** and obtained

$$\boldsymbol{R} = \begin{bmatrix} \dfrac{1}{\sqrt{2}} & \dfrac{1}{\sqrt{2}} \\ -\dfrac{1}{\sqrt{2}} & \dfrac{1}{\sqrt{2}} \end{bmatrix}.$$ Is this a plausible rotation tensor? Justify your response.

Explain why **U**, the stretch tensor, must be positive definite from a physical perspective (that is not a purely mathematical proof).

9. You are given a 2×2 tensor $\boldsymbol{C} = \boldsymbol{F}^\mathsf{T}\boldsymbol{F}$. Let $\lambda_1{}^2$ and $\lambda_2{}^2$ be the eigenvalues of $\boldsymbol{C}$, with $|\lambda_1| > |\lambda_2|$. Prove that λ_1 is the maximum stretch for an arbitrarily oriented

cell undergoing deformation described by **F**, and that λ_2 is the minimum stretch under the same conditions.

10. Use dimensional analysis to obtain an expression for the drag force on a sphere under laminar flow. You must identify the parameters that govern the force and then obtain a dimensionless expression. Next, show that the obtained expression is consistent with Stokes' law (Equation 6.1).

11. Consider the case in which you are devising a shear-based technique to estimate the adhesion force of cells to a substrate. You have an inlet into a cylindrically symmetric outflow as shown below.

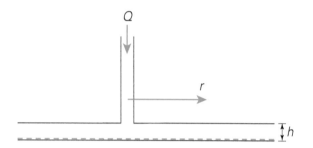

The dashed line represents the cell monolayer. The flow into the inlet is at volume flow rate Q (m³/s) across cross-sectional area A. The height of the outflow region is h, and you may assume that $h < r$ and that the flow is laminar. At the inlet, you pump fluid at a constant Q, and at the outlet, collect all the cells that detach and exit the outflow region around the perimeter. You then count the cells that have detached. Relate the number of detached cells to the maximum adhesion force of the cell, assuming that the adhesion force per cell is uniform, that cell–cell interactions can be neglected, and that each cell is 10 μm in radius.

12. Consider the case in which you are given a stretch device consisting of a membrane that is 10 cm in diameter. The membrane is stretched by a piston pushing on its bottom surface, as depicted in Figure 6.21. You may assume the piston is nearly the same diameter as the membrane. As the piston moves up and down, the media in which the cells are immersed will also flow back and forth across the cells. Assume the media has viscosity μ and density ρ and that the cells are linearly elastic with elastic modulus 10 kPa undergoing 5% stretch. Determine the ratio of stretch-based stresses to that of fluid shear stresses. You may model the cells as 50 μm × 50 μm × 5 μm rectangular solids with the smallest dimension being the height. An order of magnitude ratio is sufficient.

13. Unlike the rounded tips described in this chapter, atomic force microscopes may also be equipped with sharp tips that are typically pyramidal in shape (for example, the Berkovich tip). Owing to their geometry, these tips have modified relationships between the force, E, the indentation depth, and the geometry of the tip as given by the Hertz model. To model AFM with a pyramidal tip, it can be useful to use a scaling relationship instead of trying to get an exact solution.

    (a) First, derive a scaling relationship for the force and indentation depth of a spherical tip atomic force microscope. You may assume the radius of the bead (R) is very large compared with the radius of the area being indented (a), and that the radius of the area being indented is very large compared with the indentation depth (δ). Further, you may assume only the linear elastic modulus applies (E). To solve this, use an energy balance by calculating the strain energy of the deformation and equating that with the work done by pushing the tip in. Because this is a scaling relation, you may ignore constants.

    (b) Now, assume the tip is a pyramid instead. Assume the angle of the tip is α, the indentation depth is δ, and the cell has elastic modulus E (continue to ignore shear effects). You may find it easier to start by assuming the tip is actually a cylinder with a rounded end with radius a (and assuming that δ is much less than a), and then realize that the angle α is related to the "width" of the indenter. Once you derive the scaling relationship, you should see that there is a mathematical difference between the two types of indentation.

Annotated References

Alessandrini A & Facci P (2005) AFM: a versatile tool in biophysics. *Meas. Sci. Technol.* 16, R65–R92. *This journal review article summarizes many aspects of biological studies using AFM.*

Brown TD (2000) Techniques for mechanical stimulation of cells in vitro: a review. *J. Biomech.* 33, 3–14. *This journal review article describes several different methods for applying stretching and shear to cells and tissues.*

Cao J, Donell B, Deaver DR, et al. (1998) In vitro side-view imaging technique and analysis of human T-leukemic cell adhesion to ICAM-1 in shear flow. *Microvasc. Res.* 55, 124–37. *This paper describes the use of side-view imaging to obtain quasi-3D images of cells during rolling adhesion.*

Chen CS, Mrksich M, Huang S, Whitesides GM & Ingber DE (1997) Geometric control of cell life and death. *Science* 276, 1425–1428. *This paper describes the use of micropatterning to determine whether adhesion area or spreading area is more important for maintaining cell viability.*

Chien S (2008) Effects of disturbed flow on endothelial cells. *Ann. Biomed. Eng.* 36, 554–562. *This summarizes findings of how shear flow affects endothelial biology in a way that implicates disturbed flows in the development of atherosclerosis.*

Goubko CA & Cao X (2009) Patterning multiple cell types in co-cultures: a review. *Mat. Sci. Eng. C.* 29, 1855–1868. *This review article briefly summarizes some basic, micropatterning and then* describes the more recent pattern generation for use in multiple cell patterns.

Hoffman BD & Crocker JC (2009) Cell mechanics: dissecting the physical responses of cells to force. *Annu. Rev. Biomed. Eng.* 11, 259–288. *This paper reviews rheological properties of cells with discussion of various techniques and what the results from different techniques may have in common.*

Huang H, Kamm RD & Lee RT (2004) Cell mechanics and mechanotransduction: pathways, probes, and physiology. *Am. J. Physiol. Cell Physiol.* 287, C1–C11. *This journal review article outlines a few experimental methods for studying cell mechanics and some underlying bases for the various approaches taken.*

Janmey PA & McCulloch CA (2007) Cell mechanics: integrating cell responses to mechanical stimuli. *Annu. Rev. Biomed. Eng.* 9, 1–34. *This journal review article that discusses mostly theoretical and sensory mechanisms for understanding cell mechanics and mechanotransduction, but is also notable for presenting a nice summary in Table 1 that compares the elastic modulus by different techniques and cells with extensive references.*

Kuo SC (2001) Using optics to measure biological forces and mechanics. *Traffic* 2, 757–763. *This journal review article summarizes some key findings for using optical traps and microrheology to study biological systems.*

Landau LD & Lifshitz EM (1970) Theory of Elasticity, vol. 7, 2nd ed. Pergamon Press. *This book, part of the Course of Theoretical Physics, contains many formulae that are of relevance when trying to characterize deformations in solid objects. It does have newer editions and the other books in the series cover different topics.*

Legant WR, Miller JS, Blakely BL et al. (2010) Measurement of mechanical tractions exerted by cells in three-dimensional matrices. *Nat. Meth.* 7, 969–971. *This is a recent journal paper that describes an approach for extending traction force microscope from a two-dimensional monolayer to cells imbedded in a three-dimensional hydrogel.*

Liu J, Weller GE, Zern B, Ayyaswamy PS, Eckmann DM, Muzykantov VR, Radhakrishnan R (2010) Computational model for nanocarrier binding to endothelium validated using in vivo, in vitro and atomic force microscopy experiments. *Proc. Natl. Acad. Sci. USA* 38, 16530–16535. *This article uses computational methods to calculate the free energy of binding a functionalized "nanocarriers" to endothelial cells.*

Moseley JB, O'Malley K, Petersen NJ et al. (2002) A controlled trial of arthroscopic surgery for osteoarthritis of the knee. *N. Engl. J. Med.* 347, 81–88. *This is a journal article describes a study in humans of arthroscopic knee surgery in which the controls received a sham surgery without insertion of the scope. The patients were kept unaware of whether they had the surgery. Interestingly, the control group experienced as much improvement as the surgery group.*

Munevar S, Wang Y & Dembo M (2001) Traction force microscopy of migrating normal and H-ras transformed 3T3 fibroblasts. *Biophys J.* 80 4, 1744–1757. *This article describes traction force microscopy results applied to cell analysis.*

Neuman KC & Block SM (2004) Optical trapping. *Rev. Sci. Instrum.* 75, 2787–2809. *This paper describes the state of optical trapping and the necessary components for achieving it.*

Roy P, Rajfur Z, Pomorski P, Jacobson K (2002) Microscope-based techniques to study cell adhesion and migration. *Nat. Cell Biol.* 4, E91–6. *This journal review article discusses several microscopy-based techniques to characterize cell adhesion structures and forces, including FRET and traction force microscopy.*

Sotoudeh M, Jalali S, Usami S, Shyy JY, Chien S (1998) A strain device imposing dynamic and uniform equi-biaxial strain to cultured cells. *Ann. Biomed Eng.* 2, 181–189. *This article discusses the design and implementation of a stretch device used for cell mechanics studies, with validation of the exerted strains.*

Tan JL, Tien J, Pirone DM, et al. (2003) Cells lying on a bed of microneedles: an approach to isolate mechanical force. *Proc. Natl. Acad. Sci. USA* 100, 1484–1489. *This article describes the use of micropillars (they call it microneedles) to determine traction forces using beam-bending analysis.*

Wang JH & Lin JS (2007) Cell traction force and measurement methods. *Biomech. Model. Mechanobiol.* 6, 361–371. *This paper describes traction-force microscopy and a few different techniques for analysis of the results to extract the traction field.*

Wang N, Tolić-Norrelykke IM, Chen J et al. (2002) Cell prestress. I. stiffness and prestress are closely associated in adherent contractile cells. *Am. J. Physiol. Cell Physiol.* 282, C606–C616. *This journal article characterizes cell structures (including those based on the tensegrity hypothesis) using information from a variety of techniques, including traction force and magnetic twisting.*

Weibel DB, Diluzio WR & Whitesides GM (2007) Microfabrication meets microbiology. *Nat. Rev. Microbiol.* 5, 209–218. *This journal review article summarizes different micropatterning techniques and their applications to biological systems.*

Young EW & Simmons CA (2010) Macro- and microscale fluid flow systems for endothelial cell biology. *Lab Chip* 10, 143–160. *This paper discusses shear flow design and analysis and presents principles for shear flow design at the microscale by taking keys from macroscale shear flow devices.*

PART II: PRACTICES

CHAPTER 7

Mechanics of Cellular Polymers

We now turn our attention to one of the most important structural components of cells: biopolymers. In Section 2.1, we learned that polymers are linear molecules made up of repeating structures known as subunits. Biopolymers possess a diverse range of functions within cells. For example, DNA and RNA are polymers of nucleotides whose primary function is the storage and passage of genetic information. In Chapter 8, we will learn more about the cytoskeleton, an interconnected network of polymers that plays a critical role in a variety of functions, such as maintenance of cell shape, mechanical force generation, and intracellular transport. In many cases, the mechanical behavior of these biopolymers is critical in allowing them to perform the diverse functions they serve.

In this chapter, we focus on obtaining a quantitative understanding of the mechanics of biopolymers. We first describe the molecular structure of three important cytoskeletal polymers: microfilaments, microtubules, and intermediate filaments. We follow this with a model of polymerization dynamics in microfilaments and microtubules. Next, we consider how to characterize a polymer as being flexible or stiff and discuss three different polymer models: the ideal chain, the freely joined chain (FJC), and the wormlike chain (WLC). Through this development, it is our hope that you will have a better understanding of how the mechanics of biopolymers may dictate their capacity to perform their biological functions.

7.1 BIOPOLYMER STRUCTURE

Microfilaments are polymers composed of actin monomers

Microfilaments (MFs) are polymers composed of the protein actin, a highly conserved, 42 kDa protein. In its monomeric form, it can be referred to as globular or *G-actin*. When G-actin polymerizes, it forms filamentous actin, referred to as *F-actin*. F-actin has a double-helix-like structure consisting of two strands that spiral around the axis of the polymer (Figure 7.1). This results in the appearance of repeating helix loops, with each loop repeating every 37 nm. The overall diameter of the polymer is approximately 7–9 nm.

F-actin polymerization is influenced by the molecular characteristics of G-actin

In the cell, there is a highly dynamic and regulated balance between the G and F forms of actin. This balance is under strict biochemical regulation within the cell. This regulation allows actin to adopt different morphologies and organization and perform a variety of cellular functions, such as intracellular force generation during cell motility and maintenance of cell shape.

The molecular structure of G-actin can greatly affect the dynamics of its polymerization into F-actin. Actin is a polar molecule. Unlike the charge polarity associated with water (as we will describe in Section 9.1), here polar means that the two ends of the monomer are different. The implication of this polarity is that

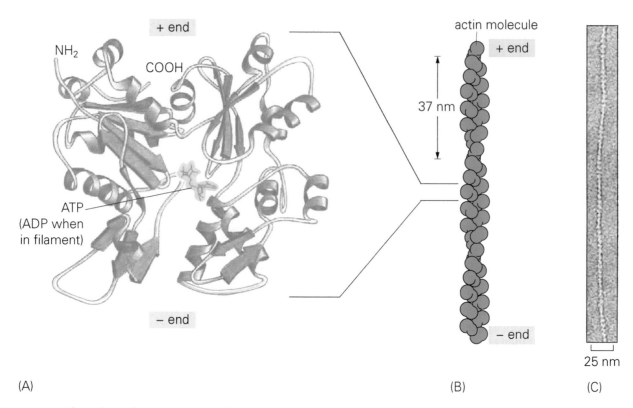

(A)   (B)   (C)

Figure 7.1 G-actin and F-actin structures. G-actin possesses a (+) and a (–) end as well as binding site for ATP. F-actin has a double-helix-like structure consisting of two strands that spiral around the axis of the polymer, with each helical loop repeating every 37 nm. (Adapted from, Alberts B et al. (2008) Molecular Biology of the Cell, 5th ed. Garland Science. Photo courtesy of Roger Craig.)

actin MFs have inherent directionality. This directionality was originally deduced from the observation that when a MF is decorated by multiple myosin molecules, the myosins tend to be angled, all in the same direction. This gave rise to the notation of a *pointed* (or " – ") and a *barbed* (or " + ") end of the polymer and the monomer (Figure 7.2). Owing to this polarity, the polymerization kinetics of actin, in other words the rates at which monomers are added or subtracted from the end of the polymer, can be very different at the (+) and (–) ends.

Another aspect of actin's molecular structure that can affect its polymerization kinetics is its capacity to bind adenosine triphosphate (ATP). Upon binding to actin, ATP hydrolyzes to adenosine diphosphate (ADP). This hydrolysis happens relatively quickly for polymeric F-actin, but is quite slow for monomeric G-actin. As a result, almost all of the G-actin in a cell is of the ATP-bound form. In contrast, the longer an actin subunit is polymerized within a MF, the more likely it will be in the ADP-bound form. In Section 7.2, we explore in detail how actin polarity and ATP/ADP binding affects its polymerization kinetics.

Figure 7.2 The two ends of a MF are denoted as the barbed, or (+), end and the pointed, or (–) end.

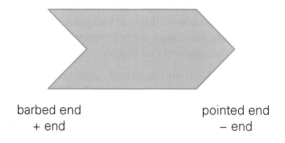

barbed end   pointed end
+ end   – end

Microtubules are polymers composed of tubulin dimers

Another class of cytoskeletal polymers is the microtubules (MTs). The subunits for microtubule polymers are heterodimers, with each subunit made up of an α and β isoform of the 55 kDa protein tubulin. The structure of the MT consists of 13 protofilaments joined together in a hollow tubular structure (Figure 7.3). The protofilaments are offset such that the dimers form a roughly helical structure, with each helical turn containing 13 subunits. The external diameter of a MT is approximately 25 nm. This is larger than that of F-actin and contributes to the increased flexural rigidity of MTs compared with MFs. Biologically, MTs are central players in several processes such as mitosis, in which they form the mitotic spindle responsible for segregation of the chromosomes (Figure 7.4). MTs also contribute to cell shape and migration and forming the structure of cilia and some flagella.

MT polymerization is affected by polarity and GTP/GDP binding

The polymerization kinetics of tubulin are dictated by its asymmetric nature and its ability to bind guanosine tiphosphate/guanosine diphosphate (GTP/GDP, analogous to ATP/ADP in actin). Like F-actin, MTs are polar, possessing a rapidly polymerizing (+) end and slowly polymerizing (−) end. Once polymerized, GTP hydrolyzes to GDP. As tubulin polymerizes, there is a region near the polymerizing end of GTP-tubulin known as the GTP cap (Figure 7.5). Only GTP tubulin tends to become polymerized because polymers of GDP tubulin are unstable and easily return to monomeric form. The GTP cap can be thought of as preventing the depolymerization of the entire MT.

Periodically, the GTP cap is lost owing to impediments to further polymerization or simple random fluctuations. When this happens, the GDP MT catastrophically depolymerizes until the GTP cap can be reestablished. Microtubules are constantly shifting from a state of slow polymerization to rapid depolymerization in a process known as *dynamic instability*. Dynamic instability results in MTs constantly extending and retracting. This is thought to be important for the ability of the MTs to find and attach to the kinetochore during mitosis. At other times, cytoplasmic MTs tend to organize into a network with their (−) ends near the nucleus and their (+) ends pointing outward toward the cell membrane.

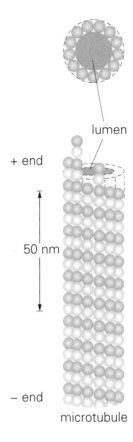

tubulin heterodimer
(= microtubule subunit)

Figure 7.3 Structure of a MT. Each MT consists of 13 protofilaments joined together in a hollow tubular structure. MT subunits consist of a heterodimer of α- and β-tubulin. (Adapted from, Alberts B et al. [2008] Molecular Biology of the Cell, 5th ed. Garland Science.)

Nota Bene

Taxol blocks dynamic instability.
The activity of certain pharmacologic therapies is derived from their effect on the dynamic instability of MTs. Taxol®, an anti-cancer drug, works by making the GDP form of tubulin stable. This blocks dynamic instability and inhibits mitosis.

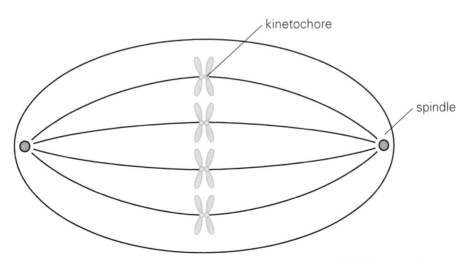

Figure 7.4 MTs during mitosis. When cells divide, MTs are responsible for segregating chromosomes and pulling them to the poles of the cell. They do this by forming two spindles and attaching to the kinetochores.

Figure 7.5 A polymerizing MT with a GTP cap. As tubulin polymerizes, there is a region of GTP + tubulin. The GTP cap arises because only GTP + tubulin tends to become polymerized as it is an unstable polymer unstable and easily returns to monomeric form.

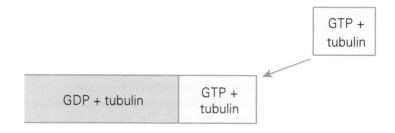

Intermediate filaments are polymers with a diverse range in composition

A third type of cytoskeletal biopolymer is the intermediate filaments. The structure and function of intermediate filaments are somewhat less well-characterized than MFs and MTs, perhaps because of their complexity. Over 70 different genetic sequences appear to code for the various intermediate-filament proteins (Table 7.1). Intermediate filaments are found in a variety of locations within the body and occur in multiple variants, depending on their location. Vimentin is a commonly studied intermediate filament in epithelial cells. Desmin is an intermediate filament that is more specific for muscle, especially the heart. Lamins are intermediate filaments that reside in the nuclear envelope, the membrane that surrounds the nuclei of cells. Keratins are another type of intermediate filament; there are over two dozen types of keratin (and likely more that have not yet been studied in detail). In humans, these intermediate filaments reside in the skin but are also expressed in hair and fingernails. In other animals, keratin is a major component of horns, hides, and scales.

Intermediate filaments possess a coiled-coil structure

Regardless of protein composition, all intermediate filaments have an α-helical structure and adopt a "coiled-coil" structure. Unlike actin and MTs, intermediate filaments are not composed of small globular subunits. Instead, the individual subunits of intermediate filaments are long α-helical regions of a protein. Two such proteins are coiled together to form a dimer. Two dimers are then staggered (in opposite orientations) to form a tetramer. Long chains of tetramers are then coiled together to form the filament (Figure 7.6). The association between filaments arises out of hydrophobic interactions of the coiled regions. Being 10 nm in diameter, intermediate filaments are larger in diameter than MFs, but smaller than MTs. Unlike MFs and MTs, intermediate filaments are not polar and do not have (+) and (−) ends. In addition, their structure does not facilitate rapid depolymerization, as the filament has to first uncoil before depolymerization can occur.

Intermediate filaments have diverse functions in cells

In general, disorders caused by dysfunction of intermediate filaments and related proteins are quite varied. As a result, a well-defined function for intermediate filaments has not yet been established. However, they are thought to provide mechanical strength to tissues in some cases. They attach to membrane plaque-like structures known as *desmosomes* that mediate cell–cell adhesions. Certain defects in desmosomes appear more pronounced in highly mechanically stressed tissues expressing them, including skin and cardiac tissue. Some intermediate filaments, such as keratin, can also attach to the extracellular matrix protein *laminin*—not to be confused with the intermediate filament lamin—by integrins using structures

Nota Bene

Naming of intermediate filaments. Although the size of intermediate filaments is between that of MFs and MTs, they actually get their name because they were first described as having a diameter between MFs and myosin fibers.

called *hemidesmosomes*. This adhesion occurs in a manner similar to focal adhesions, although the latter rely on MFs. Although hemidesmosomes have roles in mediating cell–matrix interactions, the regulatory role of hemidesmosomes is not clearly established.

Table 7.1 Intermediate filaments by type, size, chromosome, and distribution Note that there are multiple subtypes for each class; keratins, for example, consist of 20 different type 1 filaments. (From Omary MB, Coulombe PA & McLean WH. (2004) *N. Engl. J. Med.* Copyright Massachusetts Medical Society.)

Location and name	Type	Size kDa	Chromosome with associated gene	Cell or tissue distribution	Comments
Cytoplasmic					
Keratins	I	40–64	17	Epithelium (keratins 9–20); hair (keratins Ha1–Ha8)	Form obligate 1:1 heteropolymers with type II; protect from mechanical and nonmechanical forms of stress
	II	50–68	12	Epithelium (keratins 1–8); hair (keratins Hb1–Hb8)	Form obligate 1:1 heteropolymers with type I; protect from mechanical and nonmechanical forms of stress
Vimentin	III	55	10	Mesenchyme	Involved in vascular tuning and wound repair in mice
Desmin	III	53	2	All muscle	May be important for mitochondrial positioning and integrity
Glial fibrillary acidic protein	III	52	17	Astrocytes	Also found in hepatic stellate cells
Peripherin	III	54	12	Peripheral neurons	Found in enteric neurons; may be required for development of a subset of sensory neurons
Syncolin	III	54	1	Muscle (mainly skeletal and cardiac)	Interacts with α-dystrobrevin
Neurofilaments (light, medium, and heavy chains)	IV	61 (light), 90 (medium), 110 (heavy)	8 (light), 8 (medium), 22 (heavy)	Central nervous system	Form obligate 5:3:1 (light:medium: heavy) heteropolymers
α-Internexin	IV	61	10	Central nervous system	May partly compensate for peripherin in peripherin-null mice
Nestin	IV	240	1	Neuroepithilial	Is also an early developmental marker, as found in the pancreas
Synemin	IV	180 (α) and 150 (β) (two splice variants)	15	All muscle (β-isoform mainly in striated muscle)	Found at lower levels than desmin; also found in astrocytes; probably indentical to desmuslin
Nuclear					
Lamins A and C	V	62–78	1	Nuclear lamina	Arise from a single, differentially sliced gene
Lamins B1 and B2	V	62–78	5, 19	Nuclear lamina	Arise from two different genes
Other					
Phakinin (CP49)	Orphan	46	3	Lens	Forms beaded filaments; deletion in mice causes lens defect
Filensin (CP115)	Orphan	83	20	Lens	Forms beaded filaments; deletion in mice causes lens defect

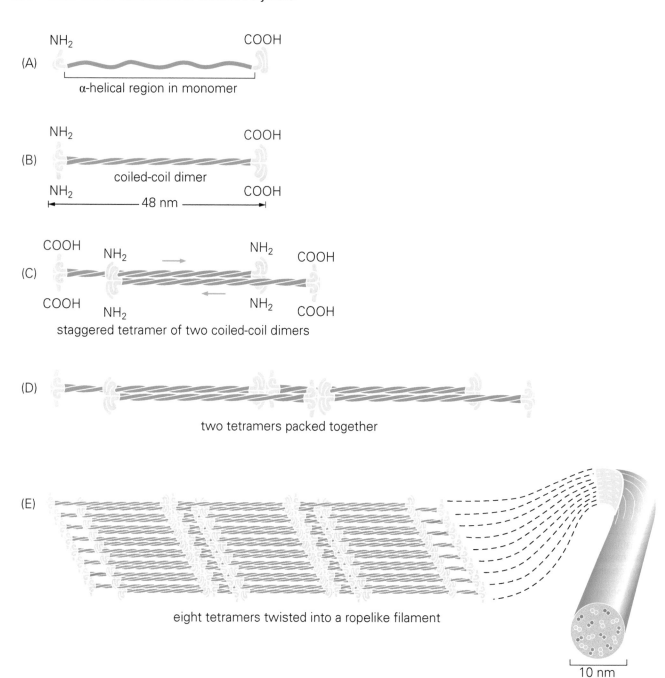

Figure 7.6 Intermediate filament structure. The monomer (A) forms a dimer with another monomer to form a coiled-coiled structure (B). Two dimers are then staggered to form a nonpolar tetramer (C). The tetramers associate in a staggered manner (D), allowing them to form the final helical filament structure (E). (Adapted from, Alberts B et al. (2008) Molecular Biology of the Cell, 5th ed. Garland Science.)

7.2 POLYMERIZATION KINETICS

In the previous section, we learned that F-actin and MT polymerization are tightly regulated intracellular processes. In addition, we learned that actin monomers and tubulin dimers share two distinct features in their molecular structure that affect their polymerization dynamics: they are both polar molecules, and they both possess the capacity to bind ATP/ADP (for actin) or GTP/GDP (for MTs). In this section, we explore how these structural features affect their polymerization kinetics.

Figure 7.7 Dynamics between the G and F forms of actin. Reversible conversion of G-actin to F-actin occurs with the addition of a monomer to a polymer at one of its free ends.

Actin and MT polymerization can be modeled as a bimolecular reaction

Consider a model polymer of some length undergoing the reversible addition of one subunit, as in Figure 7.7. If we denote the existing polymer before the subunit addition to be A_{poly}, and after the addition as A_{poly+1}, we can write the polymerization process as a chemical reaction as

$$A_{poly} + G \underset{k_{off}}{\overset{k_{on}}{\rightleftharpoons}} A_{poly+1}, \tag{7.1}$$

where k_{on} is the rate of the forward reaction (units of $s^{-1}\,M^{-1}$) and k_{off} is the rate of the reverse reaction (units of s^{-1}). We can write the rate of change of $[A_{poly+1}]$ as

$$\frac{d\left[A_{poly+1}\right]}{dt} = k_{on}\left[A_{poly}\right][G] - k_{off}\left[A_{poly+1}\right]. \tag{7.2}$$

When the polymer is not changing length, the time derivative of $[A_{poly+1}]$, $d[A_{poly+1}]/dt$, is equal to zero. When this occurs, Equation 7.2 implies that

$$k_{on}\left[A_{poly}\right][G] = k_{off}\left[A_{poly+1}\right]. \tag{7.3}$$

Solving for $[G]$ in Equation 7.3, we obtain

$$[G] = K\frac{\left[A_{poly+1}\right]}{\left[A_{poly}\right]} \quad \text{where } K = \frac{k_{off}}{k_{on}}. \tag{7.4}$$

The constant K is known as the *dissociation constant* of the reaction. It is also known as the *critical concentration*, as it is the concentration of subunit at which the polymer is neither growing nor shrinking.

The critical concentration is the only concentration at which the polymer does not change length

The critical concentration K has an interesting property for our model polymer: it is the only subunit concentration at which the polymer does not change length. To see this, we can write an expression for the elongation rate of the polymer; in other words, the change in length of the polymer per unit time. We first note that if δ is the size of the subunit, then the rate of lengthening in units of length per unit time due to polymer addition is $k_{on}[G]\delta$, and that due to subtraction is $k_{off}\delta$. The total elongation rate $\frac{dL}{dt}$ is the sum of these two rates,

$$\frac{dL}{dt} = (k_{on}[G] - k_{off})\delta. \tag{7.5}$$

A plot of Equation 7.5 can be seen in Figure 7.8. If $[G] > K$, $\frac{dL}{dt}$ is positive, the polymer is growing, whereas if $[G] < K$, $\frac{dL}{dt}$ is negative, the polymer is shrinking. The

Figure 7.8 The kinetics of polymerization as a function of monomer concentration. If the monomer concentration [G] is higher than K, the elongation rate is positive, and the polymer will grow. If it is below K, the polymer will shrink. The only concentration at which the elongation rate is zero is when [G] = K.

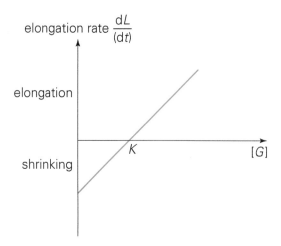

only point at which $\dfrac{dL}{dt} = 0$, in other words when the polymer is neither growing nor shrinking, occurs when [G] = K. This confirms that the critical concentration is the only subunit concentration at which the polymer does not change length.

Polarity leads to different kinetics on each end

Thus far we have assumed that the subunits were not polar and that the kinetics of polymerization were equivalent at each end of the polymer. Now we examine the implications of subunit polarity. The kinetics of the polymerization reaction are different at the (+) and (−) ends of the polymer. At the (+) end, k_{on} and k_{off} are higher than at the (−) end. This implies that if an excess of monomers were rapidly introduced, elongation would occur faster at the (+) end. Conversely, if the monomers were rapidly depleted, depolymerization would occur much faster at the (+) end. For our plot of elongation rate in Figure 7.8, we will have two different lines, one for the (+) end, and one for the (−) end. Although it is beyond the scope of this book, it can be shown by a principle called *detailed balance* that the critical concentration at both ends of a polar polymer must be the same if it is homogenously composed of a single species of subunit. If the critical concentration is the same at both the (+) and (−) ends, the lines will cross the zero axis at the same point. In Figure 7.9, we revise our plot for elongation rate to reflect that there are now two different lines, one for the (+) end, and one for the (−) end, each passing through $\dfrac{dL}{dt} v = 0$ at [G] = K.

Figure 7.9 Polymerization kinetics differ at the (+) and (−) ends. The kinetics of polymerization are different at the (+) and (−) ends of the polymer. Polymerization/depolymerization kinetics tend to be faster at the (+) end and slower at the (−) end for a given subunit concentration. However, the critical concentration is the same at both ends.

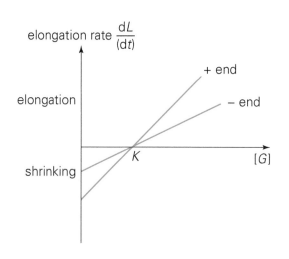

Polymerization kinetics are affected by ATP/ADP in actin and GTP/GDP binding in tubulin

To complete our discussion, we now seek to incorporate the effects of ATP/ADP binding for actin and GTP/GDP binding for tubulin. For simplicity, we will refer to ATP-bound actin and GTP-bound tubulin as T-form subunits, and ADP-bound actin and GDP-bound tubulin as D-form subunits. Recall from Section 7.1 that within T-form subunits, the ATP or GTP can undergo hydrolysis, which results in the conversion of a T-form subunit into a D-form subunit. This hydrolysis has two effects on the polymerization kinetics of our model polymer. First, recall that the primary form of the subunit in the cytoplasm is T form; however, a given polymer may contain both D- and T-form subunits. The rates of polymerization will be different depending on whether the end of the polymer that the subunit is binding contains a D-form or a T-form subunit. Second, hydrolysis has the effect of lowering the affinity of the subunits to the polymer, which increases the critical concentration for the D form. In Figure 7.10, we take these effects into account in our elongation rate plots by having one set of (+)/(−) lines for D-form actin, and one set of (+)/(−) lines for T-form actin. The critical concentration for T-form actin is denoted K_T, and the critical concentration of D-form actin K_D is denoted.

Subunit polarity and ATP hydrolysis lead to polymer treadmilling

Imagine a polymer composed entirely of D-form subunits. The polymer is quickly submerged in a bath of T-form subunits. If the subunit concentration is higher than K_T, T-form subunits will be added at both ends, with faster growth occurring at the (+) end. Therefore, at any given point in time, the T-form region on the (+) end will be longer than that on the (−) end.

Now consider what happens when we lower the subunit concentration, perhaps by diluting the solution with water. As the subunit concentration is reduced, the rate of T-form subunit addition slows. In addition, the recently polymerized T-form subunits begin undergoing hydrolysis. If the rate of hydrolysis is faster than the rate of T-form addition at the slower (−) end, then at some point in time all the T-form actin at the (−) end will have been hydrolyzed to D form. However, because there are more T-form subunits at the (+) end, there will still be some unhydrolyzed T-form subunits there. The polymer possesses two critical concentrations: K_T at the (+) end and K_D at the (−) end.

The critical concentration has interesting ramifications when the monomer concentration is lower than K_D but greater than K_T. The (−) end will be depolymerizing,

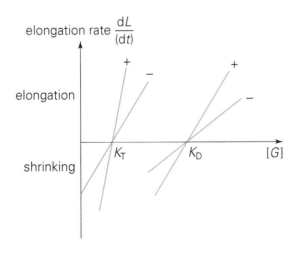

Figure 7.10 Polymerization kinetics for D- and T-form actin. The critical concentration and rates of polymerization/depolymerization are different for the D and T forms of actin.

Polymerization force. The peak force that a single fixed polymerizing filament can exert on a surface has been derived through thermodynamic considerations to be

$$F = \frac{kT}{\delta} \ln\left(\frac{k_{on}[G]}{k_{off}}\right).$$

Using this relation, it has been estimated that the peak force for F-actin is 9 pN for a typical concentration of G-actin of 50 mM. This is many times the force that one myosin motor can generate.

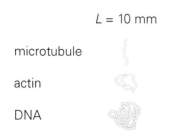

$L = 10$ mm

microtubule

actin

DNA

Figure 7.11 Representative configurations for a 10-μm segment of a MT, actin, and DNA polymer at room temperature. Under the influence of thermal forces, the tendency to take on a coiled versus straight configuration varies for each of the polymers.

but the (+) end will be polymerizing. The addition of subunits on one end and subtraction at the other is known as *treadmilling*. Once D-form subunits are lost from the shrinking (−) end, they release ADP/GDP and bind ATP/GTP to become T-form subunits and are recycled to the growing (+) end. Intracellular concentrations of G-actin and tubulin are typically within the range of subunit concentrations at which treadmilling occurs.

7.3 PERSISTENCE LENGTH

We now shift our focus from polymer structure and kinetics to mechanical behavior. In analyzing the mechanical behavior of polymers, the appropriate choice of model is determined in large part by whether the behavior is energy- or entropy-dominated. We learned in Section 7.1 that different biopolymers can have very different molecular structures. One consequence of this is that under the influence of thermal forces, these polymers will have very different tendencies to bend and coil. Consider a strand of DNA approximately 10 μm in length. If we grab it at the ends, straighten it, and then release it in into a solution at room temperature, we would see the strand of DNA begin to take on a convoluted configuration, with many undulations forming along its length. Now, we do the same for a MT 10 μm in length. The MT does not coil. In fact, it may tend to resemble a fairly straight rod (**Figure 7.11**).

The tendency for these two polymers to take on such different configurations in the presence of thermal forces can be attributed to differences in energetic and entropic influences in equilibrium. The molecular structure of MTs results in a much higher increase in potential energy during bending than that of DNA. The behavior of the MTs is energy-dominated taking on a straight configuration, as this reduces the internal energy and therefore the free energy. Conversely, the behavior of DNA is entropy-dominated. The free energy is reduced by increasing entropy, which occurs when the DNA strand takes on a coiled configuration.

Persistence length gives a measure of flexibility in a thermally fluctuating polymer

In this section, we introduce the *persistence length*, which is a characteristic length scale that gives a measure of flexibility in a thermally fluctuating polymer. Consider a continuous polymer of contour length L undergoing thermal fluctuations, as in **Figure 7.12**. We define a quantity s that runs from zero to L and gives a parameterization by which each point on the polymer can be identified. We define the orientation at each point $\theta(s)$ as the angle the polymer makes with an imaginary horizontal line. At $s = 0$, the polymer end is fixed such that $\theta(0) = 0$.

Now consider the case where we monitor $\theta(s)$ over time, and draw the probability distribution for θ. If we were to draw the distribution for small s (near the fixed

Figure 7.12 Relevant quantities in the definition of persistence length. The quantity s represents the distance along the polymer from 0 to L. The angle the polymer makes with an imaginary horizontal line at each point s is given by $\theta(s)$.

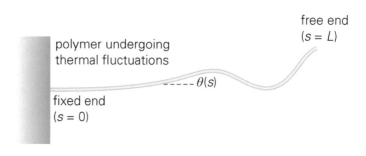

free end
$(s = L)$

polymer undergoing
thermal fluctuations

$\theta(s)$

fixed end
$(s = 0)$

end), then we would expect that the chances for the polymer to have an orientation different from $\theta = 0$ would be very small, and therefore the distribution would be very sharp, with a peak at $\theta = 0$. If we were to find the average cosine of the angle, we would find that $\langle \cos \theta \rangle \cong \langle \cos 0 \rangle \cong 1$, where the brackets indicate the time average. In contrast, at larger s, we would expect a much higher chance for the polymer to have a different orientation than $\theta = 0$, so the distribution would be wider (Figure 7.13). The distribution becomes wider and wider for larger s until, at some point, the distribution becomes essentially uniform. Beyond this point, the polymer orientation is effectively random (i.e., the orientation becomes uncorrelated from the fixed end), and so $\langle \cos \theta \rangle \cong 0$. Therefore $\langle \cos \theta \rangle$ decreases from 1 to 0 as s gets larger and larger. In fact, we now show that this decrease occurs exponentially.

Consider the following approximation for the derivative of some function $f(s)$,

$$\frac{df}{ds} \cong \frac{f(s + \Delta s) - f(s)}{\Delta s} \tag{7.6}$$

derived from the first two terms of a Taylor series approximation. Equation 7.6 is a good approximation for small values of Δs, as long as $f(s)$ is reasonably smooth. In our case, $f(s) = \langle \cos\theta'(s) \rangle$, where $\theta'(s) = \theta(s) - \theta(0)$, and

$$\frac{df}{ds} \cong \frac{\langle \cos(\theta'(s + \Delta s)) \rangle - \langle \cos(\theta'(s)) \rangle}{\Delta s}. \tag{7.7}$$

Now, if we let $\Delta\theta'(s) = \theta'(s + \Delta s) - \theta'(s)$, we can rearrange this to read $\theta'(s + \Delta s) = \Delta\theta'(s) + \theta'(s)$, and we can substitute this expression into the numerator

$$\frac{df}{ds} \cong \frac{\langle \cos(\Delta\theta'(s) + \theta'(s)) \rangle - \langle \cos(\theta'(s)) \rangle}{\Delta s}. \tag{7.8}$$

Using the fact that $\Delta\theta'(s)$ and $\theta'(s)$ are independent quantities, we can use the identity $\cos(a + b) = \cos(a)\cos(b) - \sin(a)\sin(b)$ in the above expression:

$$\frac{df}{ds} \cong \frac{\langle \cos(\Delta\theta'(s))\cos(\theta'(s)) - \sin(\Delta\theta'(s))\sin(\theta'(s)) \rangle - \langle \cos(\theta'(s)) \rangle}{\Delta s}. \tag{7.9}$$

Again, using the fact that $\Delta\theta'(s)$ and $\theta'(s)$ are independent, we can use the identity $\langle ab \rangle = \langle a \rangle \langle b \rangle$, and

$$\frac{df}{ds} \cong \frac{\langle \cos(\Delta\theta'(s)) \rangle \langle \cos(\theta'(s)) \rangle - \langle \sin(\Delta\theta'(s)) \rangle \langle \sin(\theta'(s)) \rangle - \langle \cos(\theta'(s)) \rangle}{\Delta s}. \tag{7.10}$$

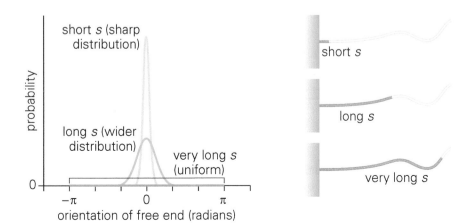

Figure 7.13 Probability distribution of orientation of the free polymer end for different values of s. For very short s, the probability for the polymer to have an orientation different from $\theta = 0$ is small, and therefore the distribution is very sharp. At large s, there is a much higher probability and so the distribution widens. At very large s, the distribution is essentially uniform.

However, because $\Delta\theta'(s)$ and $\theta'(s)$ are equally likely to be negative or positive (symmetric about zero, odd functions like sines average to zero),

$$\frac{df}{ds} \cong \frac{\langle\cos(\Delta\theta'(s))\rangle\langle\cos(\theta'(s))\rangle - \langle\cos(\theta'(s))\rangle}{\Delta s}. \tag{7.11}$$

Finally, factoring our expression for $f(s)$, we obtain

$$\frac{df}{ds} \cong \left(\frac{\langle\cos(\Delta\theta'(s))\rangle - 1}{\Delta s}\right)f(s). \tag{7.12}$$

Three-dimensional orientation correlation function. If the motion of the polymer is three-dimensional, then the orientation correlation function is

$$\langle\cos\Delta\theta(s)\rangle = e^{\left(\frac{-s}{2\ell_p}\right)}.$$

The different normalization takes into account the fact that if the polymer is allowed to move in three dimensions, it can bend in two directions orthogonal to its long axis, whereas if it is constrained to move in two dimensions, it can only bend in one direction. Therefore in two dimensions, the orientation correlation function will decay half as fast.

Notice that the term in the parentheses is a constant, not dependent on s, and in addition it will generally be negative, because $\langle\cos[\Delta\theta'(s)]\rangle$ is generally less than 1. We can rewrite the above expression as

$$\frac{df}{ds} \cong -Cf(s). \tag{7.13}$$

A solution to this equation is

$$\langle\cos\Delta\theta(s)\rangle = e^{\left(\frac{-s}{\ell_p}\right)}, \tag{7.14}$$

where for the rest of the chapter we will use the notation $\Delta\theta(s) = \theta'(s)$, $\langle\cos\Delta\theta(s)\rangle$ is called the *orientation correlation function*, $C = 1/\ell_p$, and ℓ_p is a normalization factor known as the *persistence length*. From the above expression, it is easy to see that the persistence length gives a characteristic length scale over which the orientations of a thermally undulating polymer become mostly uncorrelated.

Example 7.1: Consider a segment of actin in relation to a segment of DNA

Calculate the length of each segment such that the change in angle between the two ends of each segments is, on average, 25 degrees. Assume the persistence lengths for actin and DNA are 15 μm and 50 nm, respectively.

Using Equation 7.14 we can solve for s by taking the natural logarithm of both sides and isolating s to obtain

$$s = -\ell_p \ln(\langle\cos\Delta\theta(s)\rangle).$$

We know that a change in angle of 25 degrees corresponds to a value for the orientation correlation function, $\langle\cos\Delta\theta(s)\rangle$, of 0.9. Using this value, we obtain

$$s = 0.1\ell_p.$$

The corresponding lengths for actin and DNA are 1.5 μm and 5-nm, respectively. Physically, this means that on average, a 1.5-μm span of a thermally fluctuating actin polymer would have a difference in angle of 25 degrees from one end to the other, whereas in DNA this would occur over a 5-nm span. These dimensions make some sense considering the functions of each polymer; actin filaments must have the capacity to span significant lengths within the cell, whereas DNA must have the capacity to be tightly coiled within the nucleus.

Persistence length is related to flexural rigidity for an elastic beam

In Section 3.2, we learned that bending of an elastic beam is governed by its flexural rigidity, the product of its Young's modulus and moment of inertia EI. We now show that if we model a thermally fluctuating polymer as a curvy elastic beam, its persistence length is proportional to the flexural rigidity of the beam.

Consider an elastic three-dimensional beam with flexural rigidity EI bent 180 degrees (π radians) with a constant curvature R. In Chapter 3, you may have computed the elastic energy for this beam as a homework problem; however, for convenience we state it here as

$$Q = \frac{EI\pi}{2R}. \tag{7.15}$$

Because the beam was bent π radians, we can express Equation 7.15 in terms of a general bend angle θ as

$$Q(\theta) = \frac{EI\theta}{2R}. \tag{7.16}$$

Using the fact that the bend angle can be expressed in terms of arc length, $\theta = s/R$, we can rewrite Equation 7.16 as

$$Q(\theta) = \frac{EI\theta^2}{2s}. \tag{7.17}$$

Equation 7.17 describes the internal energy of a beam of arc length s subject to a constant curvature such that the bend angle is θ radians. We now assume we are given a polymer of length s that can be modeled as an elastic beam. If we were to immerse the polymer within a constant temperature heat bath, we could use Boltzmann's distribution to find the probability of finding a polymer with bend angle θ as

$$p(\theta) = \frac{1}{Z} e^{-Q(\theta)/k_B T}. \tag{7.18}$$

To compute the partition function Z, we must integrate the exponential term over two angles, θ and ϕ, to allow for bending in three dimensions. Specifically, Z can be computed as

$$Z = \int_0^{2\pi} \int_0^{\pi} e^{-Q(\theta)/k_B T} \, d\phi \sin\theta \, d\theta, \tag{7.19}$$

where we have taken the integral with respect to a differential element of solid angle $d\phi\sin\theta d\theta$. We now seek to quantify the average amount of polymer curvature. We can do this by computing $\langle \theta^2 \rangle$ as

$$\langle \theta^2 \rangle = \frac{1}{Z} \int_0^{2\pi} \int_0^{\pi} e^{-Q(\theta)/k_B T} \theta^2 \, d\phi \sin\theta \, d\theta. \tag{7.20}$$

For small angles, it is left to you to show that the solution to this integral is

$$\langle \theta^2 \rangle = \frac{2k_B T s}{EI}. \tag{7.21}$$

Now, we relate the above expression to the orientation correlation function. We can do this using a Maclaurin series expansion. The Maclaurin series for $\cos x$ is

$$\cos x = 1 - \frac{1}{2}x^2 + \frac{1}{24}x^4\ldots \tag{7.22}$$

Using this expansion and a small-angle assumption, we can write the orientation correlation function as

$$\langle \cos\Delta\theta(s) \rangle \approx \left\langle 1 - \frac{\Delta\theta^2(s)}{2} \right\rangle = 1 - \frac{\langle \Delta\theta^2(s) \rangle}{2}. \tag{7.23}$$

The term in brackets in the right-most relation in Equation 7.23 is the average value of the difference in angle (squared) between two points separated by a distance s along the polymer. If the polymer is assumed to be a constant curvature beam of arc length s, then we can substitute in our expression for $\langle \theta^2 \rangle$ found in Equation 7.23. In this case,

$$\langle \cos \Delta \theta(s) \rangle \approx 1 - \frac{k_B T}{EI} s. \tag{7.24}$$

Using the approximation $e^{-x} \approx 1 - x$ in the limit of small x, we can write our expression for the orientation correlation function found in Equation 7.24 for small s as

$$\langle \cos \Delta \theta(s) \rangle = e^{\frac{-s}{\ell_p}} \cong 1 - \frac{-s}{\ell_p}. \tag{7.25}$$

Relating Equations 7.24 and 7.25, we can see that the persistence length ℓ_p is related to flexural rigidity as

$$\ell_p \equiv \frac{EI}{k_B T}. \tag{7.26}$$

Equation 7.26 is a useful relation, because it allows us to estimate the effective Young's modulus of a polymer from its persistence length. We can observe the thermal fluctuations of a polymer, and calculate $\cos \Delta \theta(s)$ along the polymer length. We do this multiple times and then find the average values at each point, which gives $\langle \cos \Delta \theta(s) \rangle$. These points can be fit to an exponential to find ℓ_p. Finally, assuming we know the temperature at which the measurements were made, we can compute the flexural rigidity using Equation 7.26.

Polymers can be classified as stiff, flexible, or semi-flexible by the persistence length

The persistence length can vary dramatically for different biopolymers. For example, it is 50 nm for DNA, 15 μm for F-actin, and 6 mm for MTs. The difference in persistence lengths for these three biopolymers spans more than five orders of magnitude! Because the persistence length gives a length over which the orientations of a thermally undulating polymer become uncorrelated, it is a natural length scale for the classification of a given polymer as being stiff or flexible.

Let us revisit our example of a thermally fluctuating MT and strand of DNA, each 10 μm in length. Recall that the MT tends to take on a rod-like configuration, whereas the DNA tends to take on a more coiled configuration. For the MT, its contour length is much smaller than its persistence length, $L \ll \ell_p$. We know that the contour length of the MT is much shorter than the length required for the orientations of the polymer to become uncorrelated. Because the polymer will tend to resemble a straight rod, the polymer is considered "stiff". In contrast, for the strand of DNA, $L \gg \ell_p$. We know that the contour length is much greater than the length required for the orientations to become uncorrelated, so the DNA will tend to take on a very coiled configuration. Because it takes on a coiled configuration, the polymer is considered "flexible." We can classify polymers of length L and persistence length ℓ_p as being stiff, flexible, or semi-flexible, as in Table 7.2.

It is important to note that classification of a polymer as being flexible, semi-flexible, or stiff depends not only on its persistence length, but also its contour length.

Table 7.2 Classification of polymer flexibility based on persistence length

Type of polymer behavior	Persistence length
Stiff	$\ell_p \gg L$
Flexible	$\ell_p \ll L$
Semi-flexible	$\ell_p \approx L$

L = 10 m L = 10 mm L = 10 µm L = 10 nm

microtubule

actin

DNA

Figure 7.14 Representative configurations of MT, actin, and DNA polymers of various contour lengths at room temperature. The tendency to take on a straight versus coiled configuration depends not only on persistence length but also on contour length. A 10-nm-long DNA segment is much shorter than the persistence length, and the configuration will tend to be rod-like. In contrast, a 10-m-long MT has a contour length much greater than its persistence length and will take on a coiled configuration.

For very short DNA segments that are much less than the persistence length, the configuration will tend to be rod-like. In contrast, a very long MT that has a contour length much greater than its persistence length will take on a very coiled configuration. This is demonstrated schematically in Figure 7.14.

In the next section, we introduce several common biopolymer models and their resulting mechanical (force-extension) behavior. These different models differ substantially in their assumptions. As a result, a given model may be good for flexible polymers, but not for stiff polymers. As we will see, the appropriateness of a given model is going to depend, in large part, on the persistence length of the polymer, the length of the polymer, and the degree to which it is being extended.

7.4 IDEAL CHAIN

The ideal chain is a polymer model for flexible polymers

We start our introduction to biopolymer models with the *ideal chain*. The name for this model comes from the fact that in the ideal chain all changes in internal energy are ignored. It is often used for modeling flexible polymers whose behavior is dominated by entropy.

Consider a chain of n segments of length b that are connected by freely rotating joints as in Figure 7.15. Each segment in this model is referred to as a *Kuhn segment*, with the length of each segment called the *Kuhn length*. We will see later in this chapter that the Kuhn length is related to the persistence length.

If the chain contains n segments, it will contain $n - 1$ rotating joints, or vertices, between segments. Let $\mathbf{r}_i$ be the segment vector from vertex i to vertex $i + 1$. There are n of these vectors, and all of them have the same length b. The contour length of the chain, or the length of the chain if it were completely straight, is nb. In general, the chain will not be fully straightened out. We will describe the degree to which it is straightened out by the end-to-end vector $\mathbf{R}$, which is the sum of all bond vectors in the chain

$$\mathbf{R} = \sum_{i=1}^{N} \mathbf{r}_i. \tag{7.27}$$

We would like to say something about the average end-to-end length. However, if we average the random vector $\mathbf{R}$, we get 0 due to symmetry. In particular, there is no preferred spatial orientation for our polymer, so any vector $\mathbf{R}$ that occurs can

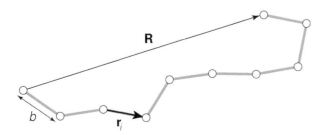

Figure 7.15 The model polymer for the ideal chain. The polymer consists of rigid links of segment length b connected by freely rotating hinges. The end-to-end vector $\mathbf{R}$ and the bond vector $\mathbf{r}_i$ for link i are depicted.

also occur with equal probability as $-\mathbf{R}$. Nonetheless, we can say something about the average magnitude of $\mathbf{R}$ by examining its square

$$\langle \mathbf{R}^2 \rangle = \langle \mathbf{R} \cdot \mathbf{R} \rangle$$

$$= \left\langle \left(\sum_{i=1}^{N} \mathbf{r}_i \right) \cdot \left(\sum_{j=1}^{N} \mathbf{r}_j \right) \right\rangle \quad (7.28)$$

$$= \sum_{i=1}^{N} \sum_{j=1}^{N} \langle \mathbf{r}_i \cdot \mathbf{r}_j \rangle.$$

We know that for the dot product of two vectors $|\mathbf{a} \cdot \mathbf{b}| = |\mathbf{a}| \, |\mathbf{b}| \cos\theta_{ab}$, where $|\mathbf{a}|$ and $|\mathbf{b}|$ are the lengths of $\mathbf{a}$ and $\mathbf{b}$, and θ_{ab} is the angle between $\mathbf{a}$ and $\mathbf{b}$. If we substitute this expression into Equation 7.28, we obtain

$$\langle \mathbf{R}^2 \rangle = \sum_{i=1}^{N} \sum_{j=1}^{N} b^2 \langle \cos\theta_{ij} \rangle \quad (7.29)$$

$$= b^2 \sum_{i=1}^{N} \sum_{j=1}^{N} \langle \cos\theta_{ij} \rangle.$$

Furthermore, we know that the direction of any given segment is independent of all of the other segments. The value of $\cos\theta_{ab}$ between any two segments will range from -1 to 1. So, the average of $\cos\theta_{ab}$ will be 0, except when computed on the same bond ($a = b$). In this case $\theta_{ab} = 0$ and $\cos 0 = 1$. An equivalent statement is $\langle \cos\theta_{ij} \rangle = 0$ for $i \neq j$, and $\langle \cos\theta_{ij} \rangle = 1$ for $i = j$, or $\langle \cos\theta_{ab} \rangle = \delta_{ij}$. Therefore,

$$\langle \mathbf{R}^2 \rangle = b^2 \sum_{i=1}^{n} \sum_{j=1}^{n} \delta_{ij} = nb^2. \quad (7.30)$$

This is the mean square end-to-end length of a three-dimensional chain composed of n segments of size b. Notice that just as in the random walk from Section 5.6, the quantity $\langle \mathbf{R}^2 \rangle$ scales linearly with n. Only now, instead of n representing the number of steps in the walk, it represents the number of links in the chain.

The probability for the chain to have different end-to-end lengths can be determined from the random walk

With our model described, we now seek to determine the probability of the chain having a specific length. This probability will allow us to determine the free energy and ultimately the force required to maintain the chain at a given length, giving us the chain's force–displacement relationship.

The probability distribution function for the ideal chain is based on the random walk. Let us return again to the one-dimensional random walk. Recall that the probability of being at position r after n steps was given by Equation 5.64

$$p_{1d}(n,r) = \frac{M(n,r)}{2^n} = \frac{1}{2^n} \frac{n!}{\left(\dfrac{n + \dfrac{r}{b}}{2} \right)! \left(\dfrac{n - \dfrac{r}{b}}{2} \right)!}. \quad (7.31)$$

and that the Gaussian approximation to this distribution was

$$p_{1d}(n,r) = \frac{1}{\sqrt{2\pi\langle r^2 \rangle}} e^{-r^2/2\langle r^2 \rangle}. \quad (7.32)$$

Nota Bene

Kronecker delta. The Kronecker delta δ_{ij} is a compact notation used to signify the following relation:

$$\delta_{ij} = \begin{cases} 1, & \text{if} \quad i = j \\ 0, & \text{if} \quad i \neq j \end{cases}$$

Equation 7.32 is for a one-dimensional random walk. We now seek to extend this relation to three dimensions. To see the connection between one and three dimensions, let us start by defining the end-to-end distance for our three-dimensional random walk in the coordinate directions, $\mathbf{R} = R_x\mathbf{e}_x + R_y\mathbf{e}_y + R_z\mathbf{e}_z$ (where $\mathbf{e}_x$, $\mathbf{e}_y$, and $\mathbf{e}_z$ are unit vectors in the x-, y-, and z-directions). The steps in each direction are independent. So, the mean square end-to-end distance is the sum of the mean square end-to-end distances along each of the three coordinate directions, or

$$\left\langle \mathbf{R}^2 \right\rangle = \left\langle R_x^2 + R_y^2 + R_z^2 \right\rangle$$

$$= \left\langle R_x^2 \right\rangle + \left\langle R_y^2 \right\rangle + \left\langle R_z^2 \right\rangle. \tag{7.33}$$

Because there is nothing special about our choice of axes, each term in Equation 7.33 must be equal and the mean square end-to-end distance along one direction is simply one-third of the total mean square end-to-end distance, in other words:

$$\left\langle R_x^2 \right\rangle = \left\langle R_y^2 \right\rangle = \left\langle R_z^2 \right\rangle = \frac{\left\langle \mathbf{R}^2 \right\rangle}{3}$$

$$= \frac{nb^2}{3}. \tag{7.34}$$

Because the three components of a three-dimensional random walk along the three coordinate directions are independent of each other, the three-dimensional probability distribution function for a random walk can be computed as the product of three one-dimensional distribution functions in each of the three directions. But, we have to be careful. We have to account properly for the fact that the one dimensional random walk is a projection of a three-dimensional random walk into one dimension. More specifically, we have to scale the mean square distance in Equation 7.32 such that $\langle r^2 \rangle$ is equal to the mean square distance in one dimension for our three-dimensional random walk,

$$\langle r^2 \rangle = \langle R_x{}^2 \rangle = \langle R_y{}^2 \rangle = \langle R_z{}^2 \rangle = \langle \mathbf{R}^2 \rangle / 3. \tag{7.35}$$

Because each of these quantities is one-third of the three-dimensional average squared end-to-end distance, $\langle r^2 \rangle = \langle \mathbf{R}^2 \rangle / 3$. Combining Equations 7.32 and 7.35, we obtain

$$p_{1d}(n, R_x) = p_{1d}(n, R_y) = p_{1d}(n, R_z) = \sqrt{\frac{3}{2\pi\langle \mathbf{R}^2 \rangle}} e^{-3R_x{}^2/2\langle \mathbf{R}^2 \rangle}. \tag{7.36}$$

The three-dimensional probability distribution is the product of three one-dimensional distributions,

$$p_{3d}(n, \mathbf{R}) = p_{1d}(n, R_x) p_{1d}(n, R_y) p_{1d}(n, R_z)$$

$$= \left(\frac{3}{2\pi\langle \mathbf{R}^2 \rangle} \right)^{3/2} e^{-3R_x{}^2/2\langle \mathbf{R}^2 \rangle} e^{-3R_y{}^2/2\langle \mathbf{R}^2 \rangle} e^{-3R_z{}^2/2\langle \mathbf{R}^2 \rangle}$$

$$= \left(\frac{3}{2\pi\langle \mathbf{R}^2 \rangle} \right)^{3/2} e^{-3(R_x^2 + R_y^2 + R_z^2)/2\langle \mathbf{R}^2 \rangle} \tag{7.37}$$

$$= \left(\frac{3}{2\pi\langle \mathbf{R}^2 \rangle} \right)^{3/2} e^{-3\mathbf{R}^2/2\langle \mathbf{R}^2 \rangle}.$$

Excluded volume interactions. In the ideal chain, we assume that segments can overlap. In other words, two segments can occupy the same space at the same time. In reality, this cannot occur. Models that restrict segment overlap are said to account for excluded volume interactions. One such model that accounts for excluded volume effects is based on the self-avoiding random walk, a walk that cannot cross a point it has traced previously.

We can express Equation 7.37 in terms of the number of segments n and the Kuhn length b by substituting the relation $\langle \mathbf{R}^2 \rangle = nb^2$. We obtain

$$p_{3d}(n,\mathbf{R}) = \left(\frac{3}{2\pi nb^2} \right)^{3/2} e^{-3\mathbf{R}^2/2nb^2}, \tag{7.38}$$

which is the three-dimensional probability distribution for an ideal chain consisting of n segments of length b to have an end-to-end vector of $\mathbf{R}$. Because the ideal chain is based on the above Gaussian distribution, it is also referred to as the *Gaussian chain*.

Advanced Material: Gaussian approximation for the random walk

Here, we show that the probability distribution for a one-dimensional random walk approaches a Gaussian distribution for large n. To begin, we first take the natural log of the exact probability given in Equation 7.31 to yield

$$\ln(p(n,R)) = -n\ln(2) + \ln(n!)$$

$$- \ln\left(\frac{n+R}{2}! \right) - \ln\left(\frac{n-R}{2}! \right). \tag{7.39}$$

Now, if a, b, and c are positive integers such that $a \geq b$, it can be shown that $((a+b)/c)!$ can be expressed as

$$\frac{a+b}{c}! = \frac{a}{c}! \prod_{s=1}^{b/c} \left(\frac{a}{c} + s \right), \tag{7.40}$$

and $((a-b)/c)!$ can be written as

$$\frac{a-b}{c}! = \frac{\frac{a}{c}!}{\prod_{s=1}^{b/c} \left(\frac{a}{c} + 1 - s \right)}. \tag{7.41}$$

The third term in Equation 7.39 can be written as

$$\ln\left(\frac{n+R}{2}! \right) = \ln\left(\frac{n!}{2} \right) \prod_{s=1}^{R/2} \left(\frac{n}{2} + s \right)$$

$$= \ln\left(\frac{n}{2}! \right) + \sum_{s=1}^{R/2} \ln\left(\frac{n}{2} + s \right) \tag{7.42}$$

and the fourth term in Equation 7.39 as

$$\ln\left(\frac{n+R}{2}! \right) = \ln\left(\frac{n}{2}! \right) - \sum_{s=1}^{R/2} \ln\left(\frac{n}{2} + 1 - s \right). \tag{7.43}$$

Combining Equations 7.39, 7.42, and 7.43, $\ln(p(n,R))$ can be expressed as

$$\ln(p(n,r)) = -n\ln(2) + \ln(n!) - 2\ln\left(\frac{n}{2}! \right)$$

$$- \sum_{s=1}^{R/2} \ln\left(\frac{n}{2} + s \right) + \sum_{s=1}^{R/2} \ln\left(\frac{n}{2} + 1 - s \right)$$

$$= -n\ln(2) + \ln(n!) - 2\ln\left(\frac{n}{2}! \right)$$

$$- \sum_{s=1}^{R/2} \ln\left(\frac{\frac{n}{2} + s}{\frac{n}{2} + 1 - s} \right)$$

$$= -n\ln(2) + \ln(n!) - 2\ln\left(\frac{n}{2}! \right) \tag{7.44}$$

$$- \sum_{s=1}^{R/2} \ln\left(\frac{1 + \frac{2s}{n}}{1 - \frac{2s}{n} + \frac{2}{n}} \right),$$

where we have divided the numerator and denominator by $n/2$ in the last term in the last line.

We now invoke our large n approximation. In particular, in the last term of Equation 7.44, any term with n in the denominator will go to zero in the limit of large n. Because $\ln(1+a) \cong a$ for $|a| \ll 1$, we can approximate the logarithm in the last term of Equation 7.44 as

$$\ln\left(\frac{1 + \frac{2s}{n}}{1 + \frac{2s}{n} + \frac{2}{n}} \right) = \ln\left(1 + \frac{2s}{n} \right) - \ln\left(1 - \frac{2s}{n} + \frac{2}{n} \right) \tag{7.45}$$

$$\cong \frac{4s}{n} - \frac{2}{n}.$$

Using Equation 7.45 and the identities

$$\sum_{s=1}^{a} s = a(a+1)/2 \qquad (7.46)$$

and

$$\sum_{s=1}^{a} 1 = a, \qquad (7.47)$$

Equation 7.44 can be expressed as

$$\ln(p(n,R)) \cong -n\ln(2) + \ln(n!) - 2\ln\left(\frac{n}{2}!\right)$$

$$- \sum_{s=1}^{R/2}\left(\frac{4s}{n} - \frac{2}{n}\right)$$

$$\cong -n\ln(2) + \ln(n!) - 2\ln\left(\frac{n}{2}!\right)$$

$$- \frac{4}{n}\sum_{s=1}^{R/2} s + \frac{2}{n}\sum_{s=1}^{R/2} 1$$

$$\cong -n\ln(2) + \ln(n!) - 2\ln\left(\frac{n}{2}!\right) \qquad (7.48)$$

$$- \frac{4\left(\dfrac{R}{2}\right)\left(\dfrac{R}{2}+1\right)}{n} + \frac{R}{n}$$

$$\cong -n\ln(2) + \ln(n!) - 2\ln\left(\frac{n}{2}!\right) - \frac{R^2}{2n}.$$

Because $\ln(a) = b$ and $a = e^b$ are equivalent statements, Equation 7.48 can be rewritten as

$$p(n,R) \cong \frac{1}{2^n}\frac{n!}{(n/2)!(n/2)!}e^{-\frac{R^2}{2n}} \cong Ce^{-\frac{R^2}{2n}}, \qquad (7.49)$$

where

$$C = \frac{1}{2^n}\frac{n!}{\dfrac{n}{2}!\dfrac{n}{2}!}. \qquad (7.50)$$

We can obtain a simplified expression for C by observing that it is a normalization constant such that

$$\int_{-\infty}^{\infty} p(n,R)\mathrm{d}R = 1. \qquad (7.51)$$

In this case, we obtain

$$C = 1\Big/\int_{-\infty}^{\infty} e^{-\frac{R^2}{2n}}\mathrm{d}R = 1/\sqrt{2\pi n}. \qquad (7.52)$$

Combining Equations 7.49 and 7.52, we obtain Equation 7.32, which shows its equivalency to Equation 7.31 in the limit of large n.

The free energy of the ideal chain can be computed from its probability distribution function

We will now use the probability distribution for the ideal chain to compute its free energy. In our model, the polymer has freely rotating joints with no capacity to store energy. The polymer has zero internal energy regardless of its conformation, and the free energy is simply $\Psi = -TS$. In Section 5.2, we learned that entropy can be calculated from the density of microstates, $\Omega(\mathbf{R})$, which is the number of configurations in which the polymer can have an end-to-end vector of $\mathbf{R}$. To calculate the density of states, we can use the probability distribution for the ideal chain obtained in Equation 7.37. More specifically, we can take advantage of the fact that the probability distribution function $p_{3d}(n,\mathbf{R})$ is proportional to the number of polymer configurations that have an end-to-end vector $\mathbf{R}$. In other words,

$$p_{3d}(n,R) \sim \Omega(n,R). \qquad (7.53)$$

This proportionality arises due to the facts that (1) the integral of $p_{3d}(n, \mathbf{R})$ over some range gives the exact probability for the polymer to have a end-to-end vector within that range, and (2) this probability is equal to the number of polymer configurations divided by the total number of all configurations. Using

Equation 7.53 we can compute the entropy as

$$S = k_B \ln \Omega(n, R) \sim k_B \ln p_{3d}(n, R). \tag{7.54}$$

Combining Equations 7.38 and 7.54, we obtain the following relation for entropy

$$S \sim k_B \ln \left(\left(\frac{3}{2\pi n b^2} \right)^{3/2} e^{-3R^2/2nb^2} \right). \tag{7.55}$$

Equation 7.55 can be simplified as

$$S(n, \mathbf{R}) = -\frac{3}{2} k_B \frac{\mathbf{R}^2}{nb^2} + S_0, \tag{7.56}$$

where we have turned the proportionality into an equivalency by accounting for any terms that do not depend on $\mathbf{R}$ in the constant S_0. Given an expression for entropy, we can calculate the free energy for a chain composed of n segments with an end-to-end vector $\mathbf{R}$ as

$$\Psi(n, \mathbf{R}) = -TS(n, \mathbf{R}) = \frac{3}{2} k_B T \frac{\mathbf{R}^2}{nb^2} + \Psi_0, \tag{7.57}$$

where $\Psi_0 = -TS_0$ does not depend on $\mathbf{R}$.

Force is the gradient of free energy in thermodynamic systems

At this point in our development we need to make a short aside to consider how force and energy are related. Recall from Section 5.1 that for mechanical systems, the principle of minimum total potential energy states that the equilibrium state is achieved when the total potential energy is minimized. Recall that in our two-spring system (Figure 5.2), forces were balanced when potential energy was minimized, in other words when the gradient of potential energy was zero. The equivalency between force and the gradient of potential energy becomes apparent.

In addition, as we discussed in Section 5.4 for thermodynamic systems, equilibrium is dictated by the principle of minimum free energy. This principle states that equilibrium is achieved when free energy is minimized, in other words forces are in balance when the gradient of free energy is zero. For thermodynamic systems, force is equivalent to the gradient of free energy. So, we can use Equation 7.57 to calculate the force necessary to extend the chain. In particular, force can be found as the gradient of free energy with respect to the end-to-end vector $\mathbf{R}$ as

$$F_x = \frac{\partial \Psi(n, \mathbf{R})}{\partial R_x} = \frac{3k_B T}{nb^2} R_x$$

$$F_y = \frac{\partial \Psi(n, \mathbf{R})}{\partial R_y} = \frac{3k_B T}{nb^2} R_y \tag{7.58}$$

$$F_z = \frac{\partial \Psi(n, \mathbf{R})}{\partial R_z} = \frac{3k_B T}{nb^2} R_z$$

or,

$$\mathbf{F} = \frac{3k_B T}{nb^2} \mathbf{R}. \tag{7.59}$$

We previously remarked that the Kuhn length b is related to the persistence length. In Section 7.6 we will show that this relation is

$$b = 2\ell_p. \tag{7.60}$$

So, we can rewrite Equation 7.59 in terms of the contour length $L = nb$ and the persistence length as

$$\mathbf{F} = \frac{3k_BT}{nb^2}\mathbf{R}$$

$$= \frac{3k_BT}{(nb)b}\mathbf{R} \tag{7.61}$$

$$= \frac{3k_BT}{2L\ell_p}\mathbf{R}.$$

Equation 7.61 gives the force–displacement relationship for an ideal chain. On inspection, a couple of interesting things can be seen. First, for any non-zero temperature and non-zero $\mathbf{R}$, the force is non-zero as well. This is perhaps somewhat counterintuitive, because the polymer consists of freely rotating segments that are unable to store energy. However, these forces are not as mysterious as they may seem. Physically, the capacity for the polymer to exert a force if the ends are separated arises out of random collisions from the thermal environment surrounding it, as well as the tendency for the polymer to take on a probabilistically favorable state. In particular, the maximum number of polymer conformations can occur with a zero end-to-end distance, and this state is the most entropically favorable. Straightening the polymer out reduces the entropy and therefore requires force.

Another interesting aspect about Equation 7.61 is that the force $\mathbf{F}$ is linear in $\mathbf{R}$, so pulling the ends apart with a distance $\mathbf{R}$ produces the same force as an elastic spring with a spring constant of $(3k_BT)/(2L\ell_p)$. This relationship has been referred to as describing an *entropic spring*. The resistance to deformation is derived entirely due to entropy and depends on temperature. We can see that the stiffness of the spring is proportional to temperature, so increasing T increases the stiffness. This is in contrast to most engineering materials such as steel, which become more compliant with increasing temperature.

The behavior of polymers tends toward that of an ideal chain in the limit of long contour length

In developing the ideal chain model, we assumed that the polymer was made up of freely rotating links with no capacity to store energy. In this model, the polymer is assumed to have zero internal energy regardless of the configuration. Of course, in reality, there will be energetic interactions within any polymer. These interactions can arise, through changes in bond angles and distances between backbone atoms, or the electrostatic interaction of backbone atoms with each other from a distance. One may ask whether the ideal chain, which has zero energy regardless of the conformation, is a good representation of any polymers in reality?

The simple answer is yes, as long as the polymer is long enough. To gain some intuition for this notion, recall our discussion of the thermally fluctuating microtubule in Section 7.3. We found that the tendency for the MT to exhibit flexible, semi-flexible, or stiff behavior depended not only on its persistence length, but also its contour length. A typical MT found in a cell has a contour length much less than the persistence length and tends to be found in rod-like configurations. However, if we were to construct a very long MT that had a contour length much greater than its persistence length, we would find that it would exhibit greatly different behavior. In particular, the behavior of the very long MT would resemble that of a flexible polymer, whose tendency to coil is an entropically driven phenomenon.

The tendency for polymers with large contour lengths to tend toward ideal chain behavior can be better understood by examining the physical meaning of the persistence length. We know that at length scales longer than the persistence length, the orientation correlation between two points of a thermally fluctuating polymer vanishes. Physically, this means that at this length scale, entropic influences become dominant over energetic influences. We also know that if a polymer is longer than its persistence length, then for every span of polymer approximately equal to this length, the orientations of the polymer become roughly uncorrelated. At very long length scales, the polymer behaves as if it were composed of many independently fluctuating chain segments, with each segment being on the order of the persistence length in size. The behavior of most polymers will be dominated by entropy and tend toward that of an ideal chain when the contour length is significantly greater than the persistence length. In modeling such cases using an ideal chain, a Kuhn segment is not necessarily a single monomer, but rather a span of polymer roughly equivalent to the persistence length in size.

7.5 FREELY JOINTED CHAIN (FJC)

The FJC model places a limit on polymer extension

Although the ideal chain is an extremely useful model, it has certain limitations. In Section 7.4 we found a relationship for the response of an ideal chain to tension. The stiffness did not depend on how much it was extended. Therefore, the model predicts that the stiffness will be constant even as the chain is extended past its contour length nb! This is clearly unphysical. Although the ideal chain is useful if the end-to-end distance is much less than the contour length, an alternative model is desirable for cases in which this condition is not met.

A model that addresses this limitation in the ideal chain is the FJC, which gives more realistic behavior at long extensions. Similar to the ideal chain, the FJC consists of n links of length b that are connected to each other by freely rotating joints. However, in the FJC model, the end-to-end length is constrained such that it cannot be longer than the contour length. We derive the force extension behavior here.

Let us revisit the chain in Figure 7.15. For the ith segment, we define the vector $\mathbf{r}_i$ as the product of b and the ith unit vector $\mathbf{u}_i$, $\mathbf{r}_i = b\mathbf{u}_i$. The end-to-end vector is given by

$$\mathbf{R} = b \sum_{i=1}^{n} \mathbf{u}_i. \tag{7.62}$$

To simplify the development (and without losing any generality) we will consider a chain that is being extended in the z-direction. We can simplify the math by expressing $\mathbf{R}$ in spherical coordinates about the z-axis with θ_i being the angle with the z-axis and ϕ_i being the azimuth (**Figure 7.16**) as

$$\mathbf{e}_x \cdot \mathbf{u}_i = \sin\theta_i \cos\phi_i$$

$$\mathbf{e}_y \cdot \mathbf{u}_i = \sin\theta_i \sin\phi_i \tag{7.63}$$

$$\mathbf{e}_z \cdot \mathbf{u}_i = \cos\theta_i,$$

where $\mathbf{e}_x$, $\mathbf{e}_y$, and $\mathbf{e}_z$ are unit vectors in the x-, y-, and z-directions.

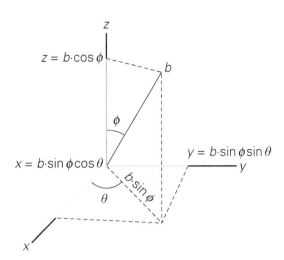

Using these relations, the end-to-end distance along the z-axis can be written as

$$R_z = \mathbf{e}_z \cdot \mathbf{R}$$

$$= b \sum_{i=1}^{n} \mathbf{e}_z \cdot \mathbf{u}_i \qquad (7.64)$$

$$= b \sum_{i=1}^{n} \cos\theta_i.$$

The force–displacement relation for the FJC can be found by the canonical ensemble

With the configurational geometry of the chain defined, we can now calculate the force it generates with extension. For the ideal chain this calculation was based on determining the change in free energy associated with a reduction in entropy with extension. For the FJC we take a different approach. In Section 5.5, we learned that for a constant temperature system at equilibrium, the probability for a given microstate is given by Boltzmann's distribution. We seek to define a system with a known potential energy such that we can use Boltzmann's distribution to find the probability of a particular polymer configuration at equilibrium.

Consider a system consisting of a FJC immersed in a constant temperature heat bath. One end of the chain is constrained such that its position is fixed but freely rotating. At the other end, we attach a small weight that pulls downward with a constant force of F_z (Figure 7.17). As the chain is extended, the potential energy of the weight (and therefore the entire system) is decreased by $F_z R_z$. We can write the internal energy as

$$Q = -F_z R_z = -F_z b \sum_{i=1}^{n} \cos\theta_i, \qquad (7.65)$$

where in the right-hand-most side we have substituted the expression for R_z found in Equation 7.64. We know that for the canonical ensemble, the probability of a microstate having internal energy Q can be found by the Boltzmann distribution. For our system, each microstate can be characterized as the set of bond angles

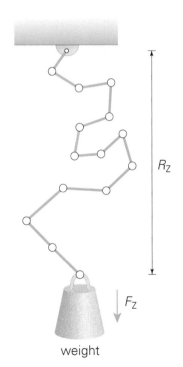

Figure 7.17 Schematic demonstrating the FJC with a weight attached to its end. The weight subjects the polymer to a constant downward force F_z. As the weight moves downward, the potential energy of the weight decreases.

$\theta_{1,2,\ldots,n}$ and $\phi_{1,2,\ldots,n}$. The Boltzmann probability associated with the microstate having energy $Q(\theta_{1,2,\ldots,n}, \phi_{1,2,\ldots,n})$ is

$$p(\theta_1, \theta_2, \ldots, \theta_n, \phi_1, \phi_2, \ldots, \phi_n) = \frac{1}{Z} e^{\kappa \sum_{i=1}^{n} \cos\theta_i}, \tag{7.66}$$

where

$$\kappa_B = \frac{F_z b}{k_B T} = \frac{2 F_z \ell_p}{k_B T}. \tag{7.67}$$

To calculate the partition function Z, we must integrate the exponential term in Equation 7.66 over all possible configurations of the chain. To do this, we integrate over all possible bond angles $\theta_{1,2,\ldots,n}$ and $\phi_{1,2,\ldots,n}$ as

$Z =$

$$\int_{\phi_1=0}^{2\pi} \int_{\phi_2=0\ldots}^{2\pi} \int_{\phi_n=0}^{2\pi} \int_{\theta_1=0}^{\pi} \int_{\theta_2=0\ldots}^{\pi} \int_{\theta_n=0}^{\pi} e^{\kappa \sum_{i=1}^{n} \cos\theta_i} \sin\theta_1 \sin\theta_2 \ldots \sin\theta_n \; d\theta_1 d\theta_2 \ldots d\theta_n \; d\phi_1 d\phi_2 \ldots d\phi_n, \tag{7.68}$$

where we have performed the integral with respect to a differential element of solid angle $\sin\theta \, d\theta \, d\phi$ because the bond angles were defined in spherical coordinates. We can rearrange the order of integration in Equation 7.68 to perform the integrals for each segment in turn as

$$Z = \int_{\phi_1=0}^{2\pi} \int_{\theta_1=0}^{\pi} e^{\kappa \cos\theta_1} \sin\theta_1 d\theta_1 d\phi_1 \int_{\phi_2=0}^{2\pi} \int_{\theta_2=0}^{\pi} e^{\kappa \cos\theta_2} \sin\theta_2 d\theta_2 d\phi_2 \ldots$$

$$\int_{\phi_n=0}^{2\pi} \int_{\theta_n=0}^{\pi} e^{\kappa \cos\theta_n} \sin\theta_n d\theta_n d\phi_n. \tag{7.69}$$

Equation 7.69 can be written in a simplified, compact form as

$$Z = \prod_{i=1}^{n} \int_{\phi_i=0}^{2\pi} \int_{\theta_i=0}^{\pi} e^{\kappa \cos\theta_i} \sin\theta_i d\theta_i d\phi_i = z^n, \tag{7.70}$$

where

$$z = \int_{\phi=0}^{2\pi} \int_{\theta=0}^{\pi} e^{\kappa \cos\theta} \sin\theta d\theta d\phi \tag{7.71}$$

is the single partition function. To compute z, we first integrate out θ as

$$z = 2\pi \int_{\theta=0}^{\pi} e^{\kappa \cos\theta} \sin\theta d\theta. \tag{7.72}$$

Next, we change variables by letting $\rho = \cos\theta$ and $d\rho = -\sin\theta \, d\theta$. Equation 7.72 can be evaluated as

$$z = -2\pi \int_{-1}^{1} e^{\kappa \rho} d\rho = 2\pi \frac{e^{\kappa} - e^{-\kappa}}{\kappa} = 4\pi \frac{\sinh\kappa}{\kappa}. \tag{7.73}$$

where $\sinh\kappa$ is the hyperbolic sine, $\sinh\kappa = (1/2)(e^{\kappa} - e^{-\kappa})$.

We now seek to use the partition function to calculate $\langle R_z \rangle$ at equilibrium. To do this, we use a similar method as in Section 5.5, in which we computed the average internal energy using the partition function in the canonical ensemble. In particular, we first start with the relation for $\langle R_z \rangle$,

$$\langle R_z \rangle = \int_\Theta p(\Theta) R_z d\Theta, \tag{7.74}$$

where for simplicity, we have used Θ in Equation 7.74 to denote integration over all possible bond angles $\theta_{1,2,\ldots,n}$ and $\phi_{1,2,\ldots,n}$. Substituting our expression for p obtained in Equation 7.68 into Equation 7.74 and using the relation

$$e^{\kappa \sum\limits_{i=1}^{n} \cos\theta_i} = e^{\frac{\kappa\left(b\sum\limits_{i=1}^{n}\cos\theta_i\right)}{b}} = e^{\frac{\kappa R_z}{b}}, \tag{7.75}$$

we obtain

$$\langle R_z \rangle = \frac{1}{Z}\int_\Theta e^{\frac{\kappa R_z}{b}} R_z d\Theta. \tag{7.76}$$

Similar to our approach in calculating internal energy in the canonical ensemble, we can rewrite Equation 7.76 as a logarithm of Z with the following manipulation

$$\begin{aligned}\langle R_z \rangle &= \frac{b}{Z}\int_\Theta \frac{\partial}{\partial\kappa}\left(e^{\frac{\kappa R_z}{b}}\right) R_z d\Theta \\ &= \frac{b}{Z}\frac{\partial}{\partial\kappa}\int_\Theta e^{\frac{\kappa R_z}{b}} R_z d\Theta \\ &= \frac{b}{Z}\frac{\partial Z}{\partial\kappa} \\ &= b\frac{\partial\ln Z}{\partial\kappa}.\end{aligned} \tag{7.77}$$

Using Equation 7.77, we can compute the mean extension as

$$\begin{aligned}\langle R_z \rangle &= b\frac{\partial\ln Z}{\partial\kappa} \\ &= bn\frac{\partial}{\partial\kappa}[\ln(\sinh\kappa) - \ln\kappa + \ln 4\pi] \\ &= bn\left(\coth\kappa - \frac{1}{\kappa}\right).\end{aligned} \tag{7.78}$$

Recall that force is embedded inside κ. Equation 7.78 gives a force-displacement relationship for the FJC. The expression is a little different than those with which we are familiar, with because it is an implicit relationship. In other words, although it does produce a force for every displacement and vice versa, we do not have an explicit expression for force. However, it is still an entirely valid relationship.

Differences between the ideal chain and the FJC emerge at large forces

We can get a sense of the force-displacement behavior predicted by the FJC model by examining some special cases. To do this, we first note that the term in

Figure 7.18 Plot of the Langevin function. The Langevin function approaches 1 for large values of x. For small values of x, the slope is approximately one-third.

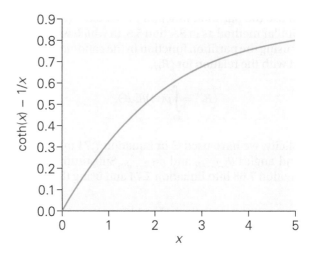

parentheses in Equation 7.78 is known as the *Langevin function*. Expanding the hyperbolic cotangent, the Langevin function $\mathcal{L}$ can be expressed as

$$\mathcal{L} = \frac{e^x + e^{-x}}{e^x - e^{-x}} - \frac{1}{x}. \tag{7.79}$$

A plot of the Langevin function can be seen in Figure 7.18. Notice that as x gets very large, $\mathcal{L}$ approaches one. This means that even as the force grows indefinitely, the end-to-end length of the polymer cannot exceed the contour length, nb. Thus, the FJC overcomes one of the most critical limitations of the ideal chain. When the value of x is small, the slope of $\mathcal{L}(X)$ approaches 1/3. In this limiting case we can approximate $\mathcal{L}(x)$ to be $x/3$ and

$$\langle R_z \rangle \approx nb\frac{\kappa}{3} = \frac{F_z nb^2}{3k_B T}, \tag{7.80}$$

which is that obtained for the ideal chain in Equation 7.61. When the extension (and force) are very small, the FJC model predicts the same force–displacement behavior as the ideal chain.

7.6 WORM-LIKE CHAIN (WLC)

The WLC incorporates energetic effects of bending

In both the ideal chain and the FJC, we modeled a polymer as rigid segments connected by freely rotating hinges. With this entirely entropic approach, we ignored energetic costs from changes in orientations between segments. In this section, we present the WLC, which incorporates both energetic and entropic effects associated with bending. In the WLC, polymers are modeled as continuous space curves rather than discrete segments.

A schematic demonstrating the relevant quantities for the WLC can be seen in Figure 7.19. The configuration of the space curve is given by the vector-valued function $\mathbf{a}(s)$. Similar to our use of s in defining the persistence length, here it runs along the polymer from 0 to the contour length L and is a parameterization by which each position on the polymer can be identified. The vector $\mathbf{a}(s)$ extends from the origin of the coordinate system and ends at some point along the polymer given by s. For simplicity, the WLC is constrained to be inextensible (that is, the contour length cannot change), which is enforced by setting

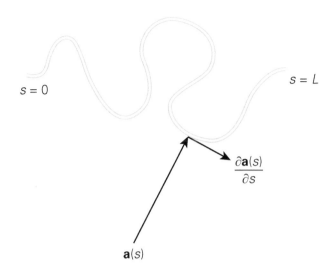

Figure 7.19 Parameterization of the space curve in the WLC. At each position s along the curve, **a**(s) is the position vector and its partial derivative is the tangent vector.

the tangent vector (the first derivative of **a**(s)) to have a unit magnitude for all values of s

$$\left(\frac{\partial \mathbf{a}(s)}{\partial s}\right)^2 = 1. \tag{7.81}$$

Alternatively, the tangent vector $\partial \mathbf{a}(s)/\partial s$ is a unit vector.

We now specify the constitutive relations for the model. In the WLC, the chain is assumed to resist bending deformation through increases in potential energy with curvature similar to an elastic beam. In Section 3.2, we found the bending energy for a beam is given by

$$Q(\mathbf{a}(s)) = \frac{EI}{2} \int_0^L \left(\frac{\partial^2 \mathbf{a}(s)}{\partial s^2}\right)^2 ds, \tag{7.82}$$

where the term in parentheses is a vector that describes the local curvature. We can express Equation 7.82 in terms of persistence length. In particular, in Section 7.3 we showed that for a curvy elastic beam subject to thermal fluctuations, its persistence length is related to its flexural rigidity as $\ell_p = EI/k_B T$. We can recast Equation 7.82 in terms of persistence length as

$$Q(\mathbf{a}(s)) = \frac{k_B T \ell_p}{2} \int_0^L \left(\frac{\partial^2 \mathbf{a}(s)}{\partial s^2}\right)^2 ds. \tag{7.83}$$

Example 7.2 DNA looping and *lac* repressor

Within the bacteria *Escherichia coli*, the genes *lacZ*, *lacY*, and *lacA* encode for enzymes that are involved in metabolizing lactose. Near the *lac* genes are two binding sites for a protein called "*lac* repressor." The binding sites are separated by a fragment of DNA with a length known as the operator distance. The binding sites must be brought together by forming a DNA loop for *lac* repressor to bind. Once bound, *lac* repressor blocks the path of RNA polymerase, preventing the transcription of *lacZ*, *lacY*, and *lacA*.

Experiments show that modifying the operator distance by inserting different-sized fragments of DNA between the *lac* repressor binding sites alters repression of the lac enzymes, with the repression of the lac enzymes peaking when the operator distance is about 70 base pairs long, and decreases when the operator distance becomes larger or smaller than this value. Calculate the probability of looping for very short and very long operator distances. Speculate why this behavior might occur.

We first calculate the probability of looping for very short operator distances. We know that the smaller the contour length relative to its persistence length, the more we must account for energetic influences. We use the WLC model. Assuming an operator distance of L loops into a circle of radius R, the energy of looping for a WLC is

$$Q_{loop} = \frac{k_B T \ell_p}{2} \int_0^L \left(\frac{1}{R}\right)^2 ds = \frac{k_B T \ell_p}{2} \frac{L}{R^2} = \frac{2\pi^2 k_B T \ell_p}{L}, \quad (7.84)$$

where in the right-most relation we made use of the fact that $R = L/2\pi$. We know from the Boltzmann distribution that the higher the value for Q_{loop}, the lower the probability of looping, p_{loop}. Thus, $p_{loop} \to 0$ as $L \to 0$.

Now let us compute the probability for looping for very long operator distances. As the contour length becomes longer and longer, the polymer begins to resemble a chain of independently fluctuating segments (each segment having length on the order of ℓ_p), and the free energy becomes dominated by the entropy. Taking this into account, for very long operator distances, we model the polymer as an ideal chain. The probability distribution for an ideal chain is given by Equation 7.38. A loop occurs when $\mathbf{R} = 0$, and so

$$p_{loop} = \left(\frac{3}{2\pi n b^2}\right)^{3/2} = \left(\frac{3}{2\pi L b}\right)^{3/2}, \quad (7.85)$$

where we made use of the fact in the right-most relation that $L = nb$. In this case, $p_{loop} \to 0$ as $L \to \infty$.

Our models predict that the probability of looping goes to zero for very short and long operator distances. In both cases, repression would be expected to decrease (causing *lac* mRNA levels to increase).

The force–displacement relation for the WLC can be found by the canonical ensemble

We now seek to determine a force–displacement relation for the WLC. We know that for a polymer immersed in a constant temperature heat bath, the probability of a microstate with a given internal energy is described by the Boltzmann distribution. To calculate the force generated by extension, we construct a system with a known internal energy such that we can use Boltzmann's distribution to find the probability of a particular polymer configuration at equilibrium.

Similar to our approach for the FJC, we consider a system consisting of a WLC immersed within a constant temperature heat bath, with one end of the chain fixed in space by a freely rotating hinge, and the other end attached to a small weight that produces a downward force of F_z. As the chain is extended, the loss in potential energy due to the decrease in height of the weight is $F_z R_z$. For the FJC, we were able to calculate R_z as a sum over chain segments. For the WLC, which is continuous curve in space, we can compute this quantity as an integral

$$R_z(\mathbf{a}(s)) = \int_0^L \frac{\partial \mathbf{a}(s)}{\partial s} \cdot \mathbf{e}_z ds, \quad (7.86)$$

where $\partial \mathbf{a}(s)/\partial s$ is the tangent vector, and the dot product with $\mathbf{e}_z$ is used to obtain the z component of the tangent vector. The total internal energy of the system for a given chain configuration of $\mathbf{a}(s)$ can be computed as

$$Q_{tot}(\mathbf{a}(s)) = Q(\mathbf{a}(s)) - F_z R_z(\mathbf{a}(s)) = \frac{k_B T \ell_p}{2} \int_0^L \left(\frac{\partial^2 \mathbf{a}(s)}{\partial s^2}\right)^2 ds - F_z \int_0^L \frac{\partial \mathbf{a}(s)}{\partial s} \cdot \mathbf{e}_z ds. \quad (7.87)$$

Now that we have the internal energy of the system for each configuration $\mathbf{a}(s)$, we can compute the probability of each curve using the Boltzmann distribution as

$$p(\mathbf{a}(s)) = \frac{1}{Z} e^{-Q_{tot}(\mathbf{a}(s))/k_B T}, \quad (7.88)$$

where

$$Z = \int_{\forall \mathbf{a}} e^{-Q_{\text{tot}}(\mathbf{a}(s))/k_B T}\, d\mathbf{a}. \qquad (7.89)$$

In principle, we can compute the average extension under the applied force as

$$\langle R_z \rangle = \int_{\forall \mathbf{a}} p(\mathbf{a}(s)) R_z \, d\mathbf{a} \qquad (7.90)$$

and relate the average extension to the partition function as

$$\langle R_z \rangle = k_B T \frac{\partial \ln Z}{\partial F_z}. \qquad (7.91)$$

However, determining analytical solutions to these integrals is a lot easier said than done. They both require the integration over all possible polymer configurations, with each configuration being a curve in space. In fact, an analytical result for the force–extension relation of the WLC does not exist, except in special limiting cases. However, it has been investigated using computational models, with the results described by the following interpolation equation

$$F_z = \frac{k_B T}{\ell_p}\left(\frac{1}{4}\left(1 - \frac{\langle R_z \rangle}{L}\right)^{-2} - \frac{1}{4} + \frac{\langle R_z \rangle}{L}\right). \qquad (7.92)$$

Differences in the WLC and FJC emerge when they are fitted to experimental data for DNA

The differences in the force–extension behavior of the FJC and the WLC can be demonstrated by fitting the two models to force–extension curves for DNA, such as in Figure 7.20. The measurements were made by attaching one end of a DNA molecule to a glass surface and the other end to a magnetic bead. The bead was pulled with a known force F_z, and the extension R_z was measured optically. The data were fitted to the WLC and FJC using the relation $b = 2\ell_p = 106$ nm. Upon inspecting the model fits, there are two noticeable trends. First, at large forces, the extension asymptotically approaches the contour length. Second, the experimental

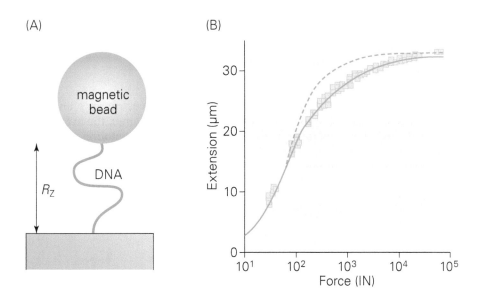

(A)

magnetic bead

DNA

R_Z

(B)

Extension (μm)

Force (IN)

Figure 7.20 Comparison of the FJC and WLC. The force–extension behavior of DNA can be obtained by applying a force to a magnetic bead tethered by DNA, and measuring the displacement of the bead optically (A). Data obtained from such an experiment can be seen in the adjacent plot (B). The experimental data (squares) were fitted to a WLC (solid line) and a FJC (dashed line) assuming a persistence length of 106 nm and a contour length of 33 μm. At large forces, the extension asymptotes at a value equal to the contour length. The experimental data are better fit by the WLC at higher forces. (B, from, Bustamante C, Marko JF, Siggia ED & Smith S (1994) *Science*. Reprinted with permission from AAAS.)

data is fit better by the WLC, especially at forces greater than approximately 0.1 pN. The WLC predicts that more force is required to extend the polymer to a given length than the FJC. The underlying reason for this can be understood by considering the thermal fluctuations in a polymer as waves of different wavelengths superimposed on top of each other. In the FJC, wave-like fluctuations are constrained to be of wavelength b or higher, because they cannot be shorter than the length of each link. This constraint is not present in the WLC. Additional force is required to smooth out these short wavelength fluctuations present in the WLC, but not the FJC.

Persistence length is related to Kuhn length

In this section, we quantify the $\langle \mathbf{R}^2 \rangle$ behavior of the WLC under thermal fluctuations. We shall see that by comparing this behavior to that of discrete chain models, we will be able to relate persistence length to Kuhn length, in other words the origins of Equation 7.60.

Let us start by defining the polymer end-to-end vector $\mathbf{R}$ for the WLC. We do this by integrating the tangent vector $\partial \mathbf{a}(s)/\partial s$ over its length as

$$\mathbf{R} = \int_0^L \frac{\partial \mathbf{a}(s)}{\partial s} ds. \tag{7.93}$$

Using Equation 7.93, we can now compute $\langle \mathbf{R}^2 \rangle$ as

$$\langle \mathbf{R}^2 \rangle = \langle \mathbf{R} \cdot \mathbf{R} \rangle = \left\langle \int_0^L \frac{\partial \mathbf{a}(s)}{\partial s} ds \cdot \int_0^L \frac{\partial \mathbf{a}(s')}{\partial s'} ds' \right\rangle$$

$$= \left\langle \int_0^L \int_0^L \frac{\partial \mathbf{a}(s)}{\partial s} ds \cdot \frac{\partial \mathbf{a}(s')}{\partial s'} ds' \right\rangle \tag{7.94}$$

$$= \int_0^L \int_0^L \left\langle \frac{\partial \mathbf{a}(s)}{\partial s} \cdot \frac{\partial \mathbf{a}(s')}{\partial s'} \right\rangle ds ds'.$$

Recall that in our parameterization of the WLC, the tangent vector $\partial \mathbf{a}(s)/\partial s$ was constrained to be a unit vector. Therefore the magnitude of the product in the brackets is one and Equation 7.94 can be written as

$$\langle \mathbf{R}^2 \rangle = \int_0^L \int_0^L \langle \cos \Delta \theta_{s-s'} \rangle ds ds', \tag{7.95}$$

where $\Delta \theta_{s-s'}$ denotes the angle between the tangent vectors at s and s'. In Section 7.3 we found that for a thermally fluctuating polymer, the average difference in angle between tangent vectors is given by the orientation correlation function, which decays exponentially such that $\langle \cos(\Delta \theta(s)) \rangle = e^{-s/\ell_\mathrm{p}}$. We assume that on average, the difference in angle between any two points on the chain will decay similarly as the distance between the two point is decreased, or in other words,

$$\langle \mathbf{R}^2 \rangle = \int_0^L \int_0^L e^{-(s-s')/\ell_\mathrm{p}} ds ds'. \tag{7.96}$$

It is left as an exercise to show that the solution to the double integral in Equation 7.96 is

$$\left\langle \mathbf{R}^2 \right\rangle = 2\ell_p^2 \left(e^{-\frac{L}{\ell_p}} - 1 + \frac{L}{\ell_p} \right). \tag{7.97}$$

Consider the case where the contour length is much greater than the persistence length, $L \gg \ell_p$ in Equation 7.98. The exponential term in the brackets goes to zero, and assuming the L/ℓ_p term is much greater than one, then

$$\left\langle \mathbf{R}^2 \right\rangle \approx 2\ell_p^2 \frac{L}{\ell_p} = 2L\ell_p. \tag{7.98}$$

If we compare this quantity to the mean squared end-to-end length of nb^2 for the FJC, we obtain a relation between Kuhn length and persistence length,

$$2L\ell_p = nb^2 = (nb)b = Lb. \tag{7.99}$$

Equation 7.100 can be simplified to

$$b = 2\ell_p, \tag{7.100}$$

which shows the origins of Equation 7.60. Equation 7.100 implies that a discrete chain with Kuhn length b has a persistence length of $b/2$ in the limit of $b \ll L$.

Key Concepts

- The cellular cytoskeleton primarily consists of three biopolymers: MFs, MTs, and intermediate filaments. The subunits for MFs and MTs are actin monomers and tubulin dimers, respectively. Intermediate filaments can be composed of different proteins depending on their location within the body.

- There is a highly dynamic balance between the G and F forms of actin. Owing to G-actin's polarity and the ability for bound ATP to undergo hydrolysis, polymerization kinetics can be very different at the (+) and (–) ends. Treadmilling occurs when D-form subunits are lost from the shrinking (–) end, become T-form subunits, and are recycled to the growing (+) end.

- The persistence length gives a characteristic length scale over which the orientations of a thermally undulating polymer become mostly uncorrelated. A polymer can be classified as being flexible, semi-flexible, or stiff depending on the relation of its persistence length to its contour length

- In the ideal chain, all conformational changes in internal energy are ignored. It is often used for modeling flexible polymers whose behavior is dominated by entropy. The probability distribution function for the ideal chain is based on the Gaussian approximation to the random walk.

- For thermodynamic systems, force is equivalent to the gradient of free energy. Separating the ends of an ideal chain reduces the entropy, and thus requires force.

- The mechanical behavior of polymers will tend toward that of an ideal chain when the contour length is significantly greater than the persistence length.

- The FJC addresses a key limitation of the ideal chain, which is unphysical behavior for large forces. The FJC is similar to the ideal chain at low extensions, but gives more realistic behavior at long extensions.

- The WLC does not have joints but rather treats the polymer like a flexible beam with elastic energy. Compared with the FJC, the WLC predicts that more force is required to extend the polymer a given length.

Problems

1. Using the one-dimensional Gaussian approximation to the random walk and step size of $b = 1$, show that the root mean square displacement $\langle R^2 \rangle^{1/2}$, is the square root of the number of steps taken. You may use integral identities.

2. Compare the root mean square end-to-end distance $\langle R^2 \rangle^{1/2}$ in three dimensions of strands of spectrin, actin, and MTs ($\ell_p = 15$, 15×10^3, and 2×10^6 nm, respectively) with 200-nm contour length (L).

3. Show that in three dimensions $\langle \cos \theta(s) \rangle = e^{\left(\frac{-s}{2\ell_p} \right)}$.

4. What is the force generated in polymers of spectrin, actin, and tubulin ($\ell_p = 15$, 15×10^3, and 2×10^6 nm, respectively) with a 100-nm contour length predicted by the ideal chain model when the polymer is extended to 50 nm? And to 150 nm? Why is this not realistic? Assume a temperature of 300 K in your calculations.

5. Assuming a persistence length for DNA of 50 nm, determine the force at which the ideal chain and the FJC differ by 10% in the extension length, assuming they are used to model the same strand of DNA with a 1-μm contour length. Which model predicts a longer polymer at this force? Assume a temperature of 300 K.

6. Show that $\langle \theta^2 \rangle = \int\limits_0^{2\pi} \int\limits_0^{\pi} \theta^2 p(U) \mathrm{d}\phi \sin\theta \mathrm{d}\theta = \dfrac{2 k_B T_s}{EI}$

   assuming that the probability follows a Boltzmann distribution and the energy is the strain energy of a beam bent to an angle θ. Hint: the integral you obtain is quite challenging. Rather than attacking it directly, try the mathematical trick we used for the FJC and WLC. Namely, show that $\langle \theta^2 \rangle$ can be expressed in terms of the derivative of $\ln(Z)$ with respect to E.

7. What is the force generated in polymers of spectrin, actin, and tubulin ($\ell_p = 15$, 15×10^3, and 2×10^6 nm, respectively) with a 100-nm contour length predicted by the FJC model when the polymer is extended to 50 nm? And to 150 nm? How is this more realistic than the Ideal chain model? Hint: the relationship between force and displacement for the FJC model is implicit (you cannot plug in a value of $\langle R \rangle$ and calculate F.) Instead, solve it numerically to two significant figures by guessing values of F and interpolating (or a more sophisticated approach if you like). You can use MATLAB or another programming approach. Alternatively, plot the force–displacement relationship and estimate the point from the graph. Comment on your results.

8. What is the slope of the Langevin function, $\mathcal{L}$, near zero?

9. What is the force generated in polymers of spectrin, actin, and tubulin ($\ell_p = 15$, 15×10^3, and 2×10^6 nm, respectively) with a 100-nm contour length predicted by the WLC model when the polymer is extended to 50 nm?

10. What is the effective (tangent) spring constant as a function of average displacement for a flexible polymer using the WLC model? What is the approximation in the limit as displacement approaches zero? And when displacement approaches the contour length? Hint: the tangent stiffness is the slope of the force–displacement curve.

11. What is the effective (secant) spring constant as a function of average displacement for a flexible polymer using the WLC model? What is the approximation in the limit as displacement approaches zero? And when displacement approaches the contour length? Hint: the secant stiffness is the slope of a line from the origin to the point on the force–displacement curve.

12. Assume a persistence length, $\ell_p = 15$, 15×10^3, and 2×10^6 nm for spectrin, actin, and MTs, respectively. For filaments 1-cm long at $T = 300$ K, what is the effective spring stiffness (tangent) at zero displacement and when fully extended?

13. Consider a 30-μm length of DNA such as might be found in a virus. What force is required to stretch the DNA to an end-to-end displacement $x = 10$, 20, and 25 μm at 300 K using first a FJC model and secondly a WLC model? Assume $\ell_p = 50$ nm. Comment on the results of the two models.

14. Conduct a numerical comparison of the FJC and WLC models. Plot the force–extension behavior for a 1-cm long polymer with a 0.1-mm persistence length with each model. Is one always higher or lower than the other? Does this make sense? Comment on how the two compare in the short and long limits. Repeat with a 1-cm persistence length and a 10-cm persistence length. How does persistence length affect how the models compare?

15. Show that the solution of $\langle \mathbf{R}^2 \rangle = \int\limits_0^L \int\limits_0^L e^{-(s-s')/\ell_p} \mathrm{d}s \mathrm{d}s'$ is

    $$2 l_p^2 \left(e^{-\frac{L}{l_p}} - 1 + \frac{L}{l_p} \right).$$

Annotated References

Boal D (2001) Mechanics of the Cell. Cambridge University Press. *An excellent text on cell mechanics with many in-depth treatments of the polymer models covered in this chapter. Parts of the relation of persistence length to Kuhn length were based on developments in this text.*

Bustamante C, Marko JF, Siggia ED & Smith S (1994) Entropic elasticity of l-phage DNA. *Science* 265, 1599– 1600. *This paper gives the force extension behavior for the WLC and shows the differences in the WLC and freely jointed chain when fit to force–extension behavior for DNA.*

Howard J (2001) Mechanics of Motor Proteins and the Cytoskeleton. Sinauer Associates. *A great introduction to cytoskeletal polymerization and mechanics. Students interested in learning more about polymerization kinetics and force generation by polymerization are referred to this text. The derivation of persistence length, the relation of persistence length to flexural rigidity, as well as some parts of the relation of persistence length to Kuhn length were based on the treatments of these topics in this text.*

Muller J, Oehler S & Muller-Hill B. (1996) Repression of lac promoter as a function of distance, phase, and quality of an auxillary lac operator. *J. Molec. Biol.* 267, 21–29. *This article shows dependence of lac promoter repression on operator distances.*

Omary MB, Coulombe PA & McLean WH (2004) Intermediate filaments and their associated diseases. *N. Engl. J. Med.* 351, 2087–2100. *A nice review of intermediate filament biology. Table 1 in this chapter was obtained from this study.*

Phillips R, Kondev J & Theriot J (2009) Physical Biology of the Cell. Garland Science. *This textbook gives a more in-depth treatment of the lac repressor example as well as many interesting applications of the polymer models developed in this chapter.*

Rubenstein M & Colby RH (2003) Polymer Physics. Oxford University Press. *An excellent introduction to polymer physics. Derivation for the Gaussian approximation to the random walk is based on the treatment developed in this text.*

Spakowtiz AJ (2008) Polymer Physics (lecture notes, from course number ChemEng 466, Stanford University, Stanford, CA). *Several of the treatments in this chapter were based on excellent course notes developed by Dr. Spakowitz for a graduate course on polymer physics at Stanford University. These treatments include developments for the ideal chain, the freely jointed chain, and the WLC, as well as the lac repressor example.*

CHAPTER 8

Polymer Networks and the Cytoskeleton

In Chapter 7, we explored the mechanics of individual biopolymers. We now turn our attention to the mechanics of biopolymer networks, which are dependent not only on the mechanical behavior of the individual biopolymers within, but also on how these biopolymers are organized into a specific microstructure or architecture. For example, the microstructure of biopolymer networks such as the cytoskeleton can vary dramatically from cell to cell in terms of how the biopolymers are oriented with respect to one another, the number of filaments per unit volume, and how they are cross-linked. This microstructure can also vary substantially between different locations within individual cells. Because of the radical effect that these microstructural variations can have on network mechanics, the relationships between biopolymer network mechanical behavior and microstructural properties has become an extremely active area of research. Indeed, several investigational approaches, including analytical, experimental, and computational approaches have been used to better understand these relationships. In this chapter, we will discuss the use of scaling approaches to relate microstructure to mechanical properties, constitutive models of a class of networks called affine networks, and the mechanics of specific cytoskeletal architectures found *in vivo* as they are related to their mechanobiological function.

8.1 POLYMER NETWORKS

Polymer networks have many degrees of freedom

We learned about approaches for modeling the mechanics of individual polymers in Chapter 7. In principle, these models could be used for mechanical modeling of polymer networks. To model a network of flexible polymers, one could examine each of the polymers within the network, determine its geometry, and explicitly account for each polymer in the network model. Each polymer could be modeled as an ideal chain with some entropic spring constant and initial end-to-end distance. In reality, however, this "discrete" approach is rarely feasible. The reason is that polymer networks (such as within a cell) generally have an enormous number of filaments and an even larger number of degrees of freedom. A typical endothelial cell contains approximately 10 mg/ml of F-actin. Assuming the average filament contains 100 subunits (and assuming an average weight of $\sim 1 \times 10^{-14}$ mg per filament), and the cell volume is on the order of 1×10^{-8} ml (for a polygonal cell 5 μm high, 50 μm long, and 40 μm wide), this would mean that a cell contains approximately 10 million actin filaments. If we use a simplified mechanical model in which the state of each filament could adequately be described by its center of mass position (x, y, z), orientation (θ, α), and end-to-end length (L), explicitly modeling each polymer would involve 60 million degrees of freedom.

Effective continuums can be used to model polymer networks

To model polymer networks with large numbers of degrees of freedom, several approaches can be taken. One could try to use high-performance computing to explicitly account for all the individual polymers in a given network. However, in most cases, the numbers of degrees of freedom in physiological networks are too large even for the world's most powerful supercomputers.

An alternative approach is to represent the network as an *effective continuum* to reduce the number of degrees of freedom. What we mean by an effective continuum is that we equate the mechanical contributions of all the discrete polymers within the network to the mechanical behavior of some equivalent continuum. Then, the mechanical state of the network can be described through the analytical framework of continuum mechanics.

Consider an imaginary cube-shaped polymer network. The length of each side of the network is L. The network is filled with randomly aligned polymers that are cross-linked to one another (Figure 8.1). We apply a very small uniaxial deformation to the network in the x-direction by fixing one side of the cube so that it does not translate in the x-direction, and subjecting the opposite side of the cube to a uniaxial normal stress σ. The lengths in the y- and z-directions are assumed to be unconstrained, such that the network can still expand in the y- and z-directions. Upon deformation, the new length in the x-direction is L_1.

Next, we replace the polymer network with an isotropic continuum with Young's modulus $E = L\sigma/(L - L_1)$. The deformation of the entire network under stress σ can be described by a single homogeneous strain of $(L - L_1)/L$. This simple example demonstrates that by assigning appropriate constitutive behavior to the continuum, we are able to capture the mechanical behavior of the network without explicitly modeling each individual filament. In the following sections, our overall focus will be to form effective continuum models for the mechanical behavior of

Figure 8.1 Effective continua can be used to model the mechanical behavior of polymer networks.
(A) An imaginary cube-shaped polymer network with undeformed side length L is subjected to a small uniaxial deformation by fixing one side and subjecting the opposite side to a uniaxial stress σ, resulting in a new length in the x-direction of L_1. The lengths in the y- and z-directions are unconstrained. (B) "Replacement" of the network with a fictitious isotropic homogeneous linear elastic continuum with Young's modulus $E = L\sigma/(L - L_1)$. The deformation to the entire network is described by a single homogeneous strain, $\varepsilon = (L - L_1)/L = E/\sigma$.

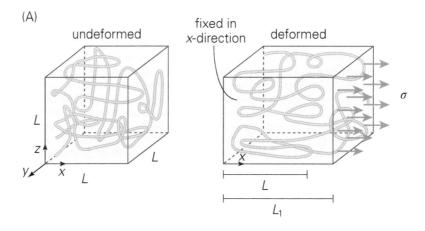

(A)

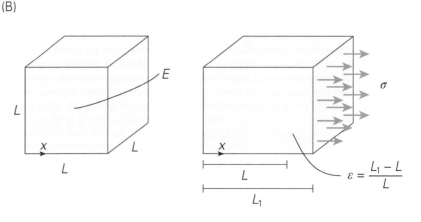

(B)

polymer networks. In doing so, we will seek to identify relationships between the effective mechanical behavior of networks and their microscopic properties, such as polymer stiffness, length, density, orientations, and other properties.

8.2 SCALING APPROACHES

We begin our discussion of effective continuum behavior for polymer networks by deriving *scaling relationships* between effective network stiffness and microstructural properties. Scaling approaches can be highly useful for making predictions of how the effective mechanical properties of a network may change with, for example, the number of filaments within the network per unit volume. However, these simplified approaches generally do not yield an explicit constitutive model. Although this may seem somewhat unsatisfying given the analytical tools we have developed thus far, such scaling relationships are easier to formulate than constitutive models, and may be useful for a variety of situations, such as interpreting experimental results during investigations, involving bead twisting/pulling, and in atomic force microscopy experiments.

Cellular solids theory implies scaling relationships between effective mechanical properties and network volume fraction

We begin our discussion of scaling approaches with the theory of *cellular solids*. In this theory (also referred to as "open foams"), one seeks to develop a model of network structure, and then use scaling arguments to find the dependence of effective elastic moduli on network volume fraction, which is the fraction of the volume of the network taken up by polymer material. For a network with a volume fraction of 0.10, 10% of its volume contains polymer, and the other 90% of its volume is empty space. In the cellular solids approach, we first construct a "unit cell" that is a representative structural unit of our network. This unit cell is assumed to possess geometric scaling such that results obtained from the analysis of the subunit are valid when scaled up to the size of an entire cell. We then find the amount of strain the unit cell would experience under an applied stress, and use this quantity to find the effective elastic modulus. In carrying out these calculations, we use scaling arguments, as we are only interested in simple relationships (if the volume fraction of the network doubles, what happens to the apparent stiffness of the network?). We will ignore all but the most basic parameters and drop constants to not vary with the dependent variables. In our analysis, we will perform calculations for two different cases: when the network deformation occurs exclusively because of polymer bending, and when the deformation occurs exclusively because of axial deformation of the polymers.

Bending-dominated deformation results in a nonlinear scaling of the elastic modulus with volume fraction

We start with a unit cell that is a representative subunit of our network and constructed such that compression of the network results in bending (Figure 8.2). The length of the beam members making up the sides of the cube in the center of the unit cell is assumed to be L, and the beams themselves are assumed to have a characteristic radius R. In this configuration, adjoining cells are coupled such that transmission of loads occurs at member mid-spans, resulting in a bending moment.

We begin our analysis by recalling Equation 3.31, which describes the deflection of a cantilevered beam,

$$w = \frac{F}{EI}\left(\frac{x^3}{6} - L\frac{x^2}{2}\right).$$

Figure 8.2 Unit cell undergoing bending deformations. (A) In the absence of force, the beam members of the unit cell have undeformed length L and radius R. (B) In the presence of force, bending of cross members occurs owing to transmission of loads at member mid-spans, resulting in a bending moment.

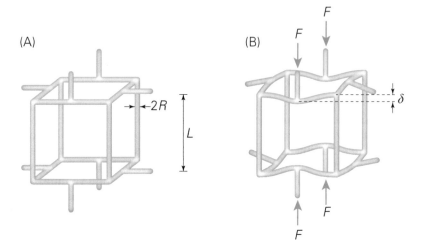

The displacement at the tip $(x = L)$ is,

$$w = \frac{-FL^3}{3EI}.$$

This relation can be used as the basis of the scaling relationship for the mid-span displacement in our unit cell, δ, because

$$\delta \sim \frac{FL^3}{E_b R^4} \sim \frac{FL^3}{E_b R^4}, \tag{8.1}$$

where F is some load applied to the subunit, E_b is the elastic modulus of the beam, and the moment of inertia, I, scales as R^4. We have omitted many constants from this expression, but those are fixed values with which we are not concerned. We want to see how the two parameters are related, not an actual magnitude. We can rearrange Equation 8.1 to solve for force as

$$F \sim \frac{E_b R^4 \delta}{L^3}. \tag{8.2}$$

We now wish to calculate the apparent stress applied to the unit cell and the apparent strain that the unit cell experiences. To do this, we consider the case in which surrounding the unit cell is a hypothetical "cube" (**Figure 8.3**). The length

Figure 8.3 Unit cell undergoing axial deformations. (A) The coupling between unit cells is modified slightly such that transmission of force results in axial deformations of beam members. (B) In the presence of force, vertical beams undergo a change in length.

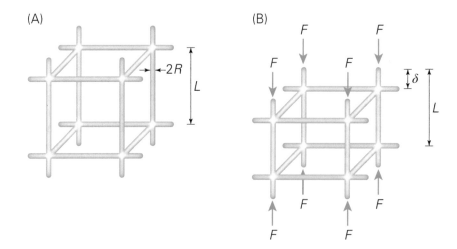

of each side of the cube is L. The apparent stress is the total force applied across a surface of the cube divided by the area L^2, and the apparent strain is the total deflection of the cube divided by the side length L. This apparent stress can be found by dividing the force in Equation 8.2 by the area of the face of the unit cell over which the force is acting,

$$\sigma \sim \frac{F}{L^2} \sim \frac{E_b R^4 \delta}{L^5}. \tag{8.3}$$

The apparent strain can be found as the displacement δ divided by the undeformed length of the unit cell L,

$$\varepsilon \sim \frac{\delta}{L}. \tag{8.4}$$

Using Equations 8.3 and 8.4, the apparent elastic modulus of the unit cell can be found by dividing the apparent stress by the apparent strain,

$$E \sim \frac{\sigma}{\varepsilon} \sim E_b \frac{R^4}{L^4}. \tag{8.5}$$

Recall that volume fraction is the fraction of the volume taken up by polymer material. We can recast Equation 8.5 in terms of this quantity, which we will denote as ρ_{vol}. For the unit cell, the volume of the beams scale as $\sim LR^2$, whereas the volume of the unit cell scales as L^3. In this case,

$$\rho_{\text{vol}} \sim \frac{LR^2}{L^3} \sim \frac{R^2}{L^2}. \tag{8.6}$$

Combining Equations 8.5 and 8.6 yields the final expression

$$E \sim E_b \rho_{\text{vol}}^2. \tag{8.7}$$

Equation 8.7 implies that when deformation of the cellular solid occurs due to bending, the apparent macroscopic modulus scales linearly with the polymer modulus but quadratically with the polymer volume fraction. This is an interesting result, as it suggests that the stiffness of the network is more dependent on changes in the polymer concentration rather than on the stiffness of the polymers themselves.

Deformation dominated by axial strain results in a linear scaling of the elastic modulus with volume fraction

We now perform a similar analysis to find out how the scaling relationship in Equation 8.7 would be altered if the deformation of the cellular solid was assumed to occur exclusively due to axial deformation of the beam members. To do this, we first slightly modify the configuration of the unit cell to that in Figure 8.3. By modifying the coupling, transmits loads from adjoining cells in such a manner that deformation occurs exclusively due to axial strain in the members. It can be shown that the force within each beam scales as

$$F \sim \frac{E_b R^2 \delta}{L}, \tag{8.8}$$

and that the apparent elastic modulus scales as

$$E \sim E_b \rho_{\text{vol}}. \tag{8.9}$$

Experimental measurements of cytoskeletal volume fraction. We have seen that the cellular solid model can predict very different scaling behaviors depending on whether the deformation is assumed to occur due to member bending or axial deformation. One may ask how these two models compare with experimental investigations of the relationship between cell stiffness and cytoskeletal polymer volume fraction, and subsequently, which model is more "correct". Unfortunately, a conclusive answer to this question has yet to be attained, owing in large part to the technical difficulties associated with measuring cytoskeletal volume fraction within a live cell.

The derivations of Equations 8.8 and 8.9 are left as student exercises. Equation 8.9 implies that when deformation of the cellular solid occurs due to axial strain, the effective elastic modulus of the network scales proportionally to both the cytoskeletal modulus and the polymer volume fraction. By assuming a different mechanism of deformation at the microstructural level, we find different dependencies of the network stiffness on the polymer volume fraction.

The stiffness of tensegrity structures scales linearly with member prestress

We conclude this section with a discussion of tensegrity structures. The term *tensegrity*, which is a contraction of tensional integrity, was originally developed by Buckminster Fuller to describe a structural principle that had been invented by Kenneth Snelson. Tensegrity structures are developed from linear structural elements that are designed to withstand either tension only (such as strings or cables) or compression only (such as rigid struts). A critical aspect of a tensegrity structure is that none of the elements experience bending moments. This lack of bending moment is typically achieved architecturally by using freely rotating joints at all connections. The resulting structure derives its stability from *preload* or *prestress* (Figure 8.4).

Donald Ingber first suggested that the cytoskeleton might function as a tensegrity structure. In this model, large-diameter microtubules act as rigid compression elements, and the microfilaments and intermediate filaments act as flexible tensile elements. There are some compelling arguments why a tensegrity structure would be advantageous. Tensegrity structures possess the capacity to deform significantly without large deformations of the individual components. That is, the tension and compression elements rearrange themselves in response to some load, but the individual elements do not need to stretch or compress by as much as the entire structure. Such elemental rearrangements also underlie the capacity for tensegrity structures to translate a deformation in one location into an equally large (or larger) displacement some distance away, termed *action at a distance*. However, a consensus of opinion on whether the cytoskeleton functions as a tensegrity structure has not yet been formed, and this topic remains a controversial one within cell mechanics.

Scaling relationships for tensegrity structures have been formed. Though its derivation is outside the scope of this book, it can be shown that in the existence of a prestress, the effective network elastic modulus scales (under certain conditions) as

Figure 8.4 Tensegrity art. Entitled *Mozart* by Kenneth Snelson, this sculpture is made up of tension-only cables and compression-only struts.

$$E \sim \frac{F_\mathrm{p}\rho_\mathrm{vol}}{r^2}, \qquad\qquad (8.10)$$

where F_p is the contractile force within the members generating the prestress.

Equation 8.10 implies that the stiffness of a tensegrity network scales linearly with the polymer volume fraction, which is perhaps not surprising, given that (1) the elements within tensegrity structures cannot sustain bending, and (2) a similar scaling occurs in the axially loaded cellular solid model. Equation 8.10 also implies that the network stiffness scales linearly with the prestress of the network.

Advanced Material: The controversy over tensegrity

One reason for the controversy over the tensegrity hypothesis is that there is evidence both in support of and against this notion. In contrast to the tensegrity model, it is generally accepted that microtubules and actin filaments cannot be classified as strictly compression- and tension-bearing elements, respectively. Extension of lamellipodia occurs by axial compression of actin fibers, and microtubules in tension are, in part, responsible for pulling chromosomes apart during mitosis. On the other hand, there is evidence supporting the tensegrity hypothesis. Nuclear motion has been observed in response to a small bead being pulled far from the nucleus when engaging certain receptors, but not others. These results are interpreted as supporting the tensegrity model's action-at-a-distance concept. Prestress in actin filaments has been demonstrated using a combination of photobleaching and laser-scalpel technology. The tensegrity hypothesis will likely continue to be controversial for many years to come.

8.3 AFFINE NETWORKS

In the previous section, we focused on deriving simple scaling relationships between effective stiffness and network density. Such relationships can be very useful for obtaining order-of-magnitudes estimates of cytoskeletal stiffness based on experimental measurements of filament density. However, because these relationships only give the scaling behavior of stiffness with polymer density, rather than an explicit expression for network stiffness, they are generally inadequate for mechanical modeling of cells. In this section, we present several formulations for constitutive models of polymer networks. As we will see, such constitutive models can be in the form of a direct relationship between stress and strain, or between strain energy and strain (from which stress/strain relationships can be calculated).

Affine deformations assume the filaments deform as if they are embedded in a continuum

In formulating relationships between network stiffness and microstructural properties, an important consideration is how the filaments deform in response to an applied load. The most common approach is to assume an *affine deformation*. An affine deformation is one that can be described by an *affine transformation*, which consists of some linear transformation (such as rotation, shear, extension, or compression) superimposed with a rigid translation. If we let the position of any point within the undeformed network be given by the vector $\mathbf{A}$, and its position within the deformed network at some time t be given by $\mathbf{a}(\mathbf{A},t)$. An affine deformation is one that can be described in the form

$$\mathbf{a}(\mathbf{A},t) = \mathbf{F}(t)\mathbf{A} + \mathbf{c}(t), \qquad\qquad (8.11)$$

where $\mathbf{F}(t)$ is the linear transformation, and $\mathbf{c}(t)$ is the rigid-body translation. Note that if $\mathbf{F}$ or $\mathbf{c}$ are functions of $\mathbf{A}$, $\mathbf{F} = \mathbf{F}(\mathbf{A}, t)$ or $\mathbf{c} = \mathbf{c}(\mathbf{A}, t)$, then the deformation in Equation 8.11 ceases to be affine. Therefore, affine transformations are sometimes referred to as homogeneous.

Under an affine deformation, the polymers deform as if they are embedded in a continuum. Consider an imaginary cubic network subjected to a uniaxial load, and assume that the filaments undergo a deformation that is affine. For all filament segments oriented similarly between two cross-link points, the stretch is the same. Similarly, at all cross-links in which the two cross-linked polymers are similarly oriented, the change in angle between each cross-linked pair will also be the same. Importantly, the affine assumption greatly simplifies making a constitutive model as it really does not matter how many filaments are in the network, because they are assumed to be all deforming in the same manner by the deformation prescribed by $\mathbf{F}$ and $\mathbf{c}$.

Flexible polymer networks can be modeled using rubber elasticity

In this section, we focus on an approach for analyzing biopolymer networks that originates from the field of rubber elasticity. The microstructures of rubbers are in some ways similar to the cytoskeleton. Rubbers are constituted of polymers that form a cross-linked network structure. The principal difference between rubber and, say, the actin cytoskeleton is that the individual rubber molecules are highly flexible. We learned in Chapter 7 that F-actin's persistence length is on the order of micrometers, similar to its contour length *in vivo*. The mechanics of actin are governed by both energetic and entropic influences. In contrast, rubber molecules have contour lengths much greater than their persistence lengths. The mechanics of individual rubber molecules are entropy-dominated. Owing to this entropy-dominated behavior, rubber-like materials have several unique mechanical properties and exhibit some very interesting behaviors. They are highly deformable and show almost complete recoverability when the load is released. In addition, the stiffness of rubber increases with temperature, which is the opposite of what occurs in most engineering materials.

To begin our analysis of rubber, consider a network of randomly oriented, cross-linked polymers. We assume that the network is shaped like a rectangle, with dimensions L_x, L_y, and L_z. Because rubber networks consist of highly flexible polymers whose mechanical behavior is entropy-dominated, we will assume that the polymers within the network can be modeled as ideal chains. Recall from Section 7.4 and Equation 7.56 that we can write the entropy of the ideal chain as

$$S(\boldsymbol{R}) = -\frac{3k_{\mathrm{B}}\boldsymbol{R}^2}{2nb^2} + S_0, \tag{8.12}$$

where S_0 is a constant.

For a single chain, the change in free energy associated with a change in initial end-to-end vector $\boldsymbol{R} = [X, Y, Z]$ to a new end-to-end vector $\boldsymbol{r} = [x, y, z]$ is

$$\Delta\Psi = -T(S(\boldsymbol{r}) - S(\boldsymbol{R})) = \frac{3k_{\mathrm{B}}T}{2nb^2} + (\boldsymbol{r}^2 - \boldsymbol{R}^2). \tag{8.13}$$

Also recall from Equation 7.30 that the mean square end-to-end length for the ideal chain is $\langle \boldsymbol{R}^2 \rangle = nb^2$. This implies that if the network was synthesized by a sequential process of polymerization followed by cross-linking, then the mean end-to-end length for each chain would be roughly nb^2, assuming that the polymers were given time to equilibrate before cross-linking, and that the density of the polymer solution was low enough that substantial inter-polymer volume exclusion interactions did not occur. Because the chains in the network are randomly oriented, this implies that the mean squared end-to-end distances in the x-, y-, and z-directions must equal each other, and therefore,

$$\langle R_x^2 \rangle = \langle R_y^2 \rangle = \langle R_z^2 \rangle = \frac{\langle \boldsymbol{R}^2 \rangle}{3} = \frac{nb^2}{3}. \tag{8.14}$$

Next, we deform the network such that the lengths in the x-, y-, and z-directions are now l_x, l_y, and l_z. The deformation of the network is given by

$$r_x = \lambda_x R_x$$
$$r_y = \lambda_y R_y \tag{8.15}$$
$$r_z = \lambda_z R_z,$$

and

$$\lambda_x = \frac{l_x}{L_x}, \quad \lambda_y = \frac{l_y}{L_y}, \quad \lambda_z = \frac{l_z}{L_z} \tag{8.16}$$

are the stretch ratios in the x-, y-, and z-directions. In this treatment, we will assume the deformations are homogeneous (and affine). This means that the stretch of the individual chains between cross-links are spatially independent, and can be described by the stretch of the bulk network (or, in other words, by the relations above). Therefore, on average, the mean squared lengths of the chains in the x-, y-, and z-directions after deformation will be

$$\langle r_x^2 \rangle = \lambda_x^2 \langle R_x^2 \rangle$$
$$\langle r_y^2 \rangle = \lambda_y^2 \langle R_y^2 \rangle \tag{8.17}$$
$$\langle r_z^2 \rangle = \lambda_z^2 \langle R_z^2 \rangle.$$

The change in free energy of a single chain will be

$$\begin{aligned}
\langle \Delta\Psi \rangle &= \frac{3k_\mathrm{B}T}{2nb^2}\left(\langle \boldsymbol{r}^2 \rangle - \langle \boldsymbol{R}^2 \rangle \right) \\
&= \frac{3k_\mathrm{B}T}{2nb^2}\left(\langle r_x^2 \rangle + \langle r_y^2 \rangle + \langle r_z^2 \rangle - \langle R_x^2 \rangle + \langle R_y^2 \rangle + \langle R_z^2 \rangle \right) \\
&= \frac{3k_\mathrm{B}T}{2nb^2}\left(\lambda_x^2 \langle R_x^2 \rangle + \lambda_y^2 \langle R_y^2 \rangle + \lambda_z^2 \langle R_z^2 \rangle - \langle R_x^2 \rangle - \langle R_y^2 \rangle - \langle R_z^2 \rangle \right) \\
&= \frac{k_\mathrm{B}T}{2}\left(\lambda_x^2 + \lambda_y^2 + \lambda_z^2 - 3 \right),
\end{aligned} \tag{8.18}$$

where we have used the relation $\langle X^2 \rangle = \langle Y^2 \rangle = \langle Z^2 \rangle = nb^2/3$. The total average change in free energy for the entire network $\langle \Delta\Psi_\mathrm{net} \rangle$ can be computed as the product of the change in free energy for a single polymer, the number of polymers per unit volume ρ_n (which is also called the number density), and the total volume of the network $V = L_x L_y L_z$ as

$$\langle \Delta\Psi_\mathrm{net} \rangle = \frac{\rho_n V k_\mathrm{B}T}{2}\left(\lambda_x^2 + \lambda_y^2 + \lambda_z^2 - 3 \right). \tag{8.19}$$

Now, consider the case of uniaxial stretching in the x-direction. We fix one end, and load the opposite end such that the deformed length in the x-direction is

$$l_x = \lambda L_x. \tag{8.20}$$

In the y- and z-directions, there is contraction due to Poisson's effect. Now, for simplicity, assume that the network as a whole is incompressible. We can calculate the stretch in the y- and z-directions. In particular, if we equate the volumes of the undeformed and deformed networks,

$$L_x L_y L_z = \lambda L_x \lambda_y L_y \lambda_z L_z. \tag{8.21}$$

For Equation 8.21 to hold,

$$\lambda_y = \lambda_z = \frac{1}{\sqrt{\lambda}}. \tag{8.22}$$

Therefore,

$$\Delta \Psi_{\text{net}} = \frac{\rho_n V k_B T}{2} - \left(\lambda^2 + \frac{2}{\lambda} - 3 \right). \tag{8.23}$$

We can find the force in the x-direction as

$$f_x = \frac{\partial \Delta \Psi_{\text{net}}}{\partial l_x} = \rho_n A k_B T \left(\lambda - \frac{1}{\lambda^2} \right), \tag{8.24}$$

where $A = L_y L_z$ is the undeformed cross-sectional area (the proof of this is left as an exercise). Now, we can find the axial stress in the x-direction as

$$\sigma = \frac{f_x}{A} = \rho_n k_B T \left(\lambda - \frac{1}{\lambda^2} \right). \tag{8.25}$$

In the limit of small strains, it can be shown that

$$\left(\lambda - \frac{1}{\lambda^2} \right) \approx 3\varepsilon. \tag{8.26}$$

Combining Equations 8.25 and 8.26, we obtain

$$\sigma = 3\rho_n k_B T \varepsilon. \tag{8.27}$$

The results of our analysis show that if we assume small deformations and that the network is linearly elastic, isotropic, and incompressible, then the Young's modulus of our network is $E = 3\rho_n k_B T$. There are several things worth noting. First, the stiffness of the network scales linearly with the density of filaments, ρ_n. This scaling of stiffness with density is similar to the scaling obtained in Section 8.2, in which we assumed that all deformations in our lattice of beams were axial, but not similar to the scaling obtained when we assumed the deformations occurred because of bending. Perhaps this is not unexpected, because in rubber-like networks, the ideal chains behave as entropic springs, which do not generate force when bent. Second, we can see that the stiffness of the network increases with increasing temperature, unlike most engineering materials. This occurs because in a network of entropic springs, the stiffness of the network is derived entirely from entropic costs of straightening out the individual polymers. In particular, the polymers become straighter and have fewer configurations available. Increasing the temperature increases the cost of this reduction of entropy, and as such the material's elastic modulus effectively increases.

Anisotropic affine networks can be modeled using strain energy approaches

In the last section, we derived the effective Young's modulus for a rubber-like network of flexible polymers. In the derivation, we assumed that the filaments were isotropically aligned (that is, they had no preferred orientation). Further, we assumed that the network was incompressible. However, in some cases, we may wish to relax these assumptions.

An isotropic, linearly elastic material has two independent elastic constants. The Young's modulus and Poisson's ratio. In Section 3.2 we found that, in general form, Hooke's law can be expressed as a matrix equation such that

$$\boldsymbol{\sigma} = \boldsymbol{C}\boldsymbol{\varepsilon} \tag{8.28}$$

where $\boldsymbol{\sigma}$ and $\boldsymbol{\varepsilon}$ are 6×1 vectors. Therefore, in general, $\boldsymbol{C}$ is a 6×6 matrix. However, it turns out that $\boldsymbol{C}$ must also be symmetric, so even for a fully anisotropic linearly elastic material there are only 21 independent elastic moduli,

$$\boldsymbol{C} = \begin{pmatrix} C_{11} & C_{21} & C_{13} & C_{14} & C_{15} & C_{16} \\ C_{21} & C_{22} & C_{23} & C_{24} & C_{25} & C_{26} \\ C_{31} & C_{32} & C_{33} & C_{31} & C_{32} & C_{33} \\ C_{41} & C_{42} & C_{43} & C_{44} & C_{56} & C_{46} \\ C_{51} & C_{52} & C_{53} & C_{54} & C_{55} & C_{56} \\ C_{61} & C_{62} & C_{63} & C_{64} & C_{65} & C_{66} \end{pmatrix}, \tag{8.29}$$

where $C_{ij} = C_{ji}$. In this section, we demonstrate a calculation of these 21 elastic moduli for a network of anisotropically aligned elastic rods undergoing an affine deformation. However, before we do this, we will first discuss computation of elastic moduli from the strain energy density, because this approach will be useful in several developments in following sections.

Elastic moduli can be computed from strain energy density

For linearly elastic materials, the elastic moduli relate stress and strain. For a rod composed of an isotropic linearly elastic material, the Young's modulus E gives the stiffness of an object in resisting a uniaxial stress, and relates the stress $\boldsymbol{\sigma}$ that results from a uniaxial strain $\boldsymbol{\varepsilon}$,

$$\boldsymbol{\sigma} = E\boldsymbol{\varepsilon}. \tag{8.30}$$

In Section 5.1 we introduced the notion of strain energy. For convenience, we denoted the strain energy density as $\mathrm{d}W$, but here we will reserve W for the total strain energy and use a lower-case w for strain energy density. In Equation 5.6 we showed that the strain energy density, or the strain energy per unit undeformed volume, is one-half the component-wise product of stress and strain,

$$w = \frac{\boldsymbol{\sigma}\boldsymbol{\varepsilon}}{2}. \tag{8.31}$$

Combining Equations 8.30 and 8.31, we can see that the strain energy density w can be formulated in terms of E and $\boldsymbol{\varepsilon}$ as

$$w = \frac{1}{2}E\boldsymbol{\varepsilon}^2. \tag{8.32}$$

Equation 8.32 implies that if we know the relationship between strain energy density and strain, we can compute the Young's modulus E for the rod. In particular, we see that E can be calculated by differentiating the strain energy density as

$$\frac{\partial^2 w}{\partial \boldsymbol{\varepsilon}^2} = \frac{\partial^2 \; \tfrac{1}{2} E \boldsymbol{\varepsilon}^2}{\partial \boldsymbol{\varepsilon}^2}$$

$$= \frac{\partial E \boldsymbol{\varepsilon}}{\partial \boldsymbol{\varepsilon}} \tag{8.33}$$

$$= E.$$

In this simple example, we assumed that the rod was composed of an isotropic linearly elastic material. However, it can be shown that this relation holds for fully anisotropic linearly elastic materials as well. Consider a continuum that is subjected to some infinitesimal, homogeneous deformation described by the strain tensor. This deformation results in a stress state described by the corresponding stress tensor. Similar to the approach of determining the Young's modulus by differentiating the strain energy density with respect to strain, it can be shown that, more generally, all 21 independent elastic moduli of $\boldsymbol{C}$ can be computed as

$$\boldsymbol{C} = \begin{pmatrix}
\dfrac{\partial^2 w}{\partial \varepsilon_{xx} \partial \varepsilon_{xx}} & \dfrac{\partial^2 w}{\partial \varepsilon_{xx} \partial \varepsilon_{yy}} & \dfrac{\partial^2 w}{\partial \varepsilon_{xx} \partial \varepsilon_{zz}} & \dfrac{\partial^2 w}{\partial \varepsilon_{xx} \partial \varepsilon_{xy}} & \dfrac{\partial^2 w}{\partial \varepsilon_{xx} \partial \varepsilon_{yz}} & \dfrac{\partial^2 w}{\partial \varepsilon_{xx} \partial \varepsilon_{xz}} \\[2ex]
\dfrac{\partial^2 w}{\partial \varepsilon_{yy} \partial \varepsilon_{xx}} & \dfrac{\partial^2 w}{\partial \varepsilon_{yy} \partial \varepsilon_{yy}} & \dfrac{\partial^2 w}{\partial \varepsilon_{yy} \partial \varepsilon_{zz}} & \dfrac{\partial^2 w}{\partial \varepsilon_{yy} \partial \varepsilon_{xy}} & \dfrac{\partial^2 w}{\partial \varepsilon_{yy} \partial \varepsilon_{yz}} & \dfrac{\partial^2 w}{\partial \varepsilon_{yy} \partial \varepsilon_{xz}} \\[2ex]
\dfrac{\partial^2 w}{\partial \varepsilon_{zz} \partial \varepsilon_{xx}} & \dfrac{\partial^2 w}{\partial \varepsilon_{zz} \partial \varepsilon_{yy}} & \dfrac{\partial^2 w}{\partial \varepsilon_{zz} \partial \varepsilon_{zz}} & \dfrac{\partial^2 w}{\partial \varepsilon_{zz} \partial \varepsilon_{xy}} & \dfrac{\partial^2 w}{\partial \varepsilon_{zz} \partial \varepsilon_{yz}} & \dfrac{\partial^2 w}{\partial \varepsilon_{zz} \partial \varepsilon_{xz}} \\[2ex]
\dfrac{\partial^2 w}{\partial \varepsilon_{xy} \partial \varepsilon_{xx}} & \dfrac{\partial^2 w}{\partial \varepsilon_{xy} \partial \varepsilon_{yy}} & \dfrac{\partial^2 w}{\partial \varepsilon_{xy} \partial \varepsilon_{zz}} & \dfrac{\partial^2 w}{\partial \varepsilon_{xy} \partial \varepsilon_{xy}} & \dfrac{\partial^2 w}{\partial \varepsilon_{xy} \partial \varepsilon_{yz}} & \dfrac{\partial^2 w}{\partial \varepsilon_{xy} \partial \varepsilon_{xz}} \\[2ex]
\dfrac{\partial^2 w}{\partial \varepsilon_{yz} \partial \varepsilon_{xx}} & \dfrac{\partial^2 w}{\partial \varepsilon_{yz} \partial \varepsilon_{yy}} & \dfrac{\partial^2 w}{\partial \varepsilon_{yz} \partial \varepsilon_{zz}} & \dfrac{\partial^2 w}{\partial \varepsilon_{yz} \partial \varepsilon_{xy}} & \dfrac{\partial^2 w}{\partial \varepsilon_{yz} \partial \varepsilon_{yz}} & \dfrac{\partial^2 w}{\partial \varepsilon_{yz} \partial \varepsilon_{xz}} \\[2ex]
\dfrac{\partial^2 w}{\partial \varepsilon_{xz} \partial \varepsilon_{xx}} & \dfrac{\partial^2 w}{\partial \varepsilon_{xy} \partial \varepsilon_{yy}} & \dfrac{\partial^2 w}{\partial \varepsilon_{xz} \partial \varepsilon_{zz}} & \dfrac{\partial^2 w}{\partial \varepsilon_{xz} \partial \varepsilon_{xy}} & \dfrac{\partial^2 w}{\partial \varepsilon_{xz} \partial \varepsilon_{yz}} & \dfrac{\partial^2 w}{\partial \varepsilon_{xz} \partial \varepsilon_{xz}}
\end{pmatrix}. \tag{8.34}$$

Equation 8.34 is a very useful result, as it implies that we can determine the effective elastic moduli of the network as long as we can formulate the functional dependence of strain energy density with strain.

Example 8.1: The cytoskeletal modulus determined from strain energy density

If you examine the cytoplasm of a very thin cell, in a region far from the nucleus, you might see the major filaments of the cytoskeleton lying horizontally, in loose layers. If you are given the strain energy density of such a section as,

$$w = 0.5\boldsymbol{C}_{11}(\varepsilon_{xx}^2 + \varepsilon_{yy}^2) + 0.5\boldsymbol{C}_{33}(\varepsilon_{zz}^2) + \boldsymbol{C}_{12}\varepsilon_{xx}\varepsilon_{yy} + \boldsymbol{C}_{13}\varepsilon_{zz}(\varepsilon_{xx} + \varepsilon_{yy}) + 0.5\boldsymbol{C}_{44}(\varepsilon_{xy}^2 + \varepsilon_{yz}^2) + 0.25(\boldsymbol{C}_{11} - \boldsymbol{C}_{12})\,\varepsilon_{xz}^2,$$

calculate the stress–strain relationship and determine the general type of symmetry, if any, that exists in this model.

Using the tensor from Equation 8.35, we can fill in the components as,

$$\boldsymbol{C} = \begin{pmatrix}
C_{11} & C_{12} & C_{13} & 0 & 0 & 0 \\
C_{12} & C_{11} & C_{13} & 0 & 0 & 0 \\
C_{13} & C_{13} & C_{33} & 0 & 0 & 0 \\
0 & 0 & 0 & C_{44} & 0 & 0 \\
0 & 0 & 0 & 0 & C_{44} & 0 \\
0 & 0 & 0 & 0 & 0 & \dfrac{C_{11} - C_{12}}{2}
\end{pmatrix}$$

This form of a material is associated with *transverse isotropy*, which means that the material has one set of properties in one longitudinal direction and another set in any transverse direction within a plane perpendicular to the longitudinal direction.

Elastic moduli of affine anisotropic networks can be calculated from appropriate strain energy density and angular distribution functions

After discussing the relationship between strain energy density and elastic moduli, we now return our attention to computing the effective elastic moduli for an anisotropic network undergoing an affine deformation. Consider a network of cross-linked elastic rods. For simplicity, we assume that these rods are all cylindrical and of the same length and cross-sectional area. The network is contained within a cube of side length L, with total volume $V_{\text{net}} = L^3$. Cross-links are assumed to be freely rotating (namely that there is no energetic cost for changes in angle between filaments), and deformations are assumed to be affine. Assume that each rod has an orientation that is specified by a unit vector $\boldsymbol{n}$ with components

$$\boldsymbol{n} = \begin{pmatrix} n_x \\ n_y \\ n_z \end{pmatrix}. \tag{8.35}$$

We can construct an angular probability density function, $\omega(\boldsymbol{n})$, which gives the distribution of filaments over the unit sphere (which we will annotate as S^2 based on convention). The angular probability density function has the property that

$$\int_{S^2} \omega(\boldsymbol{n}) \mathrm{d}S = 1, \tag{8.36}$$

which normalizes the total probability to be one. Next, consider our network under an imposed load. Recall that the deformation is assumed to be affine. The rods will freely rotate relative to one another and undergo some degree of stretch. For an elastic rod with Young's modulus E_{rod} and volume V_{rod}, the total strain energy in deforming the rod by stretching it by a factor of λ is

$$W_{\text{rod}}(\lambda) = \frac{V_{\text{rod}} E_{\text{rod}}}{2} (\lambda - 1)^2. \tag{8.37}$$

The degree to which each filament undergoes axial deformation will depend on the network deformation and the orientation of the filament. In the case of small deformations, the stretch for a filament oriented in the direction $\boldsymbol{n}$ under an imposed strain of $\boldsymbol{\varepsilon}$ is approximately

$$\lambda(\boldsymbol{\varepsilon}, \boldsymbol{n}) = \boldsymbol{n}^T \boldsymbol{\varepsilon} \boldsymbol{n}. \tag{8.38}$$

For a network with rods oriented with angular probability density $\omega(\boldsymbol{n})$, the total strain energy of the network is

$$W_{\text{net}}(\boldsymbol{\varepsilon}) = \int_{S^2} W_{\text{rod}}(\lambda(\boldsymbol{\varepsilon}, \boldsymbol{n})) \omega(\boldsymbol{n}) N \mathrm{d}S = \frac{V_{\text{rod}} E_{\text{rod}} N}{2} \int_{S^2} (\boldsymbol{n}^T \boldsymbol{\varepsilon} \boldsymbol{n} - 1)^2 \omega(\boldsymbol{n}) \mathrm{d}S, \tag{8.39}$$

where N is the total number of rods. The strain energy density can be calculated by dividing the strain energy by the undeformed volume

$$W_{\text{net}}(\boldsymbol{\varepsilon}) = \frac{\rho_{\text{vol}} E_{\text{rod}}}{2} \int_{S^2} (\boldsymbol{n}^T \boldsymbol{\varepsilon} \boldsymbol{n} - 1)^2 \omega(\boldsymbol{n}) \mathrm{d}S, \tag{8.40}$$

where $\rho_{\mathrm{vol}} = N_{\mathrm{rod}} V_{\mathrm{rod}} / V_{\mathrm{net}}$ is the volume fraction of rods within the network. Equation 8.40 gives the dependence of strain energy density with strain. Using this expression, each of the 21 independent elastic moduli can be computed using Equation 8.34. For most angular distributions, the integral and derivatives in the above expression are difficult to solve analytically but can be solved computationally using numerical methods. Upon inspection of Equation 8.40, it is noteworthy that the elastic moduli will have a linear dependence on polymer density, similar to the linear dependence on density observed in our analysis of rubber-like networks and our analysis of cellular solids in which we assumed deformation was stretching-dominated.

8.4 BIOMECHANICAL FUNCTION AND CYTOSKELETAL STRUCTURE

Up to this point, our focus has been on the formulation of relationships between effective mechanical behavior and microstructural network properties. In this section, we shift our focus to analyzing the mechanics of specific cytoskeletal structures, in particular, the red blood cell's cytoskeleton and the cross-linked actin bundles within filopodia. Although the role of the distinct architectures found in these structures in facilitating or regulating biological function is an active area of research, in this section we seek to use relatively simple analyses to explore whether the architectures found in these structures are optimally suited to perform their biological roles.

Filopodia are cross-linked bundles of actin filaments involved in cell motility

We begin our analysis of structures with a discussion of filopodia. During cell migration (a process that will be discussed in detail in Chapter 10), several cytoskeletal alterations occur, particularly within two distinct subregions of the cell: the leading edge (the part of the cell that is moving forward) and the trailing edge (the part which follows the main bulk of the cell). Filopodia are dynamic, cross-linked bundles of actin filaments at the leading edge of crawling cells. Morphologically, they resemble little fingers, and they are thought to act as "feelers" as the cell moves (Figure 8.5). These bundles of actin quickly grow out from the leading edge, pushing on the membrane, typically at rates on the order of ~0.1 μm/s to form rod-like structures, and then retract. The total time for extension and retraction is ~100 s. Filopodia are on the order of ~0.2 μm in diameter and 1–5 μm in length. Within a single filopodium, 20–30 actin filaments are aligned parallel with their barbed ends facing the membrane and cross-linked by the protein *fascin*. At the ends of the filaments are capping proteins, so named because they cap the actin filaments to prevent further growth.

The way in which cells are able to form filopodia and regulate their structure is an active area of investigation. Even fundamental questions about their structure, such as how densely cross-linked the actin filaments are, are unknown. In this section, we explore the role of mechanics in regulating their structure. In particular, we seek to determine the maximum length a filopodium may be before it undergoes mechanical failure. The motivation for this analysis comes from two observations: filopodia are slender structures; and they are generally straight (see Figure 8.5). This implies that filopodia may be structured in a specific manner to resist buckling. We will perform this analysis under two extreme conditions: (1) when no cross-linking is present, or (2) when a high degree of cross-linking is present.

Actin filaments within filopodia can be modeled as elastic beams undergoing buckling

To begin, we first consider whether to treat filipodium actin filaments as energetically or entropically dominated polymers (or both). In Section 7.3, we learned

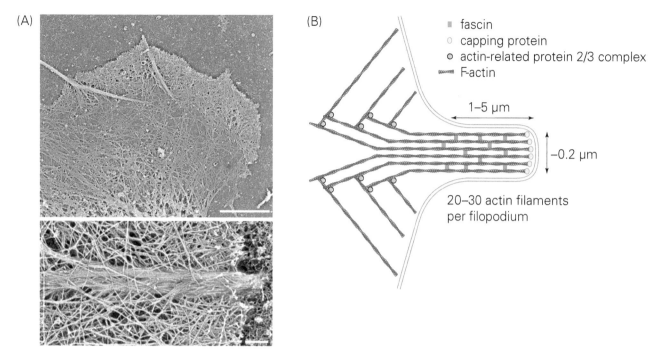

Figure 8.5 Cytoskeletal structure within filopodia. (A, top) Electron micrograph of the actin cytoskeleton within a lamellipodium of a cell. A dense network of actin filaments can be seen. Several filopodia can be seen protruding from the edge. (A, bottom) Close-up of a filopodium. The aligned bundle of actin filaments within the filopodium can be seen. (B) Schematic of several important proteins within filopodium. Note that the actin filaments are bundled together by the cross-linking protein fascin. (A, from, Mejillano MR et al. (2004). *Cell.*)

that the persistence length of F-actin is on the order of ~10 μm. Filopodia are much shorter than this, ~1 μm in length. Because the persistence length of actin is an order of magnitude larger than the average length of the filaments within the filopodia, we neglect entropic contributions and analyze the filaments as elastic beams. Since we are interested in the buckling behavior of filopodia, we seek a relation between buckling loads and beam mechanical properties and geometry. Such a relation was found in Section 3.2 when we analyzed the buckling behavior of an elastic beam of length L with one end fixed, and with an axial force F applied at the free end (Figure 8.6). We found that for such a beam it will buckle if F exceeds the buckling force (Equation 3.38)

$$F_{\text{buckle}} = \frac{\pi^2 EI}{4L^2},\qquad (8.41)$$

where E is the Young's modulus of the beam, and I is the moment of inertia.

Nota Bene

Image-based modeling of cytoskeletal networks. The angular distributions of filaments for several cases of cytoskeletal networks have been quantified using image-processing analysis. Such angular distributions could be incorporated into the model described in Equation 8.40 to simulate the mechanical behavior of such networks based on realistic microstructures obtained from imaging.

Advanced Material: Cross-links and "dangling ends"

Within polymer networks such as the cytoskeleton, an individual polymer will typically have many cross-links along its span. However, near the ends of the polymer, the ends will be "dangling": in other words, there will be a free segment near the ends of the polymer that is not bounded on both sides by a cross-link. This free segment contributes to the density of the network but in general, not its strain energy. The reason is that the load transmission occurs through the cross-links such that bending and stretching of polymers occurs between cross-links only. Because the free end is "dangling", it will likely not undergo stretch or bending. In Equation 8.40, for simplicity we assume that the free ends satisfy the affine deformation assumption. Though this is likely unphysiological, the contributions of these free ends to the total strain energy is small.

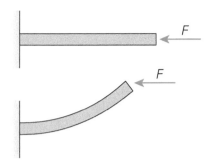

Figure 8.6 Schematic demonstrating buckling of beam with one end fixed and a force *F* applied at the free end. The beam will buckle if *F* exceeds F_{buckle}.

Figure 8.7 Protrusion of membrane by a cylindrical bundle. The force exerted on the protrusion by the membrane is proportional to the radius of the cylinder.

The membrane imparts force on the ends of filopodia

With our relation for buckling force identified, we need an estimate of the force imparted on the ends of filopodia. During filopodium formation, actin bundles exert a force against the membrane to form a local protrusion. We idealize the actin bundle as a cylinder of radius r (Figure 8.7). The force exerted by the membrane, F_{mem}, can be approximated by making an imaginary cut through the cylinder and membrane together. The resultant force of the membrane is equal to the product of the surface tension, N, of the membrane and the circumference of the cylinder,

$$F_{mem} = 2\pi RN. \tag{8.42}$$

Let us apply Equation 8.42 to a real-life situation. For neutrophils, the surface tension as measured through micropipette aspiration is approximately 35 pN/µm, resulting in a force of 22 pN for a filopodium 100 nm in radius. However, this force is likely conservative, as it does not take into account membrane bending and breaking of membrane–cortex links, both of which will contribute resistance to protrusion. Experimentally, ~50 pN of force has been found to be necessary to form filopodium-like membrane tethers within neutrophils; therefore, we will assume that $F_{mem} = 50$ pN.

The maximum filopodium length before buckling in the absence of cross-linking is shorter than what is observed *in vivo*

Using Equation 8.42 and our estimate of F_{mem}, we can now calculate the maximum filopodium length before buckling occurs. We first assume that there is no cross-linking within the bundle. Assume that there are $n = 30$ actin filaments within the bundle, and each filament has a radius of $R_{actin} = 3.5$ nm and Young's modulus of $E_{actin} = 1.9$ GPa. Each filament will experience a compressive force F_{mem}/n, so the maximum filopodium length before buckling occurs is

$$L_{nocl} = \sqrt{\frac{\pi^2 E_{actin} I_{actin}}{4 F_{buckle}}}$$

$$= \sqrt{\frac{\pi^2 E_{actin} \dfrac{\pi R_{actin}^4}{4}}{4 \dfrac{F_{mem}}{n}}}. \tag{8.43}$$

Plugging in the appropriate values gives length before buckling occurs, L_{nocl}, is $L_{nocl} = 0.57$ µm. Therefore, for a filopodium containing 30 actin filaments that are not cross-linked, it can only grow to be 0.57 µm long before it will buckle. This length is substantially shorter than the 1–5 µm long filopodia observed *in vivo*, implying that in the absence of cross-linking, filopodia do not possess adequate mechanical integrity to extend. Next, let us examine how this length is altered when we take cross-linking into account.

Cross-linking extends the maximum length before buckling

Consider a situation in which we are performing an analogous analysis for a bundle of cylindrical steel beams. To simulate a high degree of cross-linking, one could weld all the steel beams together into an approximately cylindrical-shaped bundle. As more beams were welded together, the bundle would begin to resemble a cylinder. Its mechanical behavior would more and more resemble a single solid cylindrical steel beam with the same radius as the bundle. Motivated by this thought experiment, as a first approximation, we will assume that a highly cross-linked bundle of actin will behave as a single elastic beam of effective radius r_{bundle} (Figure 8.8). To calculate this radius, we know that for a bundle of n filaments, the total cross-sectional area of F-actin is

$$n\pi R_{actin}^2 = \pi \left(\sqrt{n}\, R_{actin} \right)^2.$$

(8.44)

The term in parenthesis on the right-hand side of Equation 8.44 yields the radius of a cylinder with the same cross-sectional area of n actin filaments, which is the effective radius of the bundle,

$$R_{bundle} = \sqrt{n}\, R_{actin}.$$

(8.45)

Now, the moment of inertia of the bundle becomes

$$I_{bundle} = \frac{\pi R_{bundle}^4}{4}$$
$$= \frac{\pi \left(\sqrt{n}\, R_{actin} \right)^4}{4}.$$

(8.46)

Finally, assuming $F_{buckle} = F_{mem}$, then

$$L_{cl} = \sqrt{\frac{\pi^2 E_{actin} I_{bundle}}{F_{buckle}}}$$
$$= \sqrt{\frac{\pi^2 E_{actin} \dfrac{\pi \left(\sqrt{n}\, R_{actin} \right)^4}{4}}{4 F_{mem}}}.$$

(8.47)

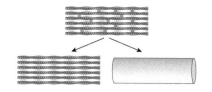

Figure 8.8 Two different approximations for analyzing the filopodium. In the first (left) we assume no cross-linking. In the second (right) we assume that the bundle is tightly cross-linked, resulting in behavior similar to that of a single, large cylinder.

Substituting the appropriate values gives $L_{cl} = 3.2$ μm. In other words, a highly cross-linked filopodium, can grow to be 3.2 μm before buckling. Comparing Equations 8.43 and 8.47, we find that the ratio of L_{cl}/L_{nocl} is $\sqrt{n}$. This indicates that the more filaments that are in a filopodium, the more pronounced the difference in the buckling lengths between the cross-linked and un-cross-linked filopodia.

What can be learned from our analysis? We have already learned that filopodia of 1–5 μm are commonly observed *in vivo*. Our analysis show that without cross-linking, filopodia will buckle before they are able to reach these lengths. However, if they are highly cross-linked, they are much more suited to resist buckling at that length. Filopodia much longer than L_{cl} (on the order of tens of micrometers) have been observed on occasion. This suggests that other mechanisms may act in concert with cross-linking to stabilize the actin bundles within filopodia, such as constraints on transverse deformations by the membrane sheath surrounding the bundle.

Is the structure of the red blood cell's cytoskeleton functionally advantageous?

We now turn our attention to another example in which cytoskeletal structure may play an important role in biological function, the highly structured red blood cell cytoskeleton. The mechanical behavior of red blood cells is extremely important to their function. To deliver oxygen to different parts of the body, these cells

(A)

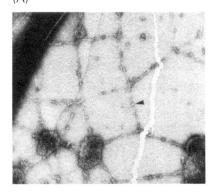

(B)

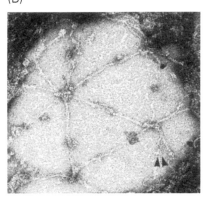

Figure 8.9 Electron micrographs of the red blood cell's cytoskeleton with different connectivities.
(A) Fourfold connectivity. (B) Sixfold connectivity. (From, Byers TJ et al. (1985). *Proc. Natl. Acad. Sci. U. S. A.*)

must be able to squeeze through tiny capillaries (many of which have diameters that are smaller than the cells themselves), and then return to their original shape. This behavior depends in large part on the cytoskeletal mechanics of these cells.

In red blood cells, the cytoskeletal network is bound to the membrane in a two-dimensional network. The main constituent of the network is a polymer called *spectrin*. The spectrin polymers are bound together at distinct vertices, or "junctions". At each vertex is a junctional complex consisting of several proteins (including F-actin) that serve to cross-link the spectrin polymers together, as well as anchor the cytoskeletal network to the membrane. There are also protein complexes along the length of the spectrin polymer that anchor it directly to the membrane. Typically, six spectrin polymers radiate from each junction. Such a junction is considered to exhibit *sixfold connectivity*. Fourfold connectivity, or four spectrin polymers radiating from each junction, has also been observed (Figure 8.9).

The question we seek to address is what are the structural and functional implications of these different connectivities? As the analytical and numerical tools for understanding the mechanics of the red blood cell's cytoskeleton become more sophisticated, the impact of such distinct microstructures on the mechanical behavior of red blood cells is becoming better understood. However, we can also learn a lot from simple analyses. In the following section, we estimate the mechanical properties of a model red blood cell's cytoskeleton from the mechanical behavior of its polymers and its microstructure. By seeing how these properties change with sixfold and fourfold connectivity, we also can see why, for red blood cells, sixfold connectivity may be functionally advantageous.

Thin structures can be analyzed using the two-dimensional shear modulus and the areal strain energy density

The red blood cell's cytoskeleton is thin, only ~100 nm thick, which is much smaller than its micrometer-sized lateral dimensions. For such structures, we usually are not interested in how quantities such as stress and strain change with depth. It is mathematically convenient to treat them as two-dimensional by integrating out the depth. To do this, consider a three-dimensional block with length l, height h, depth d, and shear modulus G undergoing shear (Figure 8.10). Recall from Section 3.2 that when we can shear the block by displacing the top surface by a small amount δ, which results in the sides of the block making an angle of γ with an imaginary vertical line. We know that

$$\gamma = \frac{\partial u}{\partial y} + \frac{\partial v}{\partial y}$$

$$= \frac{\delta}{h} \tag{8.48}$$

$$= \tan\gamma$$

$$\approx \gamma,$$

where the last line is the small angle approximation. Equation 8.48 implies that for small shear deformations, the (engineering) shear strain is simply the angle formed with the vertical, γ. The shear modulus G relates the shear stress τ to (engineering) shear strain γ,

$$\tau = G\gamma. \tag{8.49}$$

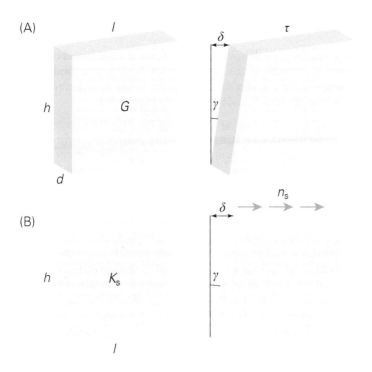

Figure 8.10 Treatment of a thin three-dimensional structure subjected to shear as a two-dimensional object. (A) A thin block of height h, length l, depth d, and shear modulus G is subjected to shear strain γ by imposing a shear stress τ, resulting in the displacement of the top surface by a small amount δ. (B) For simplicity, we treat the thin structure as a two-dimensional structure by integrating out the depth. The structure possesses a shear modulus per unit depth of K_s, and undergoes a shear strain γ under a shear force per unit length n_s.

In a manner analogous to Equation 8.33, the strain energy density for an isotropic linearly elastic material under shear strain γ is

$$w = \frac{1}{2}G\gamma^2, \tag{8.50}$$

implying that

$$G = \frac{\partial^2 w}{\partial \gamma^2}. \tag{8.51}$$

If we assume that the depth d is very small, we can treat the block as a two-dimensional structure. Instead of a shear stress (with units of force per area), we can define a shear force per unit length n_s such that

$$n_s = \tau d, \tag{8.52}$$

assuming that the shear stress is constant through the depth. The shear force per unit length can then be related to the engineering shear strain as

$$n_s = K_s \gamma, \tag{8.53}$$

where K_s is the shear modulus per unit depth,

$$K_s = Gd. \tag{8.54}$$

Similar to Equation 8.50 the areal strain energy density, or the strain energy per unit undeformed area (as opposed to volume) is

$$W_a = \frac{1}{2}K_s\gamma^2 \tag{8.55}$$

and the areal strain energy density can be differentiated to obtain the shear modulus as

$$K_s = \frac{\partial^2 W_a}{\partial \gamma^2}. \tag{8.56}$$

(A)

$k_{sp}/2$

(B)

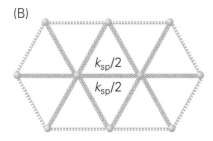

$k_{sp}/2$

$k_{sp}/2$

(C)

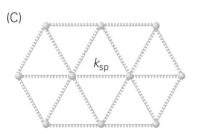

k_{sp}

Figure 8.11 Schematic demonstrating justification for choice of unit cell. For our analysis, we assume that the unit cell is an equilateral triangle with spring constant $k_{sp}/2$ (A). The spring constant is $k_{sp}/2$ when this unit cell is superimposed with similar unit cells (B). The result is a sixfold network with spring constant k_{sp} (C).

Sixfold connectivity facilitates resistance to shear

Using Equation 8.56, we will now calculate K_s for the red blood cell's cytoskeleton assuming sixfold or fourfold connectivity. Our strategy is to find the total change in strain energy W for a given shear strain γ, divide by the undeformed area to obtain w_a, and finally find K_s using the above expression.

To compute a relation for areal strain energy density, we first need to model the polymers and microstructure. In a cell, spectrin polymers have a contour length of $L = 200$ nm, and a persistence length of $\ell_p = 15$ nm. The junctions are separated by a distance of 75 nm. Because $\ell_p \ll L$ and the spectrin polymers are not fully stretched out, we can analyze the cytoskeleton as a network of entropic springs with spring constant $k_{sp} = (3k_BT)/2(L\ell_p)$ (Equation 7.60), connected at the junctions by freely rotating cross-links.

To simplify our analysis, we analyze a unit cell of the network instead of an entire network. Consider a sixfold network of springs with constant k_{sp}. If we use an equilateral triangle of springs as the unit cell of the network, then we can see in Figure 8.11 that each pair of neighboring triangles contributes a spring between each pair of junctions, resulting in two springs of stiffness k_{sp} (or, equivalently, a single spring of stiffness $2k_{sp}$) between each pair of junctions. We can correct this by making the stiffness of our springs in our equilateral triangle equal to $k_{sp}/2$.

Now, consider an equilateral triangle of springs, each with side length R_0. The triangle has a height of

$$h = \sqrt{R_0^2 - \left(\frac{R_0}{2}\right)^2}$$

$$= \frac{\sqrt{3}R_0}{2}.$$

(8.57)

If we now displace the top vertex by a small amount δ such that the triangle undergoes shear strain γ (Figure 8.12), then

$$\tan \gamma = \frac{\delta}{h}$$

$$= \frac{2\delta}{\sqrt{3}R_0}.$$

(8.58)

But $\tan \gamma \simeq \gamma$ for small γ, so Equation 8.58 can be rewritten as

$$\delta = \frac{\sqrt{3}R_0\gamma}{2}.$$

(8.59)

Figure 8.12 A simple model of sixfold symmetry. This simple model is composed of an equilateral triangle of springs undergoing shear strain γ by displacing the top surface by a small amount δ.

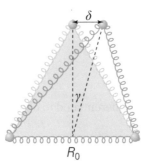

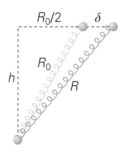

Under this deformation, the left and right diagonal springs lengthen and shorten, respectively, while the bottom spring is unchanged. The deformed length R of the left diagonal spring can be found geometrically,

$$
\begin{aligned}
R &= \sqrt{h^2 + \left(\frac{R_0}{2} + \delta\right)^2} \\
&= \sqrt{\left(\frac{\sqrt{3}R_0}{2}\right)^2 + \left(\frac{R_0}{2} + \delta\right)^2} \\
&= \sqrt{R_0^2 + R_0\delta + \delta^2} \\
&= \sqrt{R_0^2 + \frac{R_0^2\delta}{R_0} + \frac{R_0^2\delta^2}{R_0^2}} \\
&= R_0\sqrt{1 + \frac{\delta}{R_0} + \frac{\delta^2}{R_0^2}}.
\end{aligned}
\tag{8.60}
$$

Because δ is small, we ignore higher-order terms in δ, and Equation 8.60 becomes

$$
R \approx R_0\sqrt{1 + \frac{\delta}{R_0}}.
\tag{8.61}
$$

We can simplify this even further by noticing that

$$
\begin{aligned}
\left(1 + \frac{\delta}{2R_0}\right)^2 &= 1 + \frac{\delta}{R_0} + \frac{\delta^2}{4R_0^2} \\
&\approx 1 + \frac{\delta}{R_0}.
\end{aligned}
\tag{8.62}
$$

Combining Equations 8.61 and 8.62, R can be rewritten as

$$
\begin{aligned}
R &\approx R_0\sqrt{\left(1 + \frac{\delta}{2R_0}\right)^2} \\
&\approx R_0 + \frac{\delta}{2}.
\end{aligned}
\tag{8.63}
$$

This implies that, to first order, the left diagonal spring increases in by $\delta/2$. It can be shown similarly that the right diagonal spring shortens by the same amount. We know that the change in strain energy for a spring with constant k, original length R_0 and new length R is

$$
W_{\text{sp}} = \frac{k(R - R_0)^2}{2}.
\tag{8.64}
$$

The total change in strain energy is the sum of the change in strain energy for all three springs. Assuming a spring constant of $k_{\text{sp}}/2$ for our network, this expression becomes

$$W = W_{sp}^{left} + W_{sp}^{right} + W_{sp}^{bottom}$$

$$= \frac{1}{2}\left(\frac{k_{sp}}{2}\right)\left((R_0 + \delta/2) - R_0\right)^2 + \frac{1}{2}\left(\frac{k_{sp}}{2}\right)\left((R_0 - \delta/2) - R_0\right)^2 + 0 \qquad (8.65)$$

$$= \frac{k_{sp}\delta^2}{8}.$$

The areal strain energy density can be found by dividing W by the undeformed area of the triangle,

$$w_a = \frac{W^{left}}{A_{sp}}$$

$$= \frac{\dfrac{k_{sp}\delta^2}{8}}{\dfrac{1}{2}R_0 h} \qquad (8.66)$$

$$= \frac{\dfrac{k_{sp}\delta^2}{8}}{\dfrac{1}{2}R_0 \dfrac{\sqrt{3}R_0}{2}}.$$

We can write the areal strain energy density in terms of γ by substituting our relationship between δ and γ found in Equation 8.59 into Equation 8.66, which becomes

$$w_a = \frac{\sqrt{3}k_{sp}\gamma^2}{8}. \qquad (8.67)$$

Finally, the shear modulus of the network K_s is

$$K_s = \frac{\partial^2 w_a}{\partial \gamma^2}$$

$$= \frac{\sqrt{3}k_{sp}}{4}. \qquad (8.68)$$

Upon inspection of Equation 8.68, there are several interesting things to note. First, there is no explicit dependence of the shear modulus on the number, density, or volume fraction of filaments. This is because sixfold connectivity necessitates a fixed relationship between filament length L and the filament density. In particular, if we model each spectrin polymer as a rod with length L and cross-sectional area A, and we assume that the cytoskeletal network/membrane has a depth d, then the volume fraction scales as

$$\rho \sim (AL)/(L^2 d) \sim A/(Ld). \qquad (8.69)$$

In other words, the volume fraction scales linearly with $(1/L)$. Because $k_{sp} \sim 1/L$, this implies that $K_s \sim \rho$. Another interesting observation is that our prediction for K_s compares remarkably well with experiments. Substituting $b = 30$ nm, $T = 300$ K, $L = 200$ nm for spectrin yields $K_s = 0.9$ μN/m. Experimental measurements of K_s of red blood cell cytoskeletons in which the membrane was removed showed average values of the shear modulus to be $K_s = 2.4$ µN/m. This is roughly three times our predicted value, but still is in agreement to an order of magnitude. What are some potential sources of error? One may be our approximation of spectrin as an ideal chain. Remember that when R approaches L, the stiffness of a real chain will increase, while the stiffness of an ideal chain will not. Therefore, the Gaussian approximation is best when $L \gg R$, whereas for the red blood cell cytoskeleton, L is only a little more than twice that of R.

Fourfold connectivity does not sustain shear as well as sixfold

We will now calculate K_s for a fourfold network. Here, our unit cell is a square lattice of springs, each with spring constant $k_{sp}/2$ and undeformed length R_0 (Figure 8.13). If we displace the top surface of the square by a small amount δ, then the top and bottom springs do not change length, and the deformed lengths of the left and right springs are

$$R \approx \sqrt{R_0^2 + \delta^2}$$
$$\approx R_0^2,$$

(8.70)

if we ignore higher-order terms. What this means is that, to first order, the left and right springs do not change length. This implies that the change in strain energy will be effectively zero, and so $K_s = 0$. This result arises because the stiffness in shear comes from springs changing in length. However, if the springs do not change length, then the network cannot resist shear, and so it has no effective stiffness. The result of this analysis is that sixfold connectivity is advantageous for red blood cells compared with fourfold connectivity, because sixfold connectivity allows resistance to shear. Such resistance is important in allowing red blood cells to squeeze through tiny capillaries and return to their original shape.

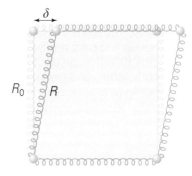

Figure 8.13 A simple model of fourfold symmetry. For fourfold symmetry we utilize a simple square network of springs undergoing shear by displacing the top surface by a small amount δ.

Example 8.2: Nonlinear fourfold connectivity

Instead of dropping the second-order term in the fourfold connectivity model, assume that we keep it. Does the analysis change?

We assume that the deformation is small compared with the side of the square; that is, $\delta \ll R_0$. Therefore, we can write

$$R^2 = R_0^2 + \delta^2, \quad \text{so } R \approx R_0 + \frac{\delta^2}{2R_0}.$$

The total strain energy is then

$$W = \frac{1}{2}\frac{k_{sp}}{2}(R - R_0)^2 + 0 + \frac{1}{2}\frac{k_{sp}}{2}(R - R_0)^2 + 0 = \frac{k_{sp}}{8}\frac{\delta^4}{R_0^2}$$

for the contributions from the left, top, right, and bottom sides (see Figure 8.13) and substituting for R. The strain energy density is the total strain energy divided by the area, or

$$w_s = \frac{1}{R_0^2}\frac{k_{sp}}{8}\frac{\delta^4}{R_0^2} = \frac{k_{sp}}{8}\gamma^4,$$

where γ is the shear strain and is very small compared with one. The shear modulus is then

$$K_s = \frac{3}{2}k_{sp}\gamma^2.$$

If you compare this expression to the expression for the sixfold network, you will see that fourfold connectivity stiffness is tiny, in part because the γ^2 term is very small. The modulus here is not constant, because it is greater as you shear the network more, but its magnitude is essentially negligible.

Key Concepts

- Polymer network models aim to capture the aggregate behavior of large numbers of individual filaments. They can be used to model cell behavior under different underlying assumptions.

- Even if explicit relationships between mechanical properties and network parameters cannot be found, scaling relationships can be identified. The cellular solids model predicts a squared modulus–density relationship for bending dominated filament loading and a linear one for axial loading.

- Tensegrity models have no bending and rely on tensile and compressive elements forming a stable structure. They predict linear scaling of modulus with both density and pre-load. They also predict the action-at-a-distance phenomenon.

- Filament cross-linking can have a substantial impact on the mechanical behavior of a network independent of filament properties and density.

- Strain energy can be used to calculate the moduli for affine materials, even if they are not isotropic. Anisotropic network behavior can be estimated from filament orientation.

- The geometric arrangement of polymers can alter the resistance of the macrostructure to deformation. Sixfold symmetric (triangular) networks tend to be stiffer in shear than fourfold (square) symmetric networks.

Problems

1. For a rubber-like network of flexible polymers loaded in uniaxial tension show that the force required to stretch the material by l_x is $\rho A k T (\lambda - \frac{1}{\lambda^2})$.

2. Show that in the small strain limit $\lambda = 1 + \varepsilon$ and $(\lambda - \frac{1}{\lambda^2}) \approx 3\varepsilon$.

3. Compute the buckling length of a primary cilium. The primary cilium is made up of nine microtubules arranged in a ring as shown. Assume that each microtubule is not linked to its neighbors and therefore buckles independently. Be sure to state what values you are assuming for any constants such as Young's modulus, membrane tension, radius and moment of inertia of the microtubules.

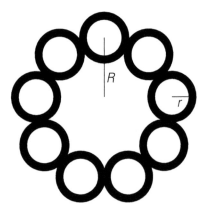

4. For the example above, use the parallel axis theorem to show that the bending moment of inertia of a collection of n microtubules is $\frac{nR^4}{4}\left(n + \frac{2n^3}{\pi^2}\right)$ assuming each

microtubule is a solid rod of radius r and they are all tightly cross-linked to one another. Assume that there are enough microtubules that the circumference of the primary cilia can be approximated by the sum of the individual microtubule diameters.

5. Use your result in Problem 4 to calculate the buckling length of a primary cilium.

6. Consider a polymer network made up of a cubic lattice of repeating junction points that are connected by polymer filaments.

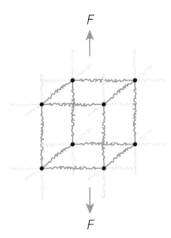

Approximate the behavior of each polymer as an ideal chain with n links of Kuhn length b at temperature T. Further assume that the equilibrium length of each molecule is R_0, such that $R_0 = b\sqrt{2n/3}$.

(a) What would the macroscopic Young's modulus, E and Poisson ratio, ν, be?

(b) What happens to the Young's modulus if we increase the temperature? Why?

(c) Now, repeat your answer to (a) and (b), but this time assume that the polymers have a long persistence length compared with their contour length. Model them as a network of rods with E of 1 GPA and diameter of 10 nm. Assume that the junctions are freely rotating pin joints.

(d) For your answer in (c), how would you expect the Young's modulus of the material to scale with the modulus of the polymer and volume fraction of the polymer. Only present a verbal argument. No calculations are required.

(e) How would you expect the shear modulus of the material to scale with the modulus of the polymer and volume fraction if the junctions were not pinned (they are bound such that the angles the polymers make with each other is fixed).

7. Examine the network response of fourfold versus sixfold connectivity for a normal stress, instead of shear stress. Is there an advantage to one versus the other under normal stress? You may assume the normal stress results in similar strains for both cases.

8. Discuss the following statements as they support or oppose tensegrity as a model.

(a) The notion of action at a distance applies to a simple beam embedded in a wall. Push down on the beam at the free end and there must be a reaction at the embedded end. If the wall is flexible enough, you will see substantial deformations at considerable distance from the applied load.

(b) If you sever a single actin filament within a living cell, the filament clearly retracts, and the local structure of the cell changes within a short time. However, within a fixed cell, severing an actin filament does not result in any dramatic changes.

9. Assuming a linearly elastic isotropic material with three-dimensional deformations (small deformations), write the strain energy density in terms of strains and constants E and Poisson's ratio.

10. Using an approach similar to that in the bending-dominated cellular solids model, derive the relations in Equations 8.8 and 8.9.

Annotated References

Boal D (2001) Mechanics of the Cell. Cambridge University Press. *An excellent source for detailed analyses of red blood cell cytoskeletal mechanics. The analysis of sixfold and fourfold connectivity was based on developments in this text.*

Byers TJ & Branton D (1985) Visualization of the protein associations in the erythrocyte membrane skeleton. *Proc. Natl. Acad. Sci. USA* 82, 6153–6157. *This references contains excellent images of the erythrocyte spectrin cytoskeleton, including fourfold and sixfold connectivity.*

Evans E & Yeung A (1989) Apparent viscosity and cortical tension of blood granulocytes determined by micropipette aspiration. *Biophys. J.* 56, 151–160. *Gives membrane tension estimates for use in filopodia buckling analysis.*

Kamm RD (2005) Molecular, Cellular, and Tissue Biomechanics (lecture notes from course number 20.310, Massachusetts Institute of Technology, Cambridge, MA). *Several of the treatments in this chapter were based on course notes developed by Dr. Kamm for this graduate course. Specific material includes the scaling analysis of cellular solids and several problems.*

Kwon RY, Lew AJ & Jacobs CR (2008) A microstructurally informed model for the mechanical response of three-dimensional actin network. *Comput. Meth. Biomech. Biomed. Eng.* 11, 407–418. *Gives details of the nonanisotropic affine network model, as well as an extension of the model that accounts for some degree of non-affine deformations.*

Mejillano MR, Kojima S, Applewhite DA et al. (2004) Lamellipodial versus filopodial mode of the actin nanomachinery: pivotal role of the filament barbed end. *Cell* 118, 363–373. *This reference assesses the role of various actin-binding proteins on the formation of lamellipodia and filopida and proposes a model whereby lamellipodial versus filopodial organization can be selected via regulation of these proteins.*

Mogilner A & Rubinstein B (2005) The physics of filopodial protrusion. *Biophys. J.* 89, 782–795. *Examines the mechanics and spatial-temporal dynamics of filopodia. The analysis of filopoidal buckling was based on treatments in this study.*

Rubenstein B & Colby RH (2003) Polymer Physics. Oxford University Press. *Gives a detailed treatment of the elasticity of rubber networks.*

Satcher RL & Dewey CF (1996) Theoretical estimates of mechanical properties of the endothelial cell cytoskeleton. *Biophys. J.* 71, 109–118. *Applies cellular solid theory to estimate the elastic moduli of the cytoskelton. This article details some of the scaling analysis and experimental support used in the cellular solids model.*

Shao EY & Hochmuth RM (1996) Micropipette suction for measuring piconewton forces of adhesion and tether formation from neutrophil membranes. *Biophys. J.* 71, 2892–2901. *This article gives estimates of force required to form filopodium-like membrane tethers within neutrophils.*

Stamenović D & Coughlin MF (1999) The role of prestress and architecture of the cytoskeleton and deformability of cytoskeletal filaments in mechanics of adherent cells: a quantitative analysis. *J. Theor. Biol.* 201, 63–74. *An excellent review of scaling analyses for cellular solids and tensegrity structures. The scaling relationship for tensegrity structures was obtained from this study.*

Wang N, Naruse K, Stamenović D et al. (2001) Mechanical behavior in living cells consistent with the tensegrity model. *Proc. Natl. Acad. Sci. U. S. A.* 98, 7765–7770. *This journal article details one of the most compelling experiments that supports the tensegrity hypothesis.*

Warren WE & Kraynik AM (1997) Linear elastic behavior of a low-density kelvin foam with open cells. *J. Appl. Mech.* 64, 787–794. *This journal article provides some very comprehensive mathematical analysis of cellular solids.*

CHAPTER 9

Mechanics of the Cell Membrane

The cell membrane is much more than a passive "bag" separating the cytoplasm from the environment. It is a heterogeneous, regulated barrier that allows for both active and passive transport of substances between the inside and outside of the cell. Its mechanical integrity is fundamental to its barrier function. Bending and stretching of the membrane is centrally involved in exocytosis and vesicle budding, as well as in fusion and viral invasion. It also incorporates structures for interacting with the extracellular matrix, other cells, and various compounds in solution. Proteins trapped in the membrane allow for signaling between the inside and outside of the cell. Understanding the biology and mechanics of the membrane can foster an appreciation of the complexity of this fascinating cellular component. In this chapter, we discuss the cell membrane's structural organization, how the two-dimensionality of the membrane limits diffusion, and how the barrier function can be understood. We end with a formal treatment of the cell membrane's mechanical function.

9.1 MEMBRANE BIOLOGY

Water is a polar molecule

Cells exist in an aqueous environment, and understanding the structure and function of the cell must include an understanding of how molecules, particularly the lipid bilayer, behave in aqueous conditions. Water molecules have an electric polarity, unusual for such a small molecule. The flanking hydrogens are not co-linear, but form an oblique angle with the oxygen (the H—O—H angle is roughly 105 degrees). Because oxygen is more electronegative than hydrogen, this leads to a net shift in the electric charge and the resulting molecule develops a polarity based on this incomplete separation of charges (Figure 9.1).

Water molecules can form hydrogen bonds among themselves by aligning the charges between different molecules. This gives rise to many of water's unique properties, including its high surface tension, the decrease in density when water freezes, and the hexagonal symmetry of snowflakes. It is important to note that these bonds are not permanent; water tends to rearrange itself all the time; however, at any given time, water molecule interactions are heavily influenced by hydrogen bonding. Indeed, as water freezes, the molecules are less free to re-form associations and, as a result, the molecules can no longer "fit" as well together (they tend to form fixed hexagonal crystals, which is why snowflakes tend to be six-sided). The net result is that water expands as it freezes, and the resulting decreased density is enough to make ice float on liquid water. The proclivity for the formation of hydrogen bonds allows a remarkably wide range of substances to be dissolved in water. Water is referred to as the *universal solvent* because of this property and the ubiquitous presence of this molecule on Earth.

Figure 9.1 A water molecule represented in the typical "ball and stick" model. (A) The oxygen atom has two pairs of electrons that occupy the "upper region" (not shown). The two pairs of electrons and the two hydrogen atoms are located roughly at the corners of a tetrahedron. (B) The sizes of the actual water molecule components. The "bending" of the water molecule gives it a polarity, with the negative charge closer to the oxygen atom. (From, Alberts B, Johnson A, Lewis J et al. (2008) Molecular Biology of the Cell, 5th ed. Garland Science.)

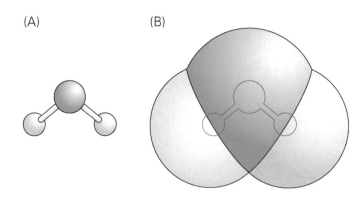

Cellular membranes form by interacting with water

Substances that are easily dissolved in water are known as *hydrophilic* (water loving). Those that repel or are repelled by water are known as *hydrophobic* (water fearing). Hydrophobic molecules tend not to be electrically polarized and thus do not readily form hydrogen bonds. When placed in an aqueous environment they tend to cluster or form bonds with themselves, as a result of being excluded by the surrounding water. In contrast, hydrophilic molecules tend to be polar, charged or easily dissociated into charged subunits. Complex molecules may exhibit regions or *domains* that are hydrophobic, while other regions are hydrophilic. These molecules are known as *amphiphilic*. This property gives rise to much of the diverse biochemical behavior that is at the root of the function of biological membranes and fundamental to cell biology.

Hydrophobic domains tend to be long hydrocarbon regions made up of repeating CH_2 units. Hydrophilic domains contain charged or polar regions. Charged regions can be *anionic* (negatively charged) like phosphates or sulfates, or they may be *cationic* (positively charged) like amines. Polar groups can be side chains such as alcohols and alkyls. Many biological membranes, including the cell membrane, are made up of phospholipids with a hydrophilic head region and a hydrophobic tail region. In an aqueous environment, the phospholipids self-assemble in structures to "isolate" the tail regions from water while keeping head regions in aqueous contact. As the concentration of phospholipids is increased, they begin to self-assemble into complex structures to protect the tail regions. The simplest of these, known as a *micelle*, is ball shaped, with the tails facing inward and away from the surrounding water. Micelles can also take on ellipsoidal or even cylindrical shapes. As concentration rises, the micelle geometry eventually becomes unstable, and a layered structure is formed with the head regions facing outward. This can take the form of a bilayer sheet or a small volume of water (and perhaps other trapped substances dissolved in the water) may be sequestered, forming a *liposome* (Figure 9.2).

Nota Bene

Micelle instability. When the lipid concentration rises to the point that lipids begin to encounter one another, they will initially fuse to form larger micelles. Eventually the water-excluding hydrophobic core becomes too large to be supported, and a bilayer results.

Figure 9.2 Lipid configurations depend on concentration. As their concentration increases, lipids begin to associate into spherical micelles, thereby hiding their hydrophobic tails from the surrounding water. At even higher concentrations, a bilayer forms.

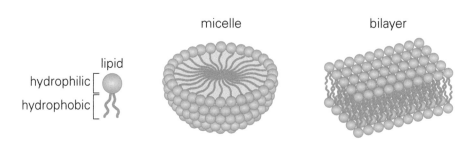

The saturation of the lipid tails determines some properties of the membrane

The phospholipid's hydrophobic (nonpolar) region is a tail consisting of two long carbon chains (Figure 9.3). Typically, one of these is saturated, meaning that each carbon is associated with two hydrogens and two carbons (one on either side). The other chain can consist of a few unsaturated carbons, meaning they form double bonds with adjacent carbons and can only accept a single hydrogen per carbon atom. Saturation leads to a more viscous behavior and a higher degree of self-association (saturated fats, like butter, tend to be solid at room temperature, whereas unsaturated fats, like vegetable oils, tend to be liquid). So it is possible to vary some of the fluid properties by controlling the amount of saturation of the lipids. In the same way that margarine is partially hydrogenated, or partially saturated, vegetable oils can be modified to become solid at higher temperatures compared.

From a mechanics point of view, the reason saturation is important has to do with steric, or spatial, properties. A fully saturated hydrocarbon chain will have some "zig-zag" in its length, but will be roughly straight (that is, the amplitude of the zig-zag is small compared with its overall length). Additionally, each bond is free to rotate, allowing for increased packing. The introduction of a double carbon bond creates a large bend in the chain. Double bonds also do not allow rotation, thereby further inhibiting packing. The result is that unsaturated phospholipids are not as tightly packed, diminishing the strength of intermolecular interactions and resulting in lower viscosity and a lower freezing point.

The cell membrane distinguishes inside and outside

The cell membrane is a lipid bilayer, also known as the *plasma membrane*, physically separating the extracellular biochemical environment from the cell's

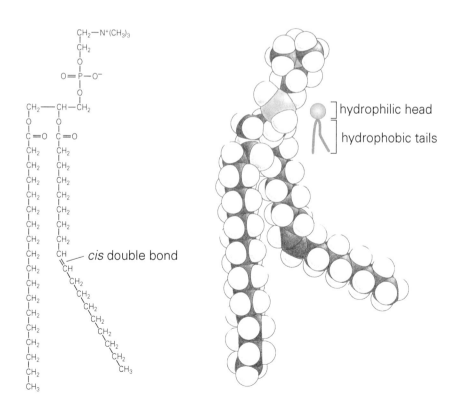

Figure 9.3 The lipids in a cell membrane consist of two fatty acid chains attached to a polar head group. Typically, one tail is fully saturated and the other consists of a varying number of carbon double bonds (making it unsaturated). The properties of the unsaturated tail are important for determining the membrane fluidity, partly because of the bend in the tail (exaggerated in this depiction). (From, Alberts, Johnson A, Lewis J et al. (2008) Molecular Biology of the Cell, 5th ed. Garland Science.)

internal fluid volume, known as the *cytoplasm*. However, the plasma membrane is not a total barrier, and certain molecules can diffuse across it. This semipermeable behavior is critical for its proper function. The barrier properties of the plasma membrane are partly a result of the thin hydrophobic layer in the interior tail region of the bilayer. Charged or polar molecules are less likely to cross this region. The plasma membrane also forms a barrier against large molecules. The barrier function of the plasma membrane is further increased by the activity of *flippases*, which are proteins that concentrate negatively charged phosphatidyl serine to the inner leaflet, forming an additional charged boundary. Permeability of the cell membrane to a given substance is hard to predict, but is generally a function of size, charge, and solubility. Water can cross the plasma membrane and does so by *osmosis*, the relatively slow transfer from the low solute concentration side (hypotonic) to the high solute concentration side (hypertonic). In general the cell's cytoplasm is hypertonic. Water movement across the plasma membrane is also facilitated by aquaporins, small protein complexes that form tiny holes in the membrane to regulate the passage of water molecules.

The fluid mosaic model of the cell membrane describes its physical properties

The molecular arrangement of the bilayer is a result of the attraction of the phospholipid head groups to water and the repulsion of the tail groups. This also produces an effective mutual attraction between the tail groups. Although an individual phospholipid can move about laterally within the bilayer, there is great resistance to its moving outside of the bilayer, which would expose the tails to water. In terms of the mechanical properties, this results in very low shear stiffness, but high resistance to *areal expansion*. Indeed, the shear stiffness of the membrane is so low; it essentially behaves as a fluid within the membrane plane. This view of the membrane as a fluid constrained to two dimensions was advanced by Singer and Nicolson in the 1970s and is termed the *fluid mosaic model*.

The practical implication of this model is that within the membrane itself, molecules such as proteins freely diffuse in the two-dimensional plane. This greatly simplifies and accelerates protein–protein interactions. There are also compartmentalizations in the membrane that can trap proteins in a "bin" or microdomain. Occasionally, the protein obtains sufficient energy (from thermal fluctuations) to break through a barrier and enter another compartment. This phenomenon was only observed with the development of small-molecule tracking techniques and fast-imaging cameras.

9.2 PHOSPHOLIPID SELF-ASSEMBLY

With our introduction to membrane biology in hand, the remainder of this chapter will be more quantitative in nature and will focus on membrane formation, mechanics, and function. We previously learned that when amphiphilic molecules such as phospholipids are placed into an aqueous environment, above a critical concentration, they self-assemble into organized structures. The concentration at which this aggregation occurs is called the *critical micelle concentration* (CMC). Experimentally, this concentration has been found to be highly sensitive to the molecular structure of the amphiphiles. The resulting shape of the self-assembled cluster (that is the spherical micelle or bilayer sheet) has also been found to be dependent on phospholipid molecular structure. In this section, we investigate the dependency of CMC and aggregate shape on various molecular attributes of phospholipids.

Nota Bene

The mechanics of the fluid mosaic model. In the sections that follow, we will formulate a mechanics description of the in-plane shear and areal behavior of the membrane. This will allow us to quantify just how "fluid" the membrane is.

Critical micelle concentration depends on amphiphile molecular structure

Experiments have demonstrated that the CMC is significantly decreased when the hydrocarbon tails are extended by adding more carbon atoms, or when the molecules are synthesized with two tails instead of one. Why would this be? One approach to understanding this phenomenon is to use a simplified model with only two idealized phases; a condensed phase in which all the amphiphiles are part of a single aggregate surrounded by water molecules and a surrounding aqueous phase. In the latter, the individual amphiphiles are dispersed throughout the solution in a dilute manner. We are able to approximate their behavior as that of an ideal gas. The advantage of this idealized treatment is its simplicity, however, some accuracy is sacrificed. In reality, the system will not be constrained to exist in these two states. There will be a spectrum of intermediary states with some molecules aggregated and others dispersed throughout the solution.

We start by considering a condensed state consisting of a spherical micelle of one-tailed phospholipids with hydrocarbon chain length l immersed in an aqueous solution. The chain length is equal to the product of the number of carbon atoms in the chain n_c and the average bond length between carbon atoms l_c, $l = l_c n_c$. We consider the condensed state to be the reference state, with zero free energy (zero free energy only occurs with zero internal energy [W] and zero entropy [S]). Next, we calculate the free energy of the aqueous state. In this state, the hydrophobic tails of the molecules are now exposed to water, and there is an increase in energy per molecule that can be approximated as the interfacial energy between the water and the exposed hydrophobic hydrocarbon region. Let γ_{int} be the interfacial energy per unit hydrocarbon chain length (this energy/length can be roughly approximated from measurements of surface tension and an estimation of the effective radius of the hydrophobic region). Then for each molecule, this energy is equal to

$$W = \gamma_{int} l = \gamma_{int} l_c n_c. \tag{9.1}$$

To calculate the entropy we assume that in this state, the amphiphiles are sufficiently dispersed to behave like an ideal gas. The entropy per molecule for an ideal gas is

$$S_{ideal} = k_B \left(\frac{5}{2} - \ln \left(\frac{\rho \hbar^3}{(2\pi m k_B T)^{3/2}} \right) \right), \tag{9.2}$$

where ρ is the number of molecules per unit volume, m is the mass of each molecule, and $\hbar$ is Planck's constant. Now, we can write the free energy per molecule in the aqueous state as

$$\Psi = W - T S_{ideal} = \gamma_{int} l_c n_c - k_B T \left(\frac{5}{2} - \ln \left(\frac{\rho \hbar^3}{(2\pi m k_B T)^{3/2}} \right) \right). \tag{9.3}$$

Remember that this free energy in the aqueous state was constructed with the condensed state defined as the reference state with zero free energy. So, the transition between the condensed and aqueous states occurs when $\Psi = 0$, or when

$$\gamma_{int} l_c n_c = k_B T \left(\frac{5}{2} - \ln \left(\frac{\rho \hbar^3}{(2\pi m k_B T)^{3/2}} \right) \right). \tag{9.4}$$

The density ρ at this transition gives the critical density of molecules at which aggregation occurs. Solving for ρ, we obtain

$$\rho = A e^{\left(5/2 - \frac{\gamma_{int} l_c n_c}{k_B T} \right)},$$ (9.5)

where

$$A = (2\pi m k_B T)^{3/2} \hbar^{-3}.$$ (9.6)

Although the mass m depends on n_c, for physiological values of γ_{int}, l_c, and n_c, the exponential term typically decreases much faster than A with n_c. Therefore, we can see that for constant values of γ_{int} and l_c, increasing the number of carbon atoms n_c decreases the aggregation density. Intuitively, this seems reasonable, as having longer tails would increase the interfacial energy of the molecule with water and make it more energetically unfavorable to remain in the aqueous state. Similarly, phospholipids with two tails have a lower aggregation density compared with their single-tailed counterparts, owing to their higher interfacial energy. What is perhaps not as intuitive is the profound impact that increasing n_c can have on lowering ρ. The predicted exponential scaling of ρ with n_c, has, in fact, been observed experimentally. Specifically, increasing n_c by x atoms decreases ρ by a factor of e^x. If we were to extend the hydrocarbon chain by just two atoms (let us say from 10 to 12), this would result in an almost tenfold decrease in the aggregation density.

Example 9.1: Lipid packing as a function of tail length

Consider a lipid having a carbon tail of length $n_c = 8$ molecules. Calculate the fold change in the critical micelle density if the length of tail was increased to $n_c = 16$ molecules. Assume $l_c = 0.1$ nm, $\gamma_{int} = 10\, k_B T/$nm, and that the mass of the 8-carbon tail lipid is 200 g/mol.

First, we must calculate the mass of the longer lipid. We know that the molecular mass of carbon is 12 g/mol. Because the longer lipid contains 8 more carbons, the mass of the longer lipid is $200 + (8 \times 12) \approx 300$ g/mol. Using Equation 9.6, we know that the fold change in A for the two different lipids is $(200^{3/2})/(300^{3/2}) \approx 0.5$.

In addition, we know that the fold change in the exponential term is

$$\frac{e^{\left(5/2 - \frac{8\gamma_{int} l_c}{k_B T} \right)}}{e^{\left(5/2 - \frac{16\gamma_{int} l_c}{k_B T} \right)}} = e^8 \approx 3000$$

We see that increasing the length of the lipid tail by 8 carbons will result in an approximately $3000 \times 0.5 = 1500$-fold decrease in the critical micelle density.

Aggregate shape can be understood from packing constraints

In the previous section, we investigated the dependency of CMC on amphiphile molecular structure. In our treatment, we did not consider the shape of the cluster. The tendency for certain amphiphiles to form certain geometries can, in principle, be investigated through free energy calculations; however, such calculations are very difficult to formulate. Nonetheless, by examining constraints associated with molecular packing, we can gain insight into the tendency for particular shapes to be preferred.

Consider a spherical micelle of radius R, each with effective head group surface area A_h. In addition, assume that the effective hydrocarbon chain volume is V_c. Because the interior of the micelle is hydrophobic, it cannot contain voids that would potentially enclose water. Therefore, the entire volume of the micelle is occupied by the hydrocarbon chains. The number of molecules n in the micelle can be calculated as either

$$n = \frac{4\pi R^2}{A_h}$$ (9.7)

from the surface area, or

$$n = \frac{4\pi R^3}{3V} \qquad (9.8)$$

from the volume. Setting Equations 9.7 and 9.8 equal, we see that for a micelle with a given radius R, V_c and A_h must be related as

$$R = 3V_c/A_h. \qquad (9.9)$$

Now, let $l = l_c n_c$ be the hydrocarbon chain length. We assume that the interior of the micelle cannot contain voids, owing to the enormous energy that creation of a vacuum would require. This implies that the radius of the micelle must be less than or equal to the hydrocarbon chain length, $R \leq l$. Equation 9.9 becomes

$$A_h/A_e \geq 3, \qquad (9.10)$$

where $A_e = V_c/l$ is the effective cross-sectional area of the hydrophobic region. Equation 9.10 implies that, based on packing constraints, a spherical micelle requires the effective area of the head group to be more than threefold greater than the hydrophobic region.

For phospholipids with progressively smaller A_h/A_e ratios, the preferred micelle shape would transition from a spherical micelle to a bilayer as this ratio approaches unity. Consider a large flat bilayer with width w, height h, and thickness t. In this case, packing requires the number of amphiphiles to be

$$n = wh/A_h \qquad (9.11)$$

based on surface area or

$$n = wht/2V_c \qquad (9.12)$$

based on volume. Setting Equations 9.11 and 9.12 equal we find

$$A_h = 2V_c/t. \qquad (9.13)$$

The thickness must be less than or equal to twice the hydrocarbon chain length, or

$$t \leq 2l. \qquad (9.14)$$

In this case,

$$V_c/l = A_0 \leq A_h. \qquad (9.15)$$

This indicates that when the effective area of the hydrophobic region approaches that of the head region, based on packing, a bilayer shape is preferred.

9.3 MEMBRANE BARRIER FUNCTION

We now turn our attention from investigation of membrane formation to that of membrane function. We previously learned that one of the primary roles of the plasma membrane is to separate the extracellular biochemical environment from the cytoplasm. However, it is only a semipermeable barrier, selectively allowing certain molecules to diffuse across it. How can we quantitatively assess the barrier properties of the membrane?

In this section, we analyze membrane barrier function using an approach based on the diffusion equation, which was derived using the random walk in

Section 5.6. The connection between random molecular processes (the random walk) and the net macroscopic or continuum-level behavior (the diffusion equation) was one of Albert Einstein's many important contributions. Owing to entropy and Brownian motion, there are many applications of random walks in biology. In this section we are going to apply the continuum level description to understand how molecular transport works at the level of the cell and the role of the membrane in that process. We begin by revisiting Fick's first and second laws of diffusion.

The diffusion equations relate concentration to flux per unit area

What does diffusion imply about the number of particles that might pass through a boundary such as the cell membrane? Using the continuum-level description the *flux* (from the Latin word *fluxus* meaning flow). J is given by

$$J = -D\frac{\partial C}{\partial x}. \tag{9.16}$$

This is Fick's first diffusion equation. It can be generalized to higher dimensions as

$$J = -D\nabla C, \tag{9.17}$$

where ∇ is the gradient operator defined as

$$\nabla(\bullet) = \left(\frac{\partial(\bullet)}{\partial x}, \frac{\partial(\bullet)}{\partial y}, \frac{\partial(\bullet)}{\partial z}\right).$$

Advanced Material: Flux and the random walk

Fick's first law can also be obtained from analysis of the flux in a discrete random walk. In one dimension, let us say we have $N(x)$ particles at location x and $N(x+d)$ particles at location $x + d$. To find the flux between these two locations, consider that in a time step Δt, half of the particles at x will move rightward (crossing the boundary) and half of the particles at $n + d$ will move leftward (crossing the boundary in the other direction). If we are interested in the number of particles moving to the right (in the positive x-direction) we get number of particles moving right per time step $= 0.5(N(x) - N(x + d))$.

To find the flux (which is a rate), we need to divide by the time step Δt as well as some area that the particles are crossing. This yields

$$J = \frac{(N(x) - N(x + d))}{2A_f\Delta t},$$

where A_f is the area through which the flux is moving. We then perform some simple manipulations,

$$J = -\frac{d^2}{2\Delta t}\frac{1}{d}\left(\frac{N(x + d) - N(x)}{A_f d}\right).$$

The difference inside the square brackets is analogous to a concentration gradient. If we take the limit as d approaches zero, we find Fick's first law (see Equation 9.17), with the diffusion constant D being $d^2/2\Delta t$.

Example 9.2: Discrete diffusion example: capture time

Consider particles that are freely diffusing until they meet a sink, wherein they become captured. How long would we need to wait for a diffusing particle to reach a given sink from some specified starting location?

Imagine a molecule that is freely diffusing along the x-axis, but with a sink at a particular location. Once the particle reaches that location, it is "captured." We want to know how long we would wait before the molecule is captured. At a position x, let the mean time of capture be $T_c(x)$. Let us start with the discrete random walk and then extend it to a differential form. Assume that I wait a small time Δt, the molecule will be at $x + b$ or $x - b$ with equal probability. We can write this recursive relationship

$$T_c(x) = \Delta t + 0.5(T_c(x + b) + T_c(x - b)).$$

Rearranging,

$$0 = 2\Delta t + (T_c(x + b) - T_c(x)) - (T_c(x) - T_c(x - b))$$

$$0 = \frac{2\Delta t}{b} + \frac{1}{b}(T_c(x + b) - T_c(x)) - \frac{1}{b}(T_c(x) - T_c(x - b)).$$

Letting b approach zero, we obtain

$$0 = \frac{2\Delta t}{b^2} + \frac{1}{b}\left(\frac{dT_c(x)}{dx} - \frac{dT_c(x - b)}{dx}\right).$$

Letting b approach zero once more

$$0 = \frac{1}{D} + \frac{d^2 T_c}{dx^2}.$$

The above equation describes the wait time for a single particle. To generalize to higher dimensions, you can replace the second term $(d^2 T_c/dx^2)$ with the Laplacian of appropriate dimension. To solve this equation one needs boundary conditions. For this example, let us assume that the absorber is a sphere of radius a located within a larger impermeable sphere of radius b ($a \ll b$). On the surface of the absorber, the wait time is obviously zero ($T_c(a) = 0$). At the second boundary, T_c has zero gradient. We leave it as an exercise to show that integrating over the known volume the mean wait time is

$$\langle T_c \rangle = \frac{b^3}{3Da}.$$

The mean capture time is

$$\langle T_c \rangle = \frac{b^2}{2D} ln\left[\frac{b}{a}\right]$$

for a particle in a two-dimensional surface with a small circular absorber of radius a. Therefore, as the distance increases, diffusion limits on kinetics increases much more slowly if the reaction is confined to a membrane than if it needed to occur in free space.

Fick's second law shows how spatial concentration changes as a function of time

In Section 5.6 we considered a collection of particles to be moving independently, in discrete time and space. We showed that if each moved either to the right or left by a distance b at each point in time, we obtained the one-dimensional diffusion equation (or Fick's second law),

$$\frac{dC}{dt} = D\frac{d^2 C}{dx^2}, \tag{9.18}$$

where C is the concentration and D is the diffusion coefficient. We can relate D to the molecular behavior of the particles and specifically that

$$D = \frac{nb^2}{2\Delta t}, \tag{9.19}$$

where Δt is the time step in which a particle moves a distance b.

We can also generalize this to higher dimensions by writing

$$\frac{\partial C}{\partial t} = D\nabla^2 C, \tag{9.20}$$

where ∇^2 is the Laplacian of C defined as

$$\nabla^2 C = \frac{\partial^2 C}{\partial x^2} + \frac{\partial^2 C}{\partial y^2} + \frac{\partial^2 C}{\partial z^2}. \tag{9.21}$$

Nota Bene

Capture time at an impermeable boundary. It may seem strange that the assumption of an impermeable boundary results in a condition on wait time gradient. One way to see this is to consider a repeating lattice of absorbers. At the midpoint between two absorbers, there will be no net flow, because particles are equally likely to end up at either one. So, this is effectively an impermeable boundary. Also, the wait time at the midpoint reaches a maximum because moving in either direction will shorten the distance to one of the absorbers and reduce the wait time. Therefore, the gradient is zero.

Fick's second law tells us how spatial concentration changes as a function of time. In essence, there is a net transport of particles from high concentration to low concentration. The rate of transport depends on the diffusion coefficient in addition to the concentration distribution. This transport is referred to as a *flux* (from the Latin word *fluxus*, meaning flow) and is usually given in terms of amount of flow per unit area.

Example 9.3: Continuous diffusion example: total flux

With Fick's first and second laws revisited, we can now begin to quantitatively assess diffusive behavior of a particular molecule through a membrane. For example, one could quantify the *current*, or integrated flux through a region. We start with the assumption that all of the space outside a given sink is filled with a collection of particles. How many particles per unit time are crossing the sink boundary when the system reaches steady state? This is analogous to the situation in which a cell with a membrane permeable to some substance is incubated in a large volume of a solution of that substance. We wish to know how much of the molecule is crossing the cell boundary, assuming that once a cell consumes the compound it is metabolized.

For simplicity, we will assume that the cell is a sphere of radius a. If we are at steady state, then we can apply Fick's second diffusion equation, with the time derivative set to zero. In spherical coordinates, this becomes

$$\frac{1}{R^2}\frac{\partial}{\partial R}\left(R^2\frac{\partial C}{\partial R}\right) = 0.$$

To model the sink we assume that the concentration of particles at the surface is zero. This provides our boundary conditions, $C(R = a) = 0$ and $C(R \to \infty) = C_0$. Fick's first diffusion law (Equation 9.17) in spherical coordinates is

$$J = -D\frac{\partial C}{\partial R},$$

which has a solution of the form

$$J(R) = -DC_0\frac{a}{R^2}.$$

The net "current" of molecules passing through the sphere is then $I = JA$, where A is the area of the boundary. At $R = a$,

$$I = -DC_0\frac{a}{R^2}4\pi R^2 = -4\pi DC_0 a.$$

Example 9.4: Fluorescence recovery after photobleaching and molecular mobility

Diffusion can be used to assist in biological microscopy experiments to assess the mobility of molecules within cells. Because it is difficult to see individual molecules, there is no simple way to determine whether, say, a particular receptor is free to move around the membrane or is anchored in place.

One type of experiment that can provide some information is fluorescence recovery after photobleaching (FRAP). In this technique, the target protein is tagged with a fluorescent molecule. After continuously exciting a small region, the fluorescent molecule will become "bleached", that is, lose its fluorescent properties. This region becomes darker compared with the surrounding regions. If the fluorescent molecule is

attached to a protein with high mobility, though, proteins attached to unbleached molecules will diffuse in, and the dark region will recover and become brighter again. Alternatively, if the protein is well-anchored (low mobility), then the dark spot will persist. This can be quantified, based on the work of Axelrod and Webb. The assumptions for this approach are that the motion of molecules is purely diffusive (there is no convection), and that the photobleached region is disk shaped.

As a function of radius, Fick's second law is

$$\frac{\partial C(R,t)}{\partial t} = D\nabla^2 C,$$

with boundary conditions $C(\infty, t) = C_\infty$ and initial conditions $C(R,0) = C_0(R)$ the initial photobleach profile. Typically,

$$C(R,0) = C_0 e^{-T\varphi(R)}$$

where T is the time interval for the bleaching and $\varphi(R)$ is a scaled excitation intensity from the laser ($\varphi = 0$ for $R > w$, where w is the disc radius, and φ = constant for $R \leq w$. The constant depends on the laser power, a scaling factor, and the characteristic size of the laser beam.)

However, what we really want is the diffusion constant D, and what we really measure is the fluorescence profile

$$F(t) = \int \alpha\varphi(R)C(R,t)\mathrm{d}A,$$

where α is a scaling factor that further corrects for imaging attenuation and the integral is over the photobleached area. The solution for $F(t)$ can be found with Fourier transforms and series analysis. If one uses the assumption that $T\varphi(0)$ is much less than unity (slight bleaching) then the solution can be expressed as

$$F(t) = (\alpha P_0 C_0)\left(1 - \frac{T\varphi(0)}{2\left(1 + \dfrac{t}{\tau_\mathrm{d}}\right)}\right),$$

where P_0 is the laser power and $\tau_\mathrm{d} = w^2/4D$ is the characteristic time for diffusion. By measuring $F(t)$, one can obtain D for the protein of interest. Alternatively, if one

has a rough idea what D is, one can design an experiment to time the FRAP.

Let us work a numerical example. Assume that $\alpha P_0 C_0 = 1$ (fluorescence units), $D = 10^{-11}$ m^2/s, and $T\varphi(0) = 0.1$. Determine the half-life of the photobleached region relative to $F(0)$, if the photobleaching region is a disk of radius 5 μm.

Because $T\varphi(0) \ll 1$, we can apply the solution for $F(t)$, which gives us

$$F(t) = \left(1 - \frac{1}{20\left(1 + \dfrac{4tD}{w^2}\right)}\right).$$

Substituting the provided numbers,

$$F(t) = \left(1 - \frac{1}{(20 + 32t)}\right),$$

with t in seconds. We can quickly get $F(0) = 1 - \dfrac{1}{20} = \dfrac{19}{20}$ (this is not much photobleaching at all). To get to half of this value, we solve

$$\frac{39}{40} = \left(1 - \frac{1}{(20 + 32t)}\right)$$

for t. One gets $t = 5/8$ s, or a little over half a second. This is very fast and unlikely to be something that would be useful using FRAP as modeled. One can either increase the intensity of illumination to photobleach out the sample more (in which case the expression no longer holds, and one needs to use a different approach) or use a smaller photobleaching area.

9.4 MEMBRANE MECHANICS I: IN-PLANE SHEAR AND TENSION

We have investigated various aspects of membrane formation (amphiphile self-assembly) and function (acting as a barrier). The rest of this chapter focuses on membrane mechanics. As we have discussed, for energetic reasons, each lipid bilayer has an inherent optimal microstructure with some optimal spacing between the lipid molecules. Any perturbation to this optimal arrangement disturbs this energetically favorable microstructure. The lipid bilayer exhibits an inherent resistance to deformation. If a portion of membrane is stretched, the lipid molecules are pulled apart, requiring energy in the presence of water. If the membrane is bent, the molecules on the outer layer are separated and those on the inner layer are compressed, again storing deformational (strain) energy (Figure 9.4). However, as we will see, the lipid molecules are relatively free to move about within the bilayer.

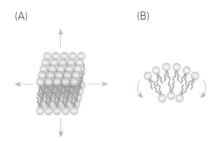

Figure 9.4 A region of membrane exposed to either tension (A) or bending (B).

With this as our motivation, the goal of this section is to develop the equations that describe the mechanics of the continuum behavior of the cell membrane. As this section is mathematics-heavy, we will consider systems of increasing levels of complexity to develop some degree of intuitive understanding before presenting the general governing equations.

Thin structures such as membranes can be treated as plates or shells

The defining characteristic of a membrane is its very thin structure relative to the overall size of the cell. The cell membrane lipid bilayer has a thickness of approximately 7 nm. The cell itself is typically on the order of micrometers. By restricting the dimensionality of the problem, certain kinematic assumptions can be made that will greatly simplify the problem from that of general three-dimensional continuum mechanics. In other words, we can approximate the membrane as a curved two-dimensional structure, or, from Chapter 3, a shell. You might be more familiar with a plate, or a flat two-dimensional structure. In this section, for simplicity, we will assume that the domain or subdomain of the problem is flat (that is, a plate). One exception is that in certain cases we will consider spherical shapes (shapes that can be described by radii in the x- and y-directions).

Advanced Material: Einstein and curved coordinate systems

The mathematics to describe behavior in curved coordinate systems was developed to a great extent by Einstein. He needed this description to cast the formulation of general relativity theory in which space is actually curved owing to the gravitational effect of mass. He developed a notational system (known as inidical notation) in which coordinates are denoted with numbers (1, 2, or 3) rather than letters (x, y, and z). In his system, indices are introduced to represent the coordinates, and when they occur in pairs, a summation is implied. The square of the magnitude of a vector v would be denoted as $v_i v_i = v_1 v_1 + v_2 v_2 + v_3 v_3 = v_x v_x + v_y v_y + v_z v_z$. This is a very powerful and compact way to represent complex vector and tensor manipulations. Any advanced treatment of continuum mechanics beyond the level used in this text will generally adopt this notation out of expediency.

As before, our continuum mechanics analysis involves three different parts, kinematics, constitutive equations, and equilibrium. For the material model we are again going to assume generalized Hookean behavior. We will consider equilibrium in distinct parts, one for the in-plane forces and one for the out-of-plane forces. Let us begin our discussion of membrane mechanics by making kinematic assumptions that take advantage of the thin shell assumption.

Kinematic assumptions help describe deformations

Similar to our approach to beams, we introduce a kinematic construct to describe the deformation. In the case of the beam, we did this in terms of planar collections of points perpendicular to the neutral axis. For shells, we are going to do this in terms of linear collections of points through the shell perpendicular to the shell's midline surface. Our kinematic assumptions are that these lines

1. remain straight,
2. do not stretch, and
3. remain normal to the midline surface.

These kinematic assumptions suggest a particular way of expressing the deformation of the shell. In particular, it makes sense to talk about the deformation due to motion in the plane of the shell (we assume this is the x–y-plane) and motion due to transverse displacements and the subsequent rotation of the normals (Figure 9.5). This is analogous to our separation of axial deformation and bending deformation when we considered a linear element. We will denote the total deformation as u^{tot} and v^{tot} in the x- and y-directions, respectively. They are simply the sum of the in-plane deformation and the displacement due to rotation of the shell's mid-surface. Our kinematic assumption amounts to just assuming that the deformations take a particular form that is a simple extension of the beam kinematics,

$$u^{\text{tot}}(x,y,z) = u(x,y) - z\frac{\mathrm{d}w}{\mathrm{d}x}$$

$$v^{\text{tot}}(x,y,z) = v(x,y) - z\frac{\mathrm{d}w}{\mathrm{d}y} \qquad (9.22)$$

$$w^{\text{tot}}(x,y,z) = w(x,y).$$

From these, we calculate strain from Equations 3.48, 3.49, and 3.52

$$\varepsilon_{xx} = \frac{\mathrm{d}u^{\text{tot}}}{\mathrm{d}x} = \frac{\mathrm{d}u}{\mathrm{d}x} - z\frac{\mathrm{d}^2w}{\mathrm{d}x^2} \quad \varepsilon_{xy} = \frac{1}{2}\left(\frac{\mathrm{d}u^{\text{tot}}}{\mathrm{d}y} + \frac{\mathrm{d}v^{\text{tot}}}{\mathrm{d}x}\right) = \frac{1}{2}\left(\frac{\mathrm{d}u}{\mathrm{d}y} + \frac{\mathrm{d}v}{\mathrm{d}x}\right) - z\frac{\mathrm{d}^2w}{\mathrm{d}x\mathrm{d}y}$$

$$\varepsilon_{yy} = \frac{\mathrm{d}v^{\text{tot}}}{\mathrm{d}y} = \frac{\mathrm{d}v}{\mathrm{d}y} - z\frac{\mathrm{d}^2w}{\mathrm{d}y^2} \quad \varepsilon_{xz} = \frac{1}{2}\left(\frac{\mathrm{d}u^{\text{tot}}}{\mathrm{d}z} + \frac{\mathrm{d}w^{\text{tot}}}{\mathrm{d}x}\right) = \frac{1}{2}\left(\frac{\mathrm{d}u}{\mathrm{d}z} - \frac{\mathrm{d}w}{\mathrm{d}x} - z\frac{\mathrm{d}^2w}{\mathrm{d}x\mathrm{d}z} + \frac{\mathrm{d}^2w}{\mathrm{d}x^2}\right)$$

$$\varepsilon_{zz} = \frac{\mathrm{d}w^{\text{tot}}}{\mathrm{d}z} = 0 \qquad\qquad \varepsilon_{yz} = \frac{1}{2}\left(\frac{\mathrm{d}v^{\text{tot}}}{\mathrm{d}z} + \frac{\mathrm{d}w^{\text{tot}}}{\mathrm{d}y}\right) = \frac{1}{2}\left(\frac{\mathrm{d}v}{\mathrm{d}z} - \frac{\mathrm{d}w}{\mathrm{d}y} - z\frac{\mathrm{d}^2w}{\mathrm{d}y\mathrm{d}z} + \frac{\mathrm{d}^2w}{\mathrm{d}y^2}\right).$$

$$(9.23)$$

At this point we have used two of our three kinematic assumptions. Namely, that the perpendiculars remain straight and that they are normal to the surface. However, we did not use the assumption that they do not stretch. This condition implies that the thickness of the shell does not change, and as a result, any strain terms involving z must be zero. We already know that $\varepsilon_{zz} = 0$, but this condition implies that $\varepsilon_{zx} = \varepsilon_{zy} = 0$ as well. The final form of our kinematics or strain-displacement relation is

$$\varepsilon_{xx} = \frac{\mathrm{d}u}{\mathrm{d}x} - z\frac{\mathrm{d}^2w}{\mathrm{d}x^2} \quad \varepsilon_{xy} = \frac{1}{2}\left(\frac{\mathrm{d}u}{\mathrm{d}y} + \frac{\mathrm{d}v}{\mathrm{d}x}\right) - z\frac{\mathrm{d}^2w}{\mathrm{d}x\mathrm{d}y}$$

$$\varepsilon_{yy} = \frac{\mathrm{d}v}{\mathrm{d}y} - z\frac{\mathrm{d}^2w}{\mathrm{d}y^2} \quad \varepsilon_{xz} = 0 \qquad\qquad (9.24)$$

$$\varepsilon_{zz} = 0 \qquad\qquad \varepsilon_{yz} = 0.$$

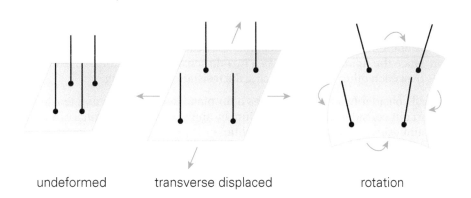

undeformed   transverse displaced   rotation

Nota Bene

The Kirchhoff hypothesis. The shell kinematic assumptions are also known as the *Kirchhoff* hypothesis. The lines perpendicular to the shell surface are known as *directors*.

Figure 9.5 The kinematic assumptions of shell mechanics. Transverse displacement within the plane causes the perpendicular lines to move but remain parallel to each other. Bending causes them to rotate.

Advanced Material: Small rotation assumption

In this development we have assumed that the deformations are small enough that our infinitesimal strain measures are adequate. However, what may not be obvious is that we have also assumed that the rotation of the normals is also infinitesimal. Often, this may not be a valid assumption, particularly for cells. One more advanced approach, known as von Kármán shell theory, assumes infinitesimal strains, but moderate rotations. This introduces an extra quadratic terms in the strains.

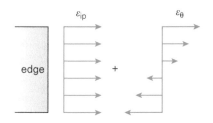

Figure 9.6 Decompositon of deformation modes. As a result of our kinematic assumption, the deformation of a shell can be separated into a component that is constant through the thickness, ε_{ip}, and a component that varies linearly through the thickness, ε_{θ}.

Notice that the in-plane strain components (ε_{xx}, ε_{yy}, ε_{xy}) contain two parts, one constant through the thickness, and one that varies linearly through the thickness. These correspond to the strains due to in-plane motion of the normal lines, ε_{ip}, and strains due to rotations of the normal lines, ε_{θ} (Figure 9.6). The in-plane strains are related to tension and shear within the plane of the shell. The linear strains are related to out-of-plane bending. Therefore, shells have three different deformational *modes*, in-plane tension (or compression), in-plane shear, and bending. The total deformation is simply the linear superposition or sum of these three modes.

A constitutive model describes material behavior

As in the past, we are going to assume a generalized Hookean material response (Equation 3.58). Owing to the simplified kinematics, it takes on a particularly simple form:

$$\sigma_{xx} = 2\mu\varepsilon_{xx} + \lambda(\varepsilon_{xx} + \varepsilon_{yy})$$

$$\sigma_{yy} = 2\mu\varepsilon_{yy} + \lambda(\varepsilon_{xx} + \varepsilon_{yy}) \tag{9.25}$$

$$\tau_{xy} = 2\mu\varepsilon_{xy}.$$

We have characterized our kinematic behavior, and given u, v, and w, we can determine strain, and through Hooke's law, stress. What remains is to determine what implications equilibrium holds for stress.

The equilibrium condition simplifies for in-plane tension and shear

When we considered kinematics, we talked about the deformation modes. Similarly, we can decompose our treatment of equilibrium by considering the different *loading* modes in turn. Specifically, it is helpful to consider in-plane loading separately from bending. Starting with in-plane loading, this mode is restricted to be within the plane of the shell locally. Although the shell may bend, we assume that there is no energy, stress, or moment associated with this bending. This is strange. What does it mean to say we will allow bending, but ignore bending stresses? Actually, this is not as unusual as you might think. Consider a thin plastic bag like you might get from the grocery store. The plastic in these bags is so thin (about 30 μm) that they bend with the slightest touch. Although there is bending, the resultant moments are so small that they can effectively be ignored. In addition, since the lipid bilayer acts as a two-dimensional fluid, the two leaflets can slide over each other further reducing the resistance to bending.

We will consider two particular cases of in-plane loading, shear and tension. Let us start our consideration of equilibrium by applying it to an imaginary free body. An infinitesimal element of the membrane subjected to in-plane forces is illustrated in Figure 9.7.

As we learned in Chapter 3, we need to apply the equilibrium conditions to forces rather than to stresses. To accomplish this, we must define resultant forces in terms of stress. The resultant forces are the equivalent force that *results*

Nota Bene

The two meanings of the term "membrane." As we briefly mentioned at the end of Section 3.4, in structural mechanics terms, a curved two-dimensional structure that can support bending moments is called a shell. A curved two-dimensional structure that cannot withstand bending is called a membrane. Unfortunately, this is a very different use of the word from its biological meaning. So be alert to the context in which these terms are used. Mechanical analysis of membranes can be thought of as a generalization of the analysis of strings to two dimensions and finds application in the analysis of rubber sheets and drums.

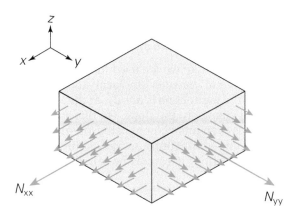

Figure 9.7 An infinitesimal element of a membrane depicting the forces N_{xx} and N_{yy}.

from the net action of the stresses acting on each edge of the element. Remember that stress is force per unit area. So, just as in the rod or column in Section 3.2, the resultant force on each edge is the integral of the stress over the area of the edge. However, there are some important differences. First, because the shell is two-dimensional, we must introduce the relevant x- and y-components denoted by subscripts. Secondly, rather than deal with the total force resultant on an edge, N, it is more convenient to divide by the edge width, b, such that $n = N/b$. The concept of resultant force per unit length (width) can be thought of as a generalization of *surface tension* that we discussed in Section 1.3. We define the resultant forces per length to be

$$n_{xx} = \int_{-h/2}^{h/2} \sigma_{xx} \mathrm{d}z \quad n_{yy} = \int_{-h/2}^{h/2} \sigma_{yy} \mathrm{d}z \quad n_{xy} = \int_{-h/2}^{h/2} \tau_{xy} \mathrm{d}z. \tag{9.26}$$

Note that sometimes (particularly in the structural mechanics literature) these are called the *stress resultants*.

With these definitions, we are now ready to derive the equilibrium conditions by constructing, as usual, a free-body diagram. If we consider an element of the membrane (Figure 9.8) that is not accelerating, equilibrium tells us that the

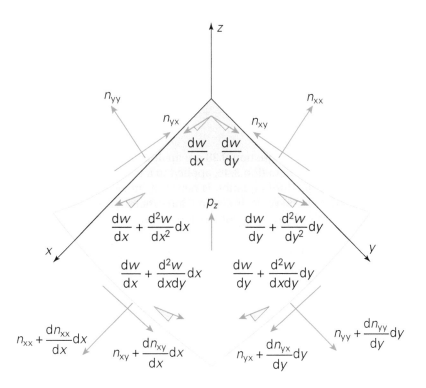

Figure 9.8 A free-body diagram for an infinitesimal element of a membrane, dx × dy. Note that because the membrane can curve, it is not restricted to remain in the x–y-plane. The slopes and rate of change of slopes are denoted on the interior of the shell.

resultant forces must sum to zero. This situation is slightly more complicated, because the membrane is not flat and the force resultants therefore change not only with respect to dx and dy, but also dz.

Figure 9.8 is more complex than we are used to. Owing to the curvature of the membrane, we now have to consider the local slopes. At the $x = 0$ edge, the slope of the surface in the y-direction is dw/dy. At the $x = $ dx edge, we must also account for the rate of change in the slope. On this edge, the slope of the surface in the y-direction is even more complicated, dw/dy + (d^2w/dy^2) dy. For the summations in the x- and y-directions, we ignore this effect, however, for the summation in the z-direction, it cannot be ignored. If you are careful, you should find,

$$\sum f_x \Rightarrow \; -n_{xx}dy + \left(n_{xx} + \frac{dn_{xx}}{dx}dx\right)dy - n_{yx}dx + \left(n_{yx} + \frac{dn_{yx}}{dy}dy\right)dx = 0$$

$$\sum f_y \Rightarrow \; -n_{yy}dx + \left(n_{yy} + \frac{dn_{yy}}{dy}dy\right)dx - n_{xy}dy + \left(n_{xy} + \frac{dn_{xy}}{dx}dx\right)dy = 0$$

$$\sum f_z \Rightarrow \qquad -n_{xx}dy\frac{dw}{dx} + \left(n_{xx} + \frac{dn_{xx}}{dx}dx\right)dy\left(\frac{dw}{dx} + \frac{d^2w}{dx^2}dx\right)$$

$$-n_{xy}dy\frac{dw}{dy} + \left(n_{xy} + \frac{dn_{xy}}{dx}dx\right)dy\left(\frac{dw}{dy} + \frac{d^2w}{dxdy}dx\right) \qquad (9.27)$$

$$-n_{yy}dx\frac{dw}{dy} + \left(n_{yy} + \frac{dn_{yy}}{dy}dy\right)dx\left(\frac{dw}{dy} + \frac{d^2w}{dy^2}dy\right)$$

$$-n_{yx}dx\frac{dw}{dx} + \left(n_{yx} + \frac{dn_{yx}}{dy}dy\right)dx\left(\frac{dw}{dx} + \frac{d^2w}{dxdy}dy\right) + p_z dxdy = 0.$$

Notice that if we divide by dx and dy, the terms simplify significantly to

$$\sum f_x = 0 \; \Rightarrow \; \frac{dn_{xx}}{dx} + \frac{dn_{xy}}{dy} = 0$$

$$\sum f_y = 0 \; \Rightarrow \; \frac{dn_{yx}}{dx} + \frac{dn_{yy}}{dy} = 0 \qquad (9.28)$$

$$\sum f_z = 0 \; \Rightarrow \; \frac{d\left(n_{xx}\frac{dw}{dx} + n_{xy}\frac{dw}{dy}\right)}{dx} + \frac{d\left(n_{xy}\frac{dw}{dx} + n_{yy}\frac{dw}{dy}\right)}{dy} + p_z = 0.$$

The first two equations of Equation 9.28 are analogous to the stress equilibrium condition we derived in Equation 3.46, applied to a two-dimensional membrane structure. However, the third equation is new and shows how the force equilibrium condition in the transverse direction relates membrane tension to pressure. We can further simplify by writing out the derivatives

$$n_{xx}\frac{d^2w}{dx^2} + 2n_{xy}\frac{d^2w}{dxdy} + n_{yy}\frac{d^2w}{dy^2} + p_z = 0. \qquad (9.29)$$

We can express Equation 9.29 in terms of curvature, just like we did for the beam, except that now there are curvatures in different directions.

$$n_{xx}\kappa_{xx} + 2n_{xy}\kappa_{xy} + n_{yy}\kappa_{yy} + p_z = 0, \qquad (9.30)$$

where

$$\kappa_{xx} = \frac{d^2w}{dx^2}, \kappa_{xy} = \frac{d^2w}{dxdy}, \kappa_{yy} = \frac{d^2w}{dy^2}. \tag{9.31}$$

In summary, equilibrium requires that the following conditions on the stress resultants be met:

$$\frac{dn_{xx}}{dx} + \frac{dn_{xy}}{dy} = 0$$

$$\frac{dn_{xy}}{dx} + \frac{dn_{yy}}{dy} = 0 \tag{9.32}$$

$$n_{xx}\frac{d^2w}{dx^2} + 2n_{xy}\frac{d^2w}{dxdy} + n_{yy}\frac{d^2w}{dy^2} + p_z = 0.$$

To gain a deeper understanding of this equation, we will take a closer look at this expression and elaborate it for two special cases, the case of planar shear and equibiaxial tension.

Equilibrium simplifies in the case of shear alone

In the case of shear loading, the applied tension in one direction, say n_{xx}, is larger than in the orthogonal direction (y-direction). The simplest case is one of "pure shear," which occurs when $n_{xx} = -n_{yy}$. In pure shear, a direction of maximum shear occurs at a 45 degree angle to the direction of load application and will have a magnitude of $(n_{xx} - n_{yy})/2$. Even if the loading is not pure shear, there is generally some amount of shear, as long as n_{xx} and n_{yy} are not exactly equal (if they are, we have equibiaxial tension, which is considered next). Recall from the constitutive equation for a general Hookean material (Equation 3.57), that the shear stresses and strains are *decoupled* from the normal stressed and strains. This means that the shear behavior (or deformational mode) is not influenced by anything else and can be analyzed on its own. Even for complex loading we can create a condition of pure shear by subtracting the equibiaxial component,

$$\tilde{n}_{xx} = n_{xx} - \left(\frac{n_{xx} - n_{yy}}{2}\right)$$

$$\tilde{n}_{yy} = n_{yy} - \left(\frac{n_{xx} + n_{yy}}{2}\right) \tag{9.33}$$

$$\tilde{n}_{xy} = n_{xy}.$$

The equilibrium condition (Equation 9.33) is not very informative. It requires that a case of pure shear can only exist if the shear resultant is constant. However, we can apply the constitutive equation to gain some insight, $\tau_{xy} = \gamma_{xy}$. We can reformulate this in terms of stress resultants. Because $n_{xy} = \tau_{xy}h$

$$n_{xy} = G\gamma_{xy}h$$

$$= K_S\gamma_{xy}. \tag{9.34}$$

We have introduced a new constant $K_S = Gh$ called the *membrane shear modulus*, which has units of force per unit length.

What are typical values of the membrane shear modulus for cells? The cell membrane of a red blood cell has a typical value of $K_S = 6 \times 10^{-6} - 9 \times 10^{-6}$ N/m = 6–9 pN/μm. Why is it so small? As we have described, the lipid bilayer is formed of molecules that are tightly bound to the layer, but easily move within the plane of

Nota Bene

Moment balance. When we applied the equilibrium equations, we did not consider the implications of the moment summation. Particularly useful is the observation that the in-plane shear resultants must be symmetric, $n_{xy} = n_{yx}$, which is a consequence of the requirement that the moments about the z-axis sum to zero.

the layer. This behavior is sometimes called a "two-dimensional liquid" because, like a liquid, if the membrane is exposed to shear it will simply flow to a new configuration until the shear is gone. For the bilayer, we can generally ignore the effects of shear. Another property of the lipids that make up the cell membrane is that they have a relatively large volume expansion or *bulk* modulus. That is, they are incompressible. As we will see in the following analysis, the in-plane bulk modulus of the lipid bilayer is also high, further supporting a two-dimensional fluid description for the membrane.

Equilibrium simplifies in the case of equibiaxial tension

In a state of equibiaxial tension the membrane experiences tensile forces that are equal in both directions, $n_{xx} = n_{yy} = n$, while the shear force vanishes, $n_{xy} = 0$. We will further assume that n is the same throughout the membrane; that it is independent of position and direction. At first glance this may appear to be a very peculiar and aggressive assumption. It would seem only to be valid for very specific artificial conditions. Surprisingly, however, this is a common and important condition for bilayer membranes. Why? As we have seen, any non-equal component in n_{xx} and n_{yy} will result in a shear in some plane. Because of the membrane's low viscosity and ability to flow under shear, it will reorganize such that the shear is again zero. As long as our boundary conditions are applied displacements, we know that n_{xy} is zero everywhere. If this is the case, equilibrium (Equation 9.32) requires that $dn_{xx}/dx = dn_{yy}/dy = 0$. The membrane tension must remain constant or the membrane will flow from regions of low tension to high tension until equilibrium is again satisfied. This is actually a very good assumption for the behavior of the lipid bilayer. It is equally applicable to other two-dimensional fluids, such as soap bubbles.

In this particular, but important, special case, we can simplify our description significantly. The question of force equilibrium in the x- and y-directions is automatically satisfied because the gradients are zero. In the transverse (z) direction, equilibrium reduces to the law of Laplace which we are familiar with from Section 1.3, but now generalized to allow curvatures in different directions,

$$n\left(\frac{d^2w}{dx^2} + \frac{d^2w}{dy^2}\right) + p_z = 0. \tag{9.35}$$

Note that the mixed term has dropped out because of the condition that $n_{xy} = 0$. Curvature is related to the local radius according to

$$\frac{d^2w}{dx^2} = \kappa_{xx} = \frac{1}{R_x}$$

$$\frac{d^2w}{dy^2} = \kappa_{yy} = \frac{1}{R_y}. \tag{9.36}$$

Therefore,

$$p_z = n\left(\frac{1}{R_x} + \frac{1}{R_{y-}}\right). \tag{9.37}$$

If the membrane is spherical in shape, we recover Equation 1.1 for the oil-drop model of the cell,

$$\frac{2n}{R} + p_z = 0. \tag{9.38}$$

Notice that the law of laplace is typically expressed as $\dfrac{2n}{R} = p_z$, because the curvature and pressure are oriented in different directions such that $p_z = -p$.

Areal strain can be a measure of biaxial deformation

In Section 1.3, we briefly introduced *areal strain* and the *areal expansion modulus* in our treatment of micropipette aspiration experiments. These quantities are particularly relevant when characterizing deformations under equibiaxial tension. We revisit them here to develop their origins more completely. In equibiaxial tension, any collection of points on the membrane will be deformed into a new state that is similar in shape to the original one, but will have its area increased or decreased. Similar to our definition of strain (the ratio of change in length to original length), we can define areal strain to be the change in area over the original area $\Delta A/A$. We can relate areal strain to linear strain by considering a small square element of dimension L on each side. Its initial area $A = L^2$ will increase to $A + \Delta A = (L + \Delta L)^2$ when it is deformed. However, this can be written in terms of linear strain, because $\Delta L = \varepsilon L$, $A + \Delta A = (1 + \varepsilon)^2 L^2$. Therefore,

$$\Delta A = (1 + \varepsilon)^2 L^2 - L^2 \tag{9.39}$$

and

$$\Delta A/A = (1 + \varepsilon)^2 - 1 = 2\varepsilon + \varepsilon^2 \simeq 2\varepsilon \tag{9.40}$$

by assuming that the term involving the square of strain will be in general much smaller than the others and can be neglected.

We can now define the material property *areal expansion modulus* for equibiaxial tension in terms of the tension and areal strain,

$$K_A = \frac{n}{(\Delta A/A)}. \tag{9.41}$$

K_A is a measure of the resistance of a membrane to in-plane biaxial stretching. It is left as an exercise for the student to show that it can be expressed as a function of Young's modulus and Poisson's ratio

$$K_A = \frac{Eh}{2(1 - v)}. \tag{9.42}$$

Typical values of the area expansion modulus for lipid bilayers are in the range of $K_A = 0.1$–1.0 N/m. The cell membrane of red blood cells, for example, has an areal expansion modulus of approximately $K_A = 0.45$ N/m ($= 450,000$ pN/μm). This value is many orders of magnitude greater than the membrane shear modulus. The bilayer is often treated as effectively inextensible. Typically, a lipid bilayer membrane can only tolerate 4–6% areal strain before rupture. The large resistance to areal change can be attributed to the changes in energy associated with exposing the hydrophobic core of the lipid bilayer to water as the spacing between the individual molecules is increased. Experimental approaches for measuring K_A are discussed in Section 9.6.

9.5 MEMBRANE MECHANICS II: BENDING

In Section 9.4 we considered a two-dimensional curved structure that does not experience bending moments (a structural membrane). Now we are going to add bending. A two-dimensional structure might be flat, in which case we are talking about a plate, or curved, in which case we are talking about a shell.

Recall that owing to its fluidity, bending resistance of the plasma bilayer is negligible for curvatures typical in a cell. It is reasonable to ask why we even worry about bending moments. Cells can, in fact, have substantial bending stiffness.

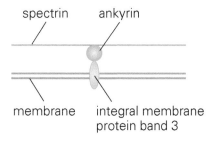

Figure 9.9 Complexes of membrane-spanning proteins and linker proteins are able to provide the cell membrane bending stiffness.

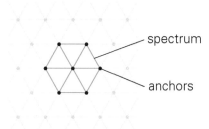

Figure 9.10 The spectrin network of the red blood cell forms a characteristic triangular network. The ankyrin and the spectrin networks are unique to red blood cells. However, other animal cells also support their bilayer with a network of membrane-associated actin. This layer of peripheral, dense actin is known as "cortical" actin and typically forms a quadrilateral repeating network.

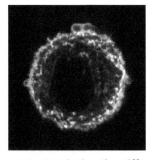

Figure 9.11 Cortical actin stiffens the cell membrane. In this fluorescent confocal micrograph, the actin forms a dense band around the periphery of the cell. See www.garlandscience for a color version of this figure. (Courtesy of Andrew Baik, from the laboratory of X. Edward Guo.)

How could the red blood cell maintain its characteristic bi-concave structure rather than acting as a flaccid bag of cytoplasm?

The answer to this paradox is that cells derive their bending stiffness from underlying cytoskeletal structures that support the bilayer. In the red blood cell, a network of the polymer *spectrin* exists directly under the bilayer. It is bound to the bilayer through periodic inclusions of *ankyrin*, a linker protein that attaches to membrane-spanning proteins (Figure 9.9). As we discussed in Section 8.4, the spectrin network takes on a characteristic triangular–symmetric form (Figure 9.10).

Another way in which cells develop bending rigidity is by covering themselves with a coating of proteoglycans known as the glycocalyx. The glycocalyx can be as thick as 0.5 μm. It is made up of long bottle-brush-like molecules that are highly bifurcated and covered with a high negative charge density. This causes intermolecular repulsion in the glycocalyx and tends to attract water molecules. The net effect is a tendency of the glycocalyx to be highly resistant to bending. In terms of bending, the bilayer tends to function primarily as a chemical barrier with the structural integrity being dominated by the cytoskeleton and glycocalyx (Figure 9.11).

In bending the kinematics are governed by membrane rotation

Analysis of bending moments in a curved shell is actually quite advanced and requires a working knowledge of differential geometry and the mathematics of curved spaces. So, we simplify our analysis by considering a flat plate as classically described by the Kirchhoff plate equation, which we will see is a fourth-order differential equation in the transverse displacement. Again, the plate equation is a result of the traditional continuum equations (the kinematics, the constitutive equations, and the equilibrium equations) combined with the definition of the stress resultants. Essentially, this is a two-dimensional extension of the development we considered previously for beam bending in Section 3.2.

Let us begin with the kinematics. We have already covered the Kirchhoff assumptions and the decomposition of strain into a part that is constant through the thickness and a part that varies linearly though the thickness. The former do not produce any bending moments, so we can drop them from Equation 9.22. For the pure bending that remains, our kinematic assumption gives us u and v in terms of the slope of the z-displacement w,

$$u = -z\frac{\mathrm{d}w}{\mathrm{d}x} \quad v = -z\frac{\mathrm{d}w}{\mathrm{d}y}. \tag{9.43}$$

From this we can calculate the strains from their continuum definitions, Equations 3.48, 3.49, and 3.52,

$$\varepsilon_{xx} = \frac{\partial u}{\partial x} = -\frac{\partial^2 w}{\partial x^2}z$$

$$\varepsilon_{yy} = \frac{\partial v}{\partial v} = -\frac{\partial^2 w}{\partial y^2}z \tag{9.44}$$

$$\varepsilon_{xy} = \frac{1}{2}\left(\frac{\partial u}{\partial y} + \frac{\partial v}{\partial x}\right) = -\frac{\partial^2 w}{\partial x \partial y}z.$$

As before, we can use our definition of curvature to obtain

$$\varepsilon_{xx} = \kappa_{xx}z \quad \varepsilon_{yy} = \kappa_{yy}z \quad \varepsilon_{xy} = \kappa_{xy}z. \tag{9.45}$$

Linear elastic behavior is assumed for the constitutive model

Next, we use the relevant constitutive equations, that is, the equations relating stress and strain. Again, we will assume a generalized Hookean material behavior and that the material properties and thickness are constant (homogeneous). In particular we need to relate the in-plane normal stresses σ_{xx} and σ_{yy} and the in-plane shear stress σ_{xy} to the corresponding strains ε or curvatures κ. Because of our kinematic assumptions, all of the z-direction strains are zero: $\varepsilon_{zz} = \varepsilon_{xz} = \varepsilon_{yz} = 0$. From Equation 3.57 we obtain,

$$\sigma_{xx} = 2\mu\varepsilon_{xx} + \lambda(\varepsilon_{xx} + \varepsilon_{yy}) = 2\mu z\kappa_{xx} + \lambda z(\kappa_{xx} + \kappa_{yy})$$

$$\sigma_{yy} = 2\mu\varepsilon_{yy} + \lambda(\varepsilon_{xx} + \varepsilon_{yy}) = 2\mu z\kappa_{yy} + \lambda z(\kappa_{xx} + \kappa_{yy}) \qquad (9.46)$$

$$\tau_{xy} = 2\mu\varepsilon_{xy} = 2\mu\kappa_{xy}z.$$

Equilibrium places conditions on resultant forces and moments

Now we are at the point at which we need to employ the equilibrium equation. However, equilibrium cannot be applied to stresses or strains directly. So, we need to consider resultant forces again. In this situation we are considering bending only, so we need to derive the *stress moment resultants* just as we did in Section 3.2 for beams. This will be a little more complex because of the two-dimensional nature of the situation, but it is quite analogous. For the beam, there was only one moment resultant. For the plate, we actually have three, denoted m_{xx}, m_{yy}, and m_{xy}. It is no surprise that they are obtained by integrating the corresponding stresses through the thickness of the plate,

$$m_{xx}(x,y) = \int_{-h/2}^{h/2} \sigma_{xx}(x,y,z)z\,\mathrm{d}z \quad m_{yy}(x,y) = \int_{-h/2}^{h/2} \sigma_{yy}(x,y,z)z\,\mathrm{d}z$$

$$m_{xy}(x,y) = \int_{-h/2}^{h/2} \sigma_{xy}(x,y,z)z\,\mathrm{d}z. \qquad (9.47)$$

With the moment resultants defined, what can equilibrium tell us? Imagining a small element exposed to moments is a little more complex than for in-plane force resultants, but it is conceptually the same. Figure 9.12 depicts the edges of a small element of the plate and the moment resultants at the edges with a double arrow symbol.

Nota Bene

The units of the moment resultants are the same as for force. Remember that for our distributed force resultants, the units were force per unit length, similar to surface tension. Normally, a moment has units of force times length. So, a distributed moment has units of force times length per unit length, or just force.

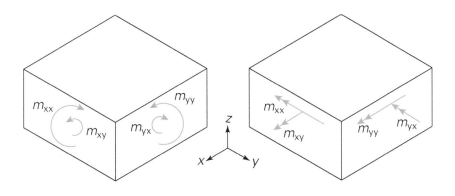

Figure 9.12 The moment resultants on the edges of the surface either directly (left) or with the double arrow symbology (right). The double arrow implies a moment around the axis of the arrow following the right-hand rule.

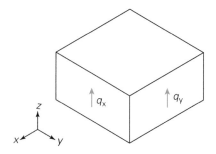

Figure 9.13 The distributed shear force resultants on the edges of the surface.

Just as with the beam, we cannot satisfy equilibrium conditions in terms of moments without considering shear. We must introduce another resultant force that we neglected before, the transverse shear force. We neglected it before because it does not play a role in membrane mechanics, but it is important in plates. The transverse shear force is the z-direction force that is produced from the x–z- and y–z-shear stresses (**Figure 9.13**).

We mathematically define the shear resultants by integrating through the thickness,

$$q_x(x,y) = \int_{-h/2}^{h/2} \sigma_{xz}(x,y,z)\,dz \quad q_y(x,y) = \int_{-h/2}^{h/2} \sigma_{yz}(x,y,z)\,dz. \qquad (9.48)$$

Now we are finally ready to apply the equilibrium equations. In the membrane analysis, we used force equilibrium to derive the governing equations and moment equilibrium about the z-axis to show a symmetry condition. For plate bending, we need to add the moment equilibrium about the x- and y-axes. Also, force equilibrium in the x- and y-directions does not apply because there are no forces in those directions. So, let us start with the force equilibrium in the z-direction. From **Figure 9.14** we can deduce the force equilibrium condition in the z-direction. Some simple manipulation (dividing by dx and dy and canceling terms) implies

$$\frac{dq_x}{dx} + \frac{dq_y}{dy} + p_z = 0. \qquad (9.49)$$

Similarly for the moments

$$\sum M_x = 0 \Rightarrow \frac{dm_{xx}}{dx} + \frac{dm_{yx}}{dy} - q_x = 0$$

$$\sum M_y = 0 \Rightarrow \frac{dm_{yy}}{dy} + \frac{dm_{xy}}{dx} - q_y = 0. \qquad (9.50)$$

Also, we can use the z-axis moment condition to show $m_{yx} = m_{xy}$.

Figure 9.14 Free-body diagram of resultants on an isolated portion of a plate depicting shear and moment resultants. Notice that the moment resultants on the edges of the surface are represented with the double arrow symbology.

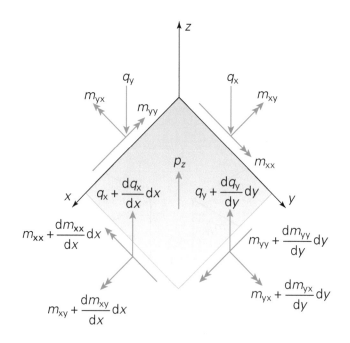

These are the equilibrium equations for a plate. They are a coupled set of first-order differential equations. You may remember that it is often possible to replace a coupled set of lower-order differential equations with a single higher-order differential equation. Indeed, this is the case here. We can eliminate the transverse shear components to find a second-order equation in the moments. It is left as an exercise for you to show that Equations 9.49 and 9.50 can be combined to yield

$$\frac{d^2 m_{xx}}{dx^2} + 2\frac{d^2 m_{xy}}{dxdy} + \frac{d^2 m_{yy}}{dy^2} + p_z = 0. \tag{9.51}$$

This is the classical equilibrium equation for a plate in terms of the distributed moments and pressure.

Now we have all the components we need to find the governing equations for plate bending. First, we combine our definition of moments (Equation 9.47) with our constitutive law (Equation 9.46). Combining the integrals yields

$$m_{xx} = \int_{-h/2}^{h/2} \sigma_{xx} z\, dz = K_B(\kappa_{xx} + \nu\kappa_{yy})$$

$$m_{yy} = \int_{-h/2}^{h/2} \sigma_{yy} z\, dz = K_B(\kappa_{yy} + \nu\kappa_{xx}) \tag{9.52}$$

$$m_{xy} = \int_{-h/2}^{h/2} \sigma_{xy} z\, dz = K_B\frac{1-\nu^2}{1+\nu}\kappa_{xy},$$

where

$$K_B = \frac{Eh^3}{12(1-\nu^2)}. \tag{9.53}$$

We can now insert this new expression for moments in terms of curvature into our equilibrium Equation 9.51 to yield

$$p_z = K_B\left(\frac{d^2\kappa_{xx}}{dx^2} + 2\frac{d^2\kappa_{xy}}{dxdy} + \frac{d^2\kappa_{yy}}{dy^2}\right) \tag{9.54}$$

or in terms of displacement

$$p_z = K_B\left(\frac{d^4 w}{dx^4} + 2\frac{d^4 w}{dx^2 dy^2} + \frac{d^4 w}{dy^4}\right). \tag{9.55}$$

This is the fourth-order differential equation describing plate bending relating the transverse displacement to applied pressure.

This completes our development and derivation of the plate-bending equation. However, let us try to get an intuitive feel for bending at a cellular level. The bending stiffness, K_B, of lipid bilayers such as the red blood cell membrane is on the order of 10^{-19} N/m. This is an exceedingly low value. Remember that K_A was on the order of 0.1 or 1 N/m. Of course, the units of these quantities are different. Nonetheless, there are about 20 orders of magnitude between typical bending and expansion forces. Bending forces in a bilayer are small even compared with shear. Recall the shear stiffness, K_S, is on the order of 10^{-6} N/m. Bending remains 13 orders smaller than this. We knew this was the case (although perhaps not how dramatically different these force regimes are), because we know resistance to bending of the cell membrane comes from membrane-associated structural proteins and not the bilayer. A support network of spectrin or cortical actin can increase the effective K_B dramatically.

Which dominates, tension or bending?

We have treated in-plane loading and bending as two separate conditions. Certainly, in reality, both occur simultaneously. Nonetheless, in many cases, one will dominate. To see this more clearly, we can formulate a description for combined loading,

$$n\left(\frac{d^2w}{dx^2} + \frac{d^2w}{dy^2}\right) + K_B\left(\frac{d^4w}{dx^4} + \frac{d^4w}{dx^2dy^2} + \frac{d^4w}{dy^4}\right) + p_z = 0. \qquad (9.56)$$

In general this equation can be difficult or impossible to solve analytically, but numerical solutions are certainly available. We can also determine which loading mode will dominate the solution and simplify in that way. We can estimate the rough relative magnitude of forces originating from each term, membrane surface tension due to n, and bending stiffness due to K_B. Assume that w is the transverse displacement that occurs over a characteristic length λ. The membrane term will scale with the quantity nw/λ^2. Similarly, the bending term will scale with K_Bw/λ^4. The ratio of these two quantities, $K_B/[n\lambda^2]$ gives us a good indication of whether tension or bending is dominant for a given situation.

$$\frac{K_B}{n\lambda^2} \lll 1 \Rightarrow \quad \text{tension dominates}$$

$$\qquad\qquad\qquad\qquad\qquad\qquad\qquad (9.57)$$

$$\frac{K_B}{n\lambda^2} \ggg 1 \Rightarrow \quad \text{bending dominates}$$

For a typical cell we might have $K_B = 10^{-18}$ N/m, $n = 5 \times 10^{-5}$ N/m, and $\lambda = 1$ μm. Therefore, $K_B/[n\lambda^2] = 0.02$, indicating that, in-plane effects are more important than bending.

9.6 MEASUREMENT OF BENDING RIGIDITY

We conclude this chapter with a discussion of the measurement of material properties of membranes. In Section 1.3, we saw that micropipette aspiration can be used to measure the areal expansion modulus K_A in lipid vesicles or in cells. Can a similar approach be used to measure the bending rigidity K_B? The answer is "yes," and conveniently, one can measure both properties simultaneously within the same experiment. To understand how bending rigidity can be ascertained in micropipette experiments, we must revisit the concept of thermal undulations.

Membranes undergo thermal undulations similar to polymers

In Section 7.3 we introduced the concept of thermal undulations in polymers and saw that entropy can play a significant role in their mechanical behavior. Membranes also undergo such fluctuations. In polymers, we characterized these thermal fluctuations using the persistence length, a characteristic length scale over which the segment of a polymer under the influence of thermal forces remains relatively straight. To review, the persistence length was derived from the orientation correlation function $f(s) = <\cos \Delta\theta(s)>$, where $\theta(s)$ is the angle the polymer makes with an imaginary horizontal line at each point s, and $\Delta\theta(s) = \theta(s) - \theta(0)$ (see Equation 7.14). In particular, we found that

$$\langle \cos \Delta\theta(s) \rangle = e^{\left(\frac{-s}{\ell_p}\right)}. \tag{9.58}$$

The orientation correlation function exponentially decreases from 1 to 0 as s gets larger and larger. In relating the persistence length to our continuum description of a beam, we found that it is related to its flexural rigidity EI as

$$\ell_p = \frac{EI}{k_B T}. \tag{9.59}$$

For surfaces such as membranes, thermal undulations can also be described using an orientation correlation function with some modifications. In particular, differences in orientations at two different points on the membrane $\mathbf{r}_1$ and $\mathbf{r}_2$ can be characterized by the dot product of the normals at these points, $\mathbf{n}(\mathbf{r}_1) \cdot \mathbf{n}(\mathbf{r}_2)$. If we let $\Delta r = \|\mathbf{r}_1 - \mathbf{r}_2\|$, then the membrane persistence length can be specified as

$$\langle \mathbf{n}(\mathbf{r}_1) \cdot \mathbf{n}(\mathbf{r}_2) \rangle = e^{-\Delta r / \ell_p}. \tag{9.60}$$

The persistence length in membranes can also be related to continuum models, in particular,

$$\ell_p \sim b e^{\frac{4\pi K_B}{3 k_B T}}, \tag{9.61}$$

where b is a characteristic length scale associated with the intermolecular spacing within the membrane. The derivation is beyond our scope here, but Boal (2001) has provided a detailed step-by-step treatment. As in a polymer, the persistence length in membranes increases with increasing resistance to bending and decreasing temperature. However, unlike in polymers, ℓ_p scales exponentially with these parameters instead of linearly.

Membranes straighten out with tension

Just like polymers, membranes are subject to thermal undulations in a manner that is dependent on bending rigidity. In polymers, we showed that "coiled" versus "straight" configurations could be characterized through differences between the end-to-end length and contour length. In addition, we found that "coiled" configurations were entropically favored; however, through the application of force, the polymer could be extended. Analogously, in membranes, "wrinkled" configurations are entropically favored, and these wrinkles can be smoothed out through the application of tension. Furthermore, a quantitative relationship exists.

Consider a surface defined through a height function over the xy-plane, $h(x,y)$. We assume that there are no overlaps in the surface, that is, h is unique at each point (x,y). Let A be the *contour area* (the true area) of the surface and A_{proj} be the projected area in the xy-plane. A_{proj} can be shown to depend on the tension, n, and bending rigidity K_B,

$$A_{\text{proj}}(n) = A - \frac{A k_B T}{8 \pi K_B} \ln\left(\frac{\pi^2 / b^2 + n / K_B}{\pi^2 / A + n / K_B} \right). \tag{9.62}$$

Note that A_{proj} goes to zero when $k_B T / K_B$ goes to zero, indicating that no wrinkles will occur in the limit of zero temperature or infinite bending rigidity. A_{proj} also goes to zero in the limit as n/K_B gets arbitrarily large.

From this we can compute the amount of tension required to increase the projected area from a tensionless reference state,

$$A_{\text{proj}}(n) - A_{\text{proj}}(0) = \frac{A k_B T}{8\pi K_B} \ln\left(\frac{1 + nA/\pi^2 K_B}{1 + nb^2 \pi^2/\pi^2 K_B} \right) \tag{9.63}$$

or

$$\Delta A/A = \frac{k_B T}{8\pi K_B} \ln(1 + nA/\pi^2 K_B), \tag{9.64}$$

where $\Delta A = A_{\text{proj}}(n) - A_{\text{proj}}(0)$, and we have assumed that $A \gg b^2$, the square of the intermolecular spacing. The above expression gives the "effective" areal strain associated with smoothing out thermal undulations under a given tension n. This expression can be superimposed with the zero temperature areal strain given by the areal expansion modulus (that is, the strain associated with increasing the intermolecular distance),

$$\Delta A/A = \frac{k_B T}{8\pi K_B} \ln(1 + nA/\pi^2 K_B) + n/K_A. \tag{9.65}$$

Example 9.5: Measurement of K_B and K_A in a single experiment

How does Equation 9.65 behave in the high and low tension limits and how might this be used to determine K_A and K_B?

In the low tension regime, Equation 9.65 becomes

$$\Delta A/A \approx \frac{k_B T}{8\pi K_B} \ln(1 + nA/\pi^2 K_B).$$

In the high tension regime, n dominates over $\ln(n)$, so, as n approaches zero,

$$\Delta A/A \approx n/K_A.$$

Consider an experiment in which a lipid vesicle is being subjected to micropipette aspiration. In the absence of tension, entropic forces induce wrinkles in the membrane. Upon application of a low tension, these entropic forces are overcome, the wrinkles are smoothed, and the response is given by the low tension expression. The areal strain is linear with $\ln(n)$; fitting the slope gives K_B. Once the membrane has been pulled taut, the stiffness dramatically increases, owing to the force associated with increasing the intermolecular spacing between the lipids in the bilayer. Here, the response is given by the high tension expression, and the areal strain is linear with n. K_A is also easily derived from the curve fit.

Advanced Material: Pipette aspiration measurement of areal expansion modulus

An alternative approach to measuring the areal expansion modulus is using a small-bore pipette. One can determine the osmotic pressure across the membrane as a function of solute concentration and temperature as

$$p_{\text{osmotic}} = k_B T \sum_i [C_i],$$

where $[C_i]$ is the molar concentration of each species present. From the law of Laplace one can relate p_{osmotic} to n, and the areal extension can be determined from the amount of membrane drawn into the pipette.

Key Concepts

- The amphiphilic nature of cell membrane phospholipids allows them to self-assemble into the bilayer. The packing behavior of the phospholipids contributes to the structures formed.

- Microdomains within the membrane enhance chemical kinetics.

- Saturated and unsaturated fatty acid chains contribute to determining membrane properties.

- The fluid mosaic model describes the two-dimensional fluid behavior of the bilayer.

- The membrane is semipermeable and has a critical barrier function that can be described with the diffusion equation.

- The continuum mechanical behavior of the membrane can be decomposed into in-plane and bending components. In-plane behavior can be further decomposed into equibiaxial and shear responses.

- Derived moduli, including K_A, K_S, and K_B, characterize the areal, shear, and bending stiffnesses-respectively.

- The quantity $K_B/n\lambda^2$ is indicative of tension-dominated or bending-dominated behavior.

- The bilayer is relatively flexible in shear and bending but can be stiff with respect to areal tension.

- Entropic membrane undulations can be described statistically and are related to continuum properties.

Problems

1. (a) In one dimension, the mean squared displacement of a particle undergoing a random walk varies with the square root of t (time). How does the mean squared displacement vary with time in two dimensions? And in three dimensions? Why?

   (b) Let us say you defined the diffusion velocity as the root mean squared displacement divided by time, for a given diffusion coefficient. If you do this, what happens as t approaches 0? How can you tell if experimental data you are analyzing are affected by this problem?

2. For typical small molecules in water, the diffusion coefficient is $D \simeq 10^{-5}$ cm^2/s. Small ions typically pass through channels diffusively. You can assume that the channels are only wide enough for a single ion to pass through at a given time.

   (a) Estimate the time it takes for a small ion to cross the channel, assuming there is no interference from other ions, and the channel has the same length as the thickness of a cell membrane.

   (b) Now assume someone is building a scaled-up macro-cell whose radius is about 1 m. How long would it take the same small ion to cross the macro-channel? Is diffusion an effective mechanism for large-scale processes?

3. Consider a hole of radius a in an infinitely large wall in space. If the concentration far away from one side of the disk is C_0, and is zero everywhere on the other side, the current flux of freely diffusing molecules passing through this hole is given by $I = 4DaC_0$. This can be a model for a single channel in a membrane (assuming the radius, a, is much smaller than the radius, b, of the cell itself). We have derived the current flux of freely diffusing molecules passing through a sphere as $I = 4\pi DC_0 b$.

   (a) If you had n channels in the cell surface, where n is small, what is the total current flux into the cell?

   (b) If the entire sphere is acting as a sink with current flux I_0, then the total current flux if you had n channels—for any n—is $I = I_0/[1 + \pi b/na]$. Is this consistent with your answer from (a)? Justify your response.

   (c) Using the information in (b), design a cell with radius $b = 10$ μm with n channels (sinks) of radius $a = 1$ nm so that its current flux through all of the channels is half of the current flux if the entire cell surface was a sink (that is, find n). You may neglect the local curvature in calculating the total area of the channels. Next, find the fraction of the cell surface area that is covered with channels for this to occur. What does this result tell you about the ability of a cell to acquire needed compounds by diffusion through channels?

4. Consider the transport of a solute through a thin blood vessel. The blood vessel has inner radius R_1, the endothelial layer increases the radius to R_2, and then the outer layer (matrix) further increases the radius to R_3. If the diffusion coefficient of the endothelial layer for the solute is D_1 and the diffusion coefficient of the matrix is D_2, find the total (steady-state) solute flux out of the blood vessel, assuming the blood concentration is effectively constant at C_0 inside the blood vessel and is zero everywhere outside (the solute is immediately consumed). Keep in mind you need to use the cylindrical form of the diffusion equation(s).

5. Using the same setup as in Problem 4, model the blood vessel as a hollow cylinder of inner radius R_1 and outer radius R_3 but consisting of only one material of diffusion coefficient D, with the same flux out of

the vessel. Find D in terms of D_1, D_2, and the given parameters.

6. You are developing a drug delivery system consisting of a small sphere of inner radius R, wall thickness h, and wall diffusion coefficient D. Assume that the internal drug concentration is constant at C and the concentration outside the sphere is zero everywhere (used up immediately) for the time interval we are interested in.

    (a) Calculate the total drug flux out of the sphere.
    (b) Next, assume you have a flat wall of the same thickness h and the same surface area as the inside of the sphere in the original problem statement. Further, assume the concentrations and diffusion coefficients are analogous to the spherical setup. Calculate the total drug flux through the wall.
    (c) Find a criterion for h and R in terms of the other parameters such that the difference between your answers for parts (a) and (b) is less than 1%. Would something the size of a typical cell with the membrane being the "wall" qualify, based on your criterion?

7. Assume the cell membrane has a fixed thickness h and a diffusion coefficient D for some solute of interest. You may model the cell membrane locally as a flat wall. The external concentration of the solute is C and is 0 everywhere inside the cell (consumed immediately).

    (a) Develop an expression for the concentration inside the membrane at steady state, making sure to identify the boundary conditions you used. Sketch this concentration profile.
    (b) At some time after this steady state has been achieved, the cell changes the diffusion coefficient to a new $D' < D$ (for example, by closing some previously open channels). You then wait for a new steady state to be achieved. List what you expect to change with the new diffusion coefficient D', assuming every other parameter remains the same.

8. You are given a finite number of identical (finite-volume) tanks separated by geometrically identical membranes. You may set the initial concentration of the solute in each tank (and assume that the concentration is always uniform inside the tank itself) as well as the diffusion coefficient of each membrane (but once you have chosen the D for a membrane, it must remain fixed forever). Configure the tanks and membranes such that the tank with the highest initial solute concentration will *receive* a flux of the solute at some point in time (before equilibrium).

9. Show that $\langle T_c \rangle = \dfrac{b^3}{3Da}$ is the solution for $\langle T_c \rangle = \dfrac{b^3}{3Da}$ for the conditions specified in Example 9.2.

10. We wish to model the cell locally as a flat wall of thickness h and diffusion coefficient D. Because we are looking very closely at the cell, we will treat it as a semi-infinite media, such that the cell starts at $x = 0$ and increases to infinity. Initially, the concentration of some solute is zero everywhere. At time $t = 0$, the concentration of the solute outside the cell suddenly increases to $C_0 > 0$. You may assume that inside the cell, at $x \to \infty$, the concentration $C = 0$ forever.

    (a) Write Fick's second law of diffusion for this setup for $t > 0$.
    (b) Define a new variable α so that

    $$\alpha = \frac{x}{\sqrt{4Dt}}.$$

    Rewrite the equation you derived in (a) in terms of α and show that the equation can be expressed as

    $$\frac{\partial^2 C}{\partial \alpha^2} = 2\alpha \frac{\partial C}{\partial \alpha} = 0$$

    You may need to use the chain rule.
    (c) Write the boundary conditions for the new equation in (b) using the same constraints you determined for (a).
    (d) Find the solution for the differential equation in (b), using the boundary conditions you found in (c). You may want to express the solution in terms of the error function (erf(x)).
    (e) What happens at the limit of the solute flux as t approaches infinity? Is it the same as the steady-state diffusive flux? Why or why not?
    (f) Someone makes the following statement: "The solution derived for the time-dependent concentration profiles in (d) suggest that if I have two materials of different diffusion coefficients, with everything else being the same as in the original problem statement, then the concentration profiles in the two materials will be identical, only shifted in time. That is, it will never be the case that for $t > 0$ one material will exhibit a concentration profile that the other will never exhibit." Support or disprove this statement.

11. Is it possible to have water climb 1 km in a capillary tube on Earth, using only surface tension? You may control the diameter of the tube (which does not have to be constant in space), the tube's angle to the vertical (but only the vertical height counts), and how deep the water the end is submerged (but only the height above the surface counts). You may also assume that water is a continuum so that you can make the diameter arbitrarily small without worrying about molecular effects (or even that the diameter may be smaller than a water molecule). If your answer is yes, what is the shape of the tube?

12. If you have a water droplet and compress it slightly, resulting in an ellipsoid (technically an oblate spheroid), and release the compressing force, show that the droplet tends to return to a spherical shape and not to distend further, using free-body diagram arguments.

13. The lipid bilayer consists of fatty acids that have varying degrees of hydrogenation. Recall that fully saturated lipid chains will be relatively straight (zig-zag, actually). Do you expect unsaturated lipid chain regions to have higher or lower viscosity compared with the saturated region? How about regions containing primarily partially hydrogenated fatty acids?

14. Show that

$$\frac{d\left(n_{xx}\dfrac{dw}{dx} + n_{xy}\dfrac{dw}{dy}\right)}{dx} + \frac{d\left(n_{xy}\dfrac{dw}{dx} + n_{yy}\dfrac{dw}{dy}\right)}{dy} + p_z =$$

$$n_{xx}\frac{d^2w}{dx^2} + 2n_{xy}\frac{d^2w}{dxdy} + n_{yy}\frac{d^2w}{dy^2} + p_z$$

15. In the text we mentioned typical strains at rupture for a bilayer. Using reasonable values given in the text, estimate the typical range of maximum membrane tension and cytosolic pressure a cell of diameter 1 μm could withstand.

16. Consider the cortex of a red blood cell. Develop a model to predict the membrane mechanics of a two-dimensional array of cross-linked spectrin polymers arranged in a simple rectangular lattice. The lattice spacing is L_e. Each one of the polymers has a persistence length ℓ_p, contour length of L, and diameter d.

   Using optical tweezers, researchers have found that the persistence length of spectrin is 10 nm. Furthermore, the contour length of each spectrin molecule is 200 nm, and L_e is approximately 70 nm.

   (a) Assuming that the deformations are small, what regime (ideal chain, semiflexible polymer, continuum mechanics) are these polymers in and why? What model would be appropriate to describe their behavior?

   (b) Now derive an expression for the membrane area expansion modulus K_A for the model.

17. Show that $\tau_{xy} = G\gamma_{xy} = E/(2(1+\nu))\gamma_{xy}$ and $K_S = Gh = Eh/(2(1+\nu))$.

18. Derive an expression for the areal expansion modulus for a membrane in terms of the membrane thickness, Young's modulus, and Poisson's ratio.

19. Show that Equations 9.49 and 9.50 can be combined to yield the equilibrium equation for a plate.

20. In the text, the claim was made that the bending stiffness, K_B, of the bilayer was much smaller than for the cell, because of the support provided by the cytoskeleton. Conduct a literature search and write a short report (no more than a quarter to half of a page) to support or refute this claim.

21. Suppose you have a 1 μm cell with a membrane thickness of 10 nm made of an incompressible material with a Young's modulus of 10^8 Pa. Estimate the change in pressure necessary to inflate the cell from an initial collapsed state in which the aspect ratio (height/diameter) is 0.2, to one in which the aspect ratio is one with the large diameter approximately equal to 10 μm. (This is roughly equivalent to the geometry change associated with a red blood cell going from its normal state to spherical.) Assume that bending stiffness dominates over tension.

22. Consider a plate experiencing a moment in one direction ($M_{xx} = M\ M_{yy} = M_{xy} = 0$).

   This is similar to the beam-bending problem we analyzed in Chapter 3. In that case we found that

   $$M = EI\frac{d^2w}{dx^2}.$$

   Does the plate theory produce the same equation for this special case? If not, why is it different? What further assumption would be needed for the plate theory to reproduce the beam theory?

23. Show that $K_B = \dfrac{Eh^3}{12(1-\nu^2)}$.

24. Show that, taken together, the moment–curvature relation (Equation 9.52) and moment equilibrium equation (Equation 9.51) result in the classical plate bending equation in terms of displacements,

   $$p_z = K_B\left(\frac{d^2\kappa_{xx}}{dx^2} + 2\frac{d^2\kappa_{xy}}{dxdy} + \frac{d^2\kappa_{yy}}{dy^2}\right)$$

Annotated References

Axelrod D, Koppel DE, Schlessinger J et al. (1976) Mobility measurement by analysis of fluorescence photobleaching recovery kinetics. *Biophys. J.* 16, 1055–1069. *This research article covers the theoretical basis underlying fluorescence recovery after photobleaching in order to characterize diffusion in 2-dimensional systems.*

Berg HC (1983) Random Walks in Biology. Princeton University Press. *Presents a wide range of biological problems that can be analyzed through the random walk. Specifically our treatment of diffusion is motivated from this book.*

Boal D (2001) Mechanics of the Cell. Cambridge University Press. *An excellent text on cell mechanics with many in-depth treatments of shell mechanics. Particularly useful for this chapter is the statistical treatment of membrane behavior.*

Evans E & Rawicz W (1990) Entropy-driven tension and bending elasticity in condensed-fluid membranes. *Phys. Rev. Lett.* 64, 2094–2097. *Gives expression for "effective" areal strain associated with smoothing out thermal undulations under a given tension.*

Helfrich W (1975) Out-of-plane fluctuations of lipid bilayers. *Z. Naturforschung. C* 30, 841–842. *Early work demonstrating that out-of-plane fluctuations of membranes lead to a decrease in effective area and alters stretching elasticity. Gives expression for projected surface area for a given tension n and bending rigidity.*

Kamm RD (2005) Molecular, Cellular, and Tissue Biomechanics (lecture notes from course number 20.310, Massachusetts Institute of Technology, Cambridge, MA). *Several of the treatments in this chapter were based on these graduate course notes. Specific material included in-plane and bending behavior of plates and shells and several problems.*

Kooppel DE, Axelrod D, Schlessinger J et al. (1976) Dynamics of fluorescence marker concentration as a probe of mobility. *Biophys J.* 16, 1315–1329. *An early description of FRAP used to measure lateral diffusion in the cell membrane.*

Simons K & van Meer G (1988) Lipid sorting in epithelial cells. *Biochemistry* 27, 6197–6202. *An early description on microdomains in lipid membranes.*

Singer SJ & Sackmann E (1972) The fluid mosaic model of the structure of cell membranes. *Science* 175, 720–731. *One of the earliest descriptions of the fluid mosaic model of the cell membrane.*

Timoshenko S & Woinowsky-Krieger S (1959) Theory of Plates and Shells. Engineering Society Monographs. *A defining text on the mathematics of plate and shell mechanics.*

CHAPTER 10

Adhesion, Migration, and Contraction

Thus far, our focus has primarily been on cell biomechanics; in other words, the study of the mechanical behavior of cells. In these treatments, we used theoretical and experimental mechanics to understand better the mechanical behavior of cells or their structural constituents. Ultimately, we are interested in using this knowledge to understand how mechanics governs cellular function. We now make a conceptual turn, shifting our focus from cell biomechanics to cellular mechanobiology; in other words, aspects of biology in which mechanical force is generated, imparted, or sensed, leading to alterations in biological function. Because mechanics plays a critical role in many different physiological and pathological processes, a complete treatment of cellular mechanobiology is more than we can address here. In this chapter, we aim to give the reader a firm foundation upon which to build by focusing on three vital cellular processes in which mechanics dictates biological function: adhesion, migration, and contraction.

10.1 ADHESION

Cells can form adhesions with the substrate

We begin this chapter with a discussion of cell–substrate adhesion, which is vital for many biological processes, including tissue cohesion, repair, inflammatory responses, and growth. Adhesion is also important for engineering applications, such as seeding new constructs with cells for implantation.

Among the best-characterized cell–substrate adhesive molecules are integrins. As we first described in Section 2.2, integrins are extracellular matrix protein receptors that are tightly bound to the plasma membrane. They are dimers composed of α and β subunits, each of which is found in one of several isoforms. Each isoform is denoted with subscripts. The different isoform dimerizations allow for the extracellular component of the integrins to bind to different extracellular matrix components such as collagen or fibronectin. For example, $\alpha_5\beta_1$-integrin is composed of an α_5 subunit and a β_1 subunit and is a fibronectin receptor, specifically targeting the arginine–glycine–aspartate (RGD) peptide sequence in fibronectin. The intracellular components of integrins are often structurally coupled to the cytoskeleton in an indirect manner through one or more intermediary molecules. The intracellular component of the integrin may also be a major site of enzymatic activity.

Integrins can be assembled into discrete adhesive plaques called focal adhesions. As they are assembled, an intracellular molecular complex is also created. This complex consists of many proteins, each of which has various roles in intracellular biochemical signaling and forms structural links with the actin cytoskeleton (Figure 10.1). It is currently unclear what, if any, advantage is associated with multiple focal attachments rather than a more distributed scheme. However, their formation and destruction are highly regulated. It has been shown that microtubules can disrupt focal adhesions locally at microtubule leading tips. It is

Figure 10.1 Focal adhesions are molecular complexes involved in cell–substrate adhesion.
Photomicrograph of focal adhesions in adherent cells. The cells have been immunostained for the focal adhesion protein talin (encircled dot). The actin cytoskeleton and nuclei have also been stained. (B) A pictorial representation of a focal adhesion demonstrating many of the key proteins comprising this molecular complex. Proteins within the complex have various roles, including functioning as intracellular signaling molecules and forming structural links with the actin cytoskeleton. (A, see www.garlandscience.com for a color version of the figure. B, adapted from, Kamm R & Lang M *Molecular, Cellular, and Tissue Biomechanics.* Massachusetts Institute of Technology)

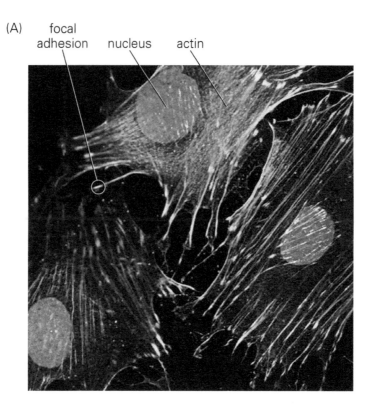

(A)

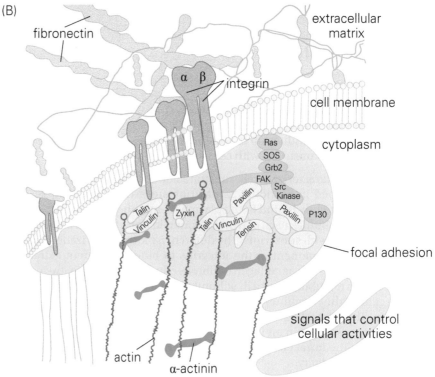

(B)

thought that that the various components of focal adhesions play different roles in adhesion, activation, and maintenance of mechanical competence, although the exact roles of each component are still being defined. Because they play important roles in both structural integrity and intracellular signaling, they are a major candidate for mechanotransduction sensors (covered in Section 11.2).

Fluid shear can be used to measure adhesion strength indirectly

It is clear that adhesions play vital roles in cellular function. Therefore, it is of great interest to quantify their adhesive strength. Adhesive strength can be assessed both directly and indirectly. Indirect measurements do not allow for the quantification of actual forces required to detach cells from a given substrate, but they do allow for relative comparisons of adhesive strength between multiple samples.

One approach for indirectly measuring adhesive strength is to measure the level of fluid shear stress required to detach adherent cells from a surface. If a large enough shear stress is applied, the cells will begin to detach. As we saw from Equation 6.12, one way to apply a spatially varying shear stress over a uniform cell population is through a centralized flow that spreads out radially from a central inlet. The farther the cell is from the center, the lower the flow rate and the lower the shear stress experienced by the cell. If cells are plated on a coverslip and placed in a radial flow chamber there will be a circular patch of cleared cells. The radius of the cleared area is inversely proportional to the adhesive forces of the cells on the coverslip.

Parallel plate flow chambers can also be used for indirectly measuring adhesive strength. Recall from Section 4.2 that we obtained the flow profile within infinitely wide parallel plates. It can be shown that the surface shear stress on the bottom of a parallel plate flow chamber is

$$\tau = \frac{\mu V_0}{h},\tag{10.1}$$

where V_0 is the velocity at a height h above the surface and μ is the fluid viscosity. For a pressure-driven parallel plate configuration, Equation 10.1 becomes

$$\tau = \frac{6\mu Q}{bh^2},\tag{10.2}$$

where Q is the volume flow rate, b is the chamber width, and h is the chamber height. Different shear stresses can be applied to the same cell population by increasing the flow rate with time, or by constructing a chamber with a spatially varying geometry, such as one whose width changes along the direction of flow.

Detachment forces can be measured through direct cellular manipulation

In contrast with the use of fluid shear stress for indirectly assessing adhesive strength, the forces required to detach adherent cells from their substrate can also be measured through direct cellular manipulation. One approach involves the use of micropipette aspiration (Figure 10.2A). Suction pressure within a micropipette is used to hold a cell and place it in contact with a surface with interest, such as a functionalized bead. After binding of the cell to the surface, the cell is pulled away from the surface, and the critical force required to separate the cell from the surface is measured with a force transducer. This setup can be modified to measure cell–cell adhesion strength by using two micropipettes instead of one (Figure 10.2B). The two cells are brought into contact with each other, intercellular adhesions are allowed to form, and the force required to separate the cells is measured. This technique is particularly useful for cells that function normally in the absence of a substrate, such as white blood cells.

For adherent cells, the lack of a substrate and a spread-out morphology may affect processes associated with adhesion. Therefore, detachment forces in a non-spread configuration may not be representative of physiological conditions. To address this problem, microplate manipulation can be used (Figure 10.2C).

Figure 10.2 Schematic depicting the use of (A) single pipette, (B) double pipette, (C) microplate, and (D) atomic force microscopy to measure adhesion strength. The arrows indicate the direction of motion to attach/detach the cell to the probe or to another cell.

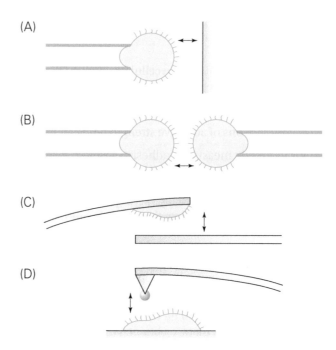

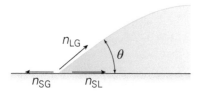

Dispase can be used to measure cell–cell dissociation. An assay unique to cell–cell adhesion is the dispase-dissociation assay. Cells are plated and allowed to grow to confluence. They are then detached from the substrate using *dispase*, which typically targets cell–substrate receptors but not cell–cell receptor interactions. The result is that the cells are released from the substrate as a *sheet*. This sheet of cells is then vigorously mixed so that shearing forces separate the cell sheet into smaller sections. The size of the aggregates is a surrogate for cell–cell adhesion.

Figure 10.3 Surface tensions for a drop partly wetting a surface. This schematic depicts a close-up of the point of contact between the droplet and the substrate. At this point, there are three surface tensions between the solid, liquid, and gas phases: one tangent to the solid–gas interface (n_{SG}), one tangent to the solid–liquid interface (n_{SL}), and one tangent to the liquid–gas interface (n_{LG}). The angle between the gas–liquid interface and the substrate is θ.

Microplate manipulation is similar to micropipette aspiration. Instead of using a vacuum within a micropipette to hold a cell in place, one uses a small island to which the cell adheres. This technique can be modified to measure cell–cell adhesion through the addition of a second microplate manipulator. However, the force required to break the cell–cell adhesion must be smaller than the force required to break the cell–matrix adhesion. Similar to microplates, atomic force microscopy has been used for measuring cell adhesion using the built-in cantilever to measure the force and functionalizing the cantilever tip to facilitate cell attachment (Figure 10.2D).

Another scheme is to use micropipette aspiration and micromanipulation to peel an edge of the adherent cell off a substrate and then pull the cell back. This has the advantage of being a more direct measurement of cell–substrate adhesive force. However, peeling is less a measure of whole-cell adhesive force than of individual molecular strength in sequence. We consider this in more quantitative detail in the following section.

The surface tension/liquid-drop model can be used to describe simple adhesion

We now shift our focus from experimentally measuring adhesive strength to mathematical models of cell–substrate adhesion. As we will see, such models can provide a great deal of insight into such mechanobiological functions as the factors that govern whole adhesion strength or how cells control their morphology through the expression of adhesive molecules. We begin this section with a discussion of the energy of adhesion of liquid drops in contact with a solid surface. We have already discussed an application of the liquid-drop model in Section 1.3, in which we found that, during micropipette aspiration, some cells behave as a liquid droplet with a particular surface tension. In this section, we will see that the liquid-drop model can also be integrated with force- or energy-based models of adhesion to gain insight into processes such as peeling behavior.

To begin our treatment, consider a liquid drop partly wetting a surface, as depicted in Figure 10.3. Recall that surface tension arises from an imbalance of molecules at the interface of two different phases. At the edge of the liquid drop, there will be three surface tensions between the solid, liquid, and gas phases: one tangent to the solid–gas interface (n_{SG}), one tangent to the solid–liquid interface (n_{SL}), and one tangent to the liquid–gas interface (n_{LG}). Assuming the drop is at equilibrium, we can perform a horizontal force balance at the liquid-drop edge,

$$-n_{SG} + n_{SL} + n_{LG} \cos\theta = 0. \tag{10.3}$$

We now seek to calculate the energy required to "lift" the drop from the surface. Physically, surface energy is the energy required to generate a new surface, and arises due to the disruption of intermolecular bonds when new surfaces are created. For a liquid, surface tension (which has units of force per unit length) and the surface energy density (surface energy per unit area, which also has units of force per unit length) are identical. The energy density of adhesion is

$$J_{LG} = n_{LG} + n_{SG} - n_{SL}. \tag{10.4}$$

Substituting into Equation 10.3, we can rewrite Equation 10.4 as

$$J = n(1 + \cos\theta), \tag{10.5}$$

where $J = J_{LG}$ and $n = n_{LG}$. Equation 10.5 is sometimes referred to as Young's equation or the Young–Dupré equation. It implies that the energy density required to "lift" the drop from its surface is dependent on only two factors, the surface tension n, and the angle that the membrane makes with the surface.

It is possible to derive a relation similar to Equation 10.5 by considering the internal force within membrane receptors directly. It can be shown that the relationship between adhesion energy density J and membrane tension n is

$$\frac{\rho_b F_R L}{2} = n(1 + \cos\theta), \tag{10.6}$$

where ρ_b is the receptor bond density, F_R is the force that would rupture the receptor–ligand bond, and L is the critical bond length at which the receptor is stretched with force F_R. Notice that this is essentially a reformulation of Young's equation (Equation 10.5).

From the form of Equation 10.6 we can make some predictions about the capacity for cells to control the degree to which they spread. Consider a cell that is at adhesive equilibrium as a half-dome on a flat surface. We can treat the cell as a liquid drop with contact angle $\pi/2$ and membrane tension $n = J_0$.

Equation 10.5 implies that, the adhesion energy density is $J = J_0$. One might suppose that to spread out more toward a flat pancake shape, the cell would be required to increase the adhesion energy density indefinitely, but this is not the case. In particular, if we increase the adhesion energy density by a factor of q, then Equation 10.6 predicts that

$$qJ_0 = n(1 + \cos\theta). \tag{10.7}$$

But because we know $n = J_0$, Equation 10.7 simplifies to

$$q = (1 + \cos\theta). \tag{10.8}$$

This implies that q is at most two, which occurs when θ is equal to zero. In other words, doubling the energy adhesion density is sufficient to spread the cell to the maximum extent allowed by the membrane tension. This example suggests that a cell may only need to adjust its adhesion energy density by small amounts to produce substantial changes in spreading (for instance, by adjusting its receptor bond density as in Equation 10.6).

Advanced Material: Young's equation can be derived using energy considerations

Equation 10.5 was derived using a combination of a force balance and energy considerations. One can also derive this relation solely using energy considerations. Consider a droplet to be a truncated sphere of radius R, parameterized by the angle θ (Figure 10.4). A "cone" with vertex angle θ will truncate the part of the sphere that we use to model the drop, resulting in a contact angle of θ as well. Using our relation for energy density of adhesion given in Equation 10.4, the energy of the drop can be written as

$$E = (n_{SL} - n_{SG})A_{SL} + N_{LG}A_{LG}, \tag{10.9}$$

where

$$A_{SL} = \pi R^2 \sin^2 \theta \tag{10.10}$$

is the area of the flat surface with which the drop is in contact and

$$A_{LG} + 2\pi R^2 (1 - \cos\theta) \tag{10.11}$$

is the total area of the "dome shaped" liquid–gas interface. To obtain Equation 10.3, we can use the fact that at equilibrium there is no change in energy with respect to A_{SL},

$$\frac{\partial E}{\partial A_{SL}} = 0. \tag{10.12}$$

Combining Equations 10.9, 10.10, and 10.11, and solving Equation 10.12, leads directly to Equation 10.3.

Figure 10.4 A droplet of water approximated as a truncated sphere. The angle of truncation, or the vertex angle of the cone, is θ. This angle is also the contact angle of the droplet on the surface. The area of the solid–liquid interface is "underneath" the droplet and the area of the liquid–gas interface is the "upper surface" of the droplet.

Adhesive peeling can be modeled using continuum mechanics

In the last section, we investigated the dependence of adhesion energy density on membrane surface tension and geometry. In our analysis, we assumed that the detachment of the cells from the surface occurs instantaneously. However, when a cell is detached from a surface, the process often does not occur all at once, but progressively. This process of peeling can greatly alter the force required to detach a cell from its substrate. Consider the free-body diagram of a small section of a membrane immediately adjacent to the attached section, as shown in Figure 10.5.

Figure 10.5 A section of membrane being peeled off a surface. The region to the left is still attached, whereas to the right it has released. The relevant balance of forces occurs between the membrane tension, n, the adhesive force per unit length, F_a, and internal bending moments.

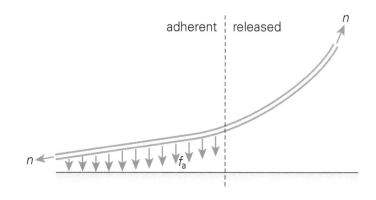

As we have done in the past, we could create a free-body diagram for Figure 10.3. However, we have already analyzed a more general case. Note that this situation is entirely analogous to a plate bending owing to a distributed pressure. Recall Equation 9.37,

$$n\left(\frac{d^2w}{dx^2} + \frac{d^2w}{dy^2}\right) + K_B\left(\frac{d^4w}{dx^4} + \frac{d^4w}{dx^2dy^2} + \frac{d^4w}{dy^4}\right) + F_a = 0, \qquad (10.13)$$

where K_B is the bending modulus and we have replaced p_z with F_a to indicate adhesion force. Assuming that the peeling is occurring in the x-direction, we have

$$n\left(\frac{d^2w}{dx^2}\right) + K_B\left(\frac{d^4w}{dx^4}\right) + F_a = 0. \qquad (10.14)$$

We can model the adhesion force with simple linear spring-like behavior such that the force in a single bond is given by

$$F_b = \begin{cases} \left(\dfrac{F_m}{l_m}\right)y & 0 < y < l_m \\ 0 & y < 0, \;\; y > l_m \end{cases}, \qquad (10.15)$$

where F_b is the force for an individual bond, F_m is the maximum force generated by the bond before bond breakage, l_m is the length of the bond at the maximum force, and w is the displacement (Figure 10.6). If $w > l_m$, then the bond has broken and is no longer capable of sustaining force. Next, we observe that the total adhesion force is simply the sum of the individual bond forces:

$$F_a = n_b F_b, \qquad (10.16)$$

where n_b is the area density of bonds attaching the cell to the surface. We can use energy density of adhesion by defining

$$J = \frac{n_b F_m l_m}{2}, \qquad (10.17)$$

the work done in extending all the bonds in a unit area from zero length to their maximum length.

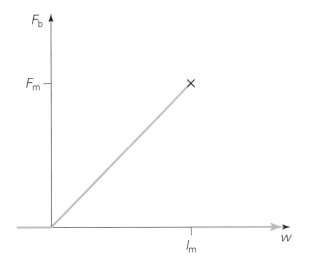

Figure 10.6 The plot of the adhesion force modeled as a linear spring. For y less than the maximum extension l_m, the bond acts as a spring with spring constant F_m/l_m. At l_m, however, the bond breaks and no restoring force is present. Unlike a classical spring, the bond cannot be compressed, so that if w is less than zero, there is also no force generated.

Advanced Material: Energy density of adhesion

To determine the energy density of adhesion, we first use the relationship that work equals force times distance and that this work goes entirely into the adhesion energy density. Force, however, is not constant. It varies with the length according to Equation 10.15. Therefore the work for a single bond is the integral of the force-displacement curve, or $W = 0.5 \times k \times x^2$ for a classical spring. For our example, the spring constant k is the slope (F_m/l_m) and x is the maximum length of the spring l_m. Substituting yields $W = 0.5 \times F_m \times l_m$. Because this is for a single bond, we can convert it to the desired energy density by multiplying by the number of bonds per unit area n_b,

$$W = 0.5 \times n_b \times F_m \times l_m = J.$$

Introducing this energy allows us to rewrite the adhesion force density as

$$F_a = n_b \left(\frac{F_m}{l_m} \right) y = \left(\frac{2J}{l_m} \right) \left(\frac{y}{l_m} \right) = \left(\frac{2J}{l_m^2} \right) y \tag{10.18}$$

by using appropriate substitutions. Substituting into Equation 10.14, we obtain an expression for adhesion in terms of y alone, with the remaining terms based on physical parameters. Although this model is fairly explicit and can be (and has been) used to model peeling, it relies on knowing parameters that are typically difficult to measure. For this reason, many studies of adhesion rely on strict comparisons from direct force measurements rather than fitting to parameterized models. Also, bond strength is known to depend on the level of prestress, which this treatment ignores.

Adhesion energy density can be obtained through consideration of strain energy

Previously, we obtained relations for adhesion energy density by considering surface tensions associated with a liquid drop, as well as peeling of a two-dimensional continuum. An alternative approach is to consider strain energy following adhesion, because the energy associated with adhesion must overcome the energy associated with cellular deformation.

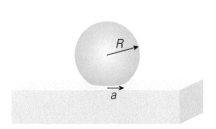

Figure 10.7 A mostly spherical cell attached to a flat surface. The radius of the cell is R. At the contact surface, the cell and substrate form a circular region of contact with base radius a, which we assume to be much less than R.

Consider a spherical cell of radius R. After adhesion, the cell retains a mostly spherical morphology, except for the base region contacting the surface (Figure 10.7). At this region, the cell and substrate form a circular region of contact with radius a, which is assumed to be much less than R. The deformation of the cell is assumed to occur solely because of the adhesion process.

Now consider the case in which we model the cell as a homogeneous isotropic linear elastic continuum with Young's modulus E. It can be shown through Hertz contact theory (described in Section 6.1) that the strain energy associated with this deformation scales as

$$W_{def} \propto \frac{Ea^5}{R^2}. \tag{10.19}$$

The adhesion energy scales as

$$W_{adh} \propto Ja^2, \tag{10.20}$$

where J is the adhesion energy density. Assuming that the strain energy is balanced by the adhesion energy, then $W_{def} \sim W_{adh}$. We can relate the right handed sides of the expression in Equations 10.19 and 10.20 as

$$\frac{Ea^5}{R^2} \propto Ja^2, \tag{10.21}$$

which simplifies to

$$a \propto \sqrt[3]{\frac{JR^2}{E}}. \tag{10.22}$$

This is only a scaling relationship, and is only valid for a relatively small cellular deformation occurring upon adhesion. Still, it allows us to predict some interesting scaling behavior. Consider what happens to adhesion contact if the cell radius decreases by a factor of two, but all other parameters remain constant. A reduction in R by a factor of two would result in an approximately 1.6-fold reduction in adhesion contact radius. If the elastic modulus increases by a factor of two the adhesion contact radius decreases by approximately 1.3-fold. We can see that the contact radius of the cell is more sensitive to changes in its radius than changes in its stiffness, a somewhat unexpected result.

Targeting of white blood cells during inflammation involves the formation of transient and stable intercellular adhesions

We conclude our discussion on adhesion by analyzing the mechanics of cell–cell adhesions. Intercellular adhesions are involved in a variety of biological processes. For example, intercellular adhesions are critical for the formation of impermeable or semipermeable cell linings within tissues. They are also necessary for the formation of structures that facilitate direct intercellular communication, such as *gap junctions*, specialized channel pairs that allow for the flow of signaling molecules from one cell to another. Cell–cell adhesion in known to be mediated by many transmembrance proteins, including cadherins, junctional adhesion molecules, some integrins, and desmosomal proteins, each of which has associations with many other proteins.

One of the best-characterized processes associated with intercellular adhesion is the interaction of white blood cells with blood vessel walls during inflammation. White blood cells are key players in the body's immune response to infection. By flowing with blood though the circulatory system, they have rapid access to almost all of the tissues and compartments of the body. However, we will see that they are able to exit the bloodstream quickly and target specific locations by processes mediated by intercellular adhesions (Figure 10.8).

One may observe the formation of two distinct adhesions in neutrophils interacting with a vessel wall (Figure 10.9). Initially the cells of the endothelium become *activated*. Endothelial activation is marked by the expression of a class of receptors called *selectins* that bind glycoproteins coating the surface of the white blood cell. Neutrophils that bump into the activated endothelial cells are held in contact with them by the selectins. However, the selectin bonds have relatively low affinity and are not strong enough to fully resist the hemodynamic drag force. Rather, the neutrophil begins to roll along the endothelial layer with selectin bonds forming and breaking transiently.

In the second step, the neutrophil becomes activated by signals released by the endothelial cells. Activation of the neutrophil is marked by expression of integrins on the neutrophil membrane. Unlike selectins, which bind to the glycoprotein layer coating the cell, integrin-mediated intercellular adhesion is much stronger and eventually halts the neutrophil rolling, resulting in a so-called *firm* adhesion. During this halting period, formation of tethers have been observed (Figure 10.10). These tethers are small elastic strands that serve to help arrest the neutrophils. After arrest, the neutrophils then migrate to the junctions between endothelial cells and perform paracellular migration, squeezing between the cells to exit the bloodstream.

Nota Bene

Selectin nomenclature. The three types of selectins are named after the cell types in which they were first characterized. E-selectin is found on the surface of endothelial cells, and L-selectin on leukocytes. P-selectin was first found on platelets, but later discovered to be expressed by endothelial cells as well. The feature they all have in common is that they all bind to sugar molecules in the glycoprotein coating of the target cell.

Figure 10.8 *In vivo* image of neutrophils passing through a blood vessel. The neutrophils are in various stages of rolling and attachment. (Courtesy of Gustavo Menezes; see www.garlandscience.com for a color version and movie of the figure.)

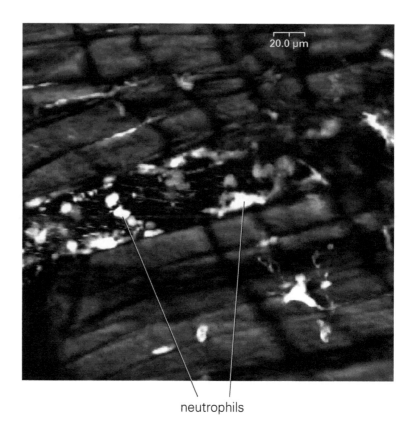

neutrophils

Mediators of firm adhesions. The integrins involved in firm adhesion have been identified as $\alpha_4\beta_1$ and $\alpha_L\beta_2$. They bind transmembrane members of the immunoglobulin superfamily (which also includes antibodies), vascular cellular adhesion molecules (VCAMs), intercellular adhesion molecules (ICAMs), and nerve cellular adhesion molecules (NCAMs).

Kinetics of receptor–ligand binding can be described with the law of mass action

Given that neutrophil adhesion to an endothelial cell layer involves two types of interaction, transient binding mediated by selectins and more stable bonds mediated by integrins, we now seek a means to characterize the difference in kinetics between these interactions. We will start by adopting a descriptive approach taken from chemical kinetics. A ligand–receptor pair can be in a bound (B) or unbound (L + R) state. We arbitrarily consider the rate at which binding is occurring to be the "forward" or " + " direction of the reaction and therefore call the binding rate constant k_+ and the unbinding rate constant k_-. The reaction can be written as analogous to Equation 7.1,

$$L + R \underset{k_-}{\overset{k_+}{\rightleftharpoons}} B. \tag{10.23}$$

Figure 10.9 Schematic demonstrating the stages of neutrophil adhesion in a blood vessel during an inflammatory reaction.

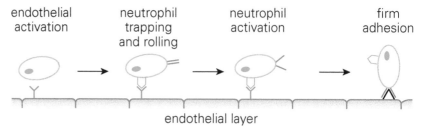

| endothelial activation | neutrophil trapping and rolling | neutrophil activation | firm adhesion |

endothelial layer

Y selectin   ‖ unactivated integrin
U selectin ligand   \/ activated integrin
∧ intercellular adhesion molecule (ICAM)

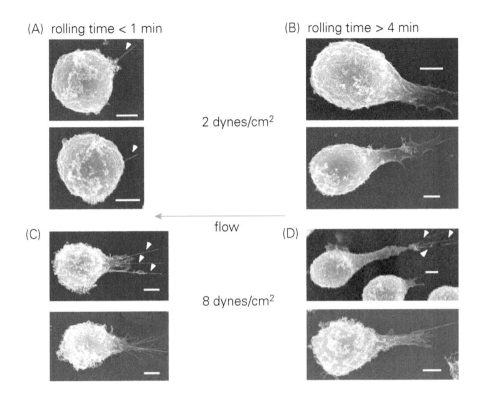

(A) rolling time < 1 min

(B) rolling time > 4 min

2 dynes/cm²

flow

(C)

(D)

8 dynes/cm²

Figure 10.10 Neutrophil tethering at various fluid shear stresses on a surface coated with P-selectin. (A) and (C) show the tether formation <1 min of attachment, and (B) and (D) show tethering after >4 min of attachment, so the tethers are more fully developed (From, Ramachandran V et al. (2004) *Proc. Natl. Acad. Sci. USA* with permission from the National Academy of Sciences.)

As we saw in Section 7.2, the kinetics of this reaction can be described quantitatively through the so-called *law of mass action*. Specifically, the assumption, which can be obtained from statistical mechanics by calculating the rate at which reactants would randomly collide, is that the rate at which a reaction occurs is linearly related to the rate constant and the concentration of the reactants. In our example of two reactants, the forward reaction occurs at a rate of

$$k_+[L][R] \tag{10.24}$$

and the reverse at

$$k_-[B], \tag{10.25}$$

where the square brackets indicate concentrations. If the reaction is at or near equilibrium, the forward and reverse reactions occur at nearly the same rate, or

$$k_+[L][R] = k_-[B]. \tag{10.26}$$

The *equilibrium constant*, or *dissociation constant*, is defined as the ratio of the two reaction constants, analogous to Equation 7.4,

$$K_d = \frac{k_-}{k_+} = \frac{[L][R]}{[B]}. \tag{10.27}$$

The dissociation constant has the unique property that when the concentration of ligand is equal to K_d, half of the receptors will be bound. This can be better understood by computing the fraction of bound receptors as

$$\%_{\text{bound}} = \frac{[B]}{[B]+[R]} = \frac{[B]\frac{[L]}{[B]}}{[B]\frac{[L]}{[B]}+[R]\frac{[L]}{[B]}} = \frac{[L]}{[L]+K_d}. \tag{10.28}$$

Figure 10.11 The percentage of bound reactants plotted as a function of ligand concentration. When the ligand concentration is K_d, exactly half of the receptors will be in the bound state.

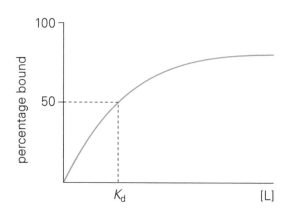

Notice that when the $[L] = K_d$, $\%_{bound} = 0.5$. A plot of Equation 10.28 can be seen in Figure 10.11.

The Bell model describes the effect of force on dissociation rate

Although the dissociation constant is a useful way to describe the behavior of ensembles of bonds on long timescales, it does not take into account the effects of force. Consider the kinetics of force-induced bond rupture between a single receptor–ligand pair. Characterization of such kinetics can be performed using experimental approaches such as micromanipulation or atomic force microscopy. After a single bond is identified, a controlled force can be applied in what is known as a force clamp configuration. This allows for the generation of force–displacement curves (Figure 10.12). At the single-molecule scale, entropic influences become extremely important. If we were to probe a single ligand–receptor interaction, even in the absence of an applied force, thermal forces will eventually cause the bond between the two molecules to dissociate. In general, the time that a bond exists decreases as the force across the bond increases.

In 1978 George Bell advanced a theory to describe how force would affect the rate of dissociation. He assumed that when a bond is subjected to loading, once it is ruptured the receptor and ligand move too far away from each other to rebind (in other words $k_+ = 0$). This means that the relevant kinetics can be described only by

Figure 10.12 A typical force–displacement curve from a single-molecule binding experiment. Notice that initially some force is required to push the receptor and ligand together. Then, as the bond begins to be pulled apart, a tension force is evident that was not present previously. If the force is the same as before contact, no bond is formed. Alternatively, if the adhesion fails in a series of steps, it suggests that more than one bond is formed. Once an isolated bond is identified, the investigator can apply a fixed force and measure the bond lifetime.

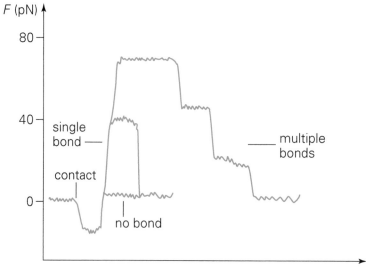

the rate of dissociation, k_-, and does not depend on k_+. Next, he assumed that force has an exponential influence on bond rupture such that the rate of dissociation in the presence of a force F across the bond is

$$k_- = k_-^0 e^{\frac{\sigma F}{k_B T}}, \qquad (10.29)$$

where k_-^0 is the rate of dissociation in the absence of force, k_B is Boltzmann's constant, T is absolute temperature, and σ is a constant that characterizes the influence of force. Notice that the form of this relationship is very similar to the Boltzmann probability derived in Section 5.5. Because the denominator in the exponent has units of energy, σF must also have units of energy, and therefore σ must have units of length. Indeed, it can be shown that σ is the bond extension that is associated with its minimum potential energy configuration. Equation 10.29 can also be expressed as

$$k_- = k_-^0 e^{\frac{F}{F_B}}, \quad F_B = \frac{k_B T}{\sigma}. \qquad (10.30)$$

F_B is a characteristic measure of bond strength, although not in the traditional sense, because bond dissociation is always occurring at any force level. Rather, it is a measure of the influence that force has on unbinding kinetics. It is noteworthy that in the Bell model, the time evolution of the concentration of bound molecules is also described by an exponential relation. Specifically, since only bond rupture is considered and rebinding is not allowed,

$$\frac{d[B]}{dt} = -k_-[B]. \qquad (10.31)$$

The solution of this differential equation is an exponential of the form

$$[B(t)] = [B(0)]e^{-k_- t}. \qquad (10.32)$$

There are actually two exponentials in Bell's model, one describing the influence of force on the rate of dissociation, and another describing the time evolution of [B].

Shear enhances neutrophil adhesion—up to a point

From what we have discussed so far, we would expect that the rate at which bonds are broken increases with the force across the bond. Indeed, for most bonds, this is the case. In other words, $k_- \propto F$. However, in experiments in which neutrophil adhesion and rolling was examined quantitatively, a surprising result was observed. As flow increased, the cells were actually observed to adhere better. This was reflected in the number of adherent cells and the number of tethers per cell increasing with flow, and rolling velocity decreasing with flow (Figure 10.13). In fact, up to a certain value of wall shear stress, adhesion seems to increase linearly. Beyond this *shear threshold*, adhesion again decreases linearly with flow, as expected.

For some time the mechanism behind the shear threshold phenomenon was unknown, but it seemed to suggest the existence of *catch bonds*. The term "catch bond" is meant to suggest that some pulling is required for the bond to engage or "catch." Without the pull, the bonds tend to float around engaging and disengaging randomly. In our everyday life, this might be similar to picking up a shopping bag with a hooked finger. Perhaps the receptor binding involves the engagement of hook-shaped molecules? Such bonds have been shown to exist experimentally at the single-molecule level. P-selectin exhibits catch behavior up to approximately 25 pN. Beyond this level, lifetime dramatically decreases in a behavior termed *slip*. Such bonds are sometimes called catch-slip, reflecting the two types of relationship between bond lifetime and force.

Figure 10.13 Plot of the number of adherent cells as a function of shear stress during neutrophil adhesion and rolling. Up to a certain shear stress, the number of adherent cells increases with increased shear.

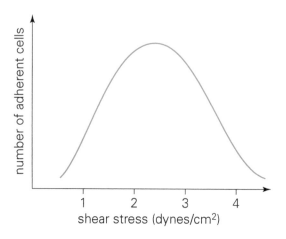

10.2 MIGRATION

Cell migration can be studied *in vitro* and *in vivo*

Cell migration is a fundamental aspect of cellular behavior that is central to processes of immunity, regeneration, repair, inflammation, and cancer, to name a few. *In vitro* assays, such as tracking motility or movement through a polymer gel, are powerful reductionist approaches. However, *in vivo* strategies are becoming more common. One elegant study was performed on heart transplant patients. Of the people who had heart failure significant enough to require transplantation, there were eight males who received hearts from female donors. After the male patients died (mostly within a year, some after nearly 2 years), their female hearts were assayed for the Y chromosome. It was found that a substantial fraction of organ cells (up to 10%, depending on the type of cell) were from the male recipient (turning the organ into a "chimeric tissue", Figure 10.14). Although the source of those cells is not known, such changes can influence other readouts, including migration.

Cell locomotion occurs in distinct steps

Before we discuss quantitative approaches for analyzing cell locomotion, we must first introduce some background biology of this process. Cell locomotion

Figure 10.14 A section of heart showing the presence of a Y chromosome. The bright spot (see arrow) is a Y chromosome detected by *in situ* hybridization in a female heart transplanted into a male patient. (From Quaini F et al. (2002). *New Engl. J. Med.* 346. With permission from the Massachusetts Medical Society.)

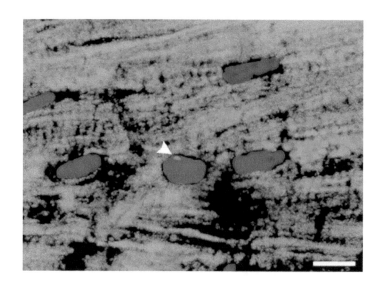

can generally be delineated as a series of four steps: protrusion, attachment, translocation, and release (Figure 10.15). *Protrusion* occurs when the cell extends actin-rich projections outward from its current location. There are several such projections that can generally be classified by their gross morphology. For example, lamellipodia are flat, broad, veil-like extensions containing highly branched actin networks at the leading edge of crawling cells. As we learned in Section 8.4, filopodia are finger-like extensions containing cross-linked actin bundles. The process of protrusion has been likened to an exploratory process by which the cell "feels" new surfaces and "determines" the direction of motion. The second step is *attachment*, which occurs when the cellular extension forms a stable adhesion with the surface such that it can be used as an anchorage point for subsequent locomotion. Although this adhesion is generally stable, it need not be permanent, as such adhesions are often observed to be transient and withdrawn after the third step, *translocation*. In this step, the cell moves in the direction of attachment. Because this process involves a substantial amount of cell motion, it is often characterized by a large degree of actin–myosin contractile activity. The fourth step is *release*, or detachment of the trailing end of the cell. This release is often associated with withdrawal of adhesions formed during attachment. However, it can sometimes be more forcible. If the cell's motion is faster than the release of the focal contacts in the trailing end, little bits of cellular material can be broken off and left behind. Though we have delineated cell locomotion as four sequential steps, it is important to note that the cell can be undergoing more than one step at any given time.

Protrusion is driven by actin polymerization

The generation of intracellular force is critical for cellular locomotion, particularly during the processes of protrusion and translocation. Translocation is accepted to be largely governed by activation of actin–myosin contraction, a process that we will discuss in great detail in Section 10.3. However, the molecular mechanisms driving cellular protrusion are less clear.

In Figure 10.16, a schematic depicts what are believed to be the primary molecular mediators of protrusion of lamellipodia. A key member of this process is the Arp2/3 complex, which consists of two primary molecules, Arp2 and Arp3, as well as several other binding molecules. Interestingly, the Arps (short for actin-related proteins) are structurally similar to actin and can serve as nucleation sites for the creation and extension of new filaments. In lamellipodia, Arp2/3 serves to initiate new actin branches on existing actin filaments, resulting in the formation of a

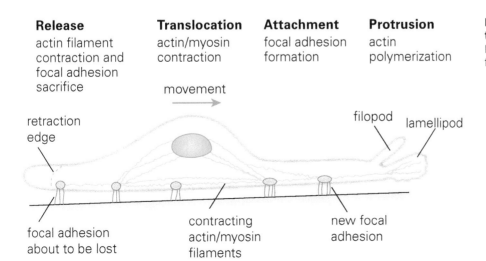

Release
actin filament contraction and focal adhesion sacrifice

Translocation
actin/myosin contraction

Attachment
focal adhesion formation

Protrusion
actin polymerization

movement

retraction edge

filopod   lamellipod

focal adhesion about to be lost

contracting actin/myosin filaments

new focal adhesion

Figure 10.15 Schematic showing the major steps for cell migration. In this depiction, the cell is migrating toward the right.

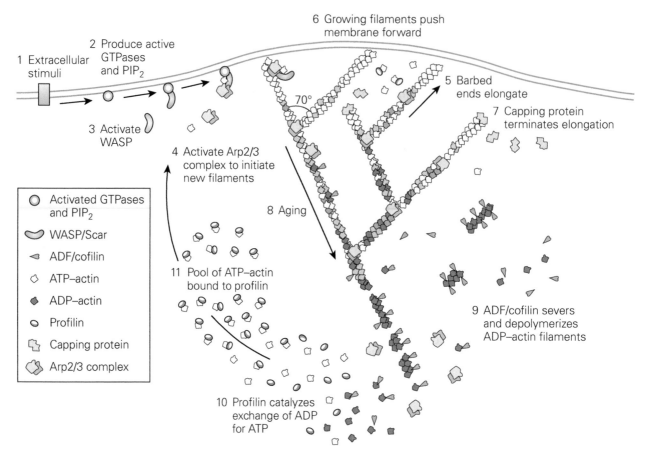

Figure 10.16 Stages for protrusion of the cell membrane in lamellipodia. In this model, external stimuli (1) result in the activation of GTPases (2), which activate Wiskott–Aldrich syndrome protein (WASP). (3). This complex activates the Arp2/3 complex (4), resulting in the formation of a new actin branch on a pre-existing actin filament. This newly formed actin branch grows by polymerization at the barbed end of the filament (5) from a pool of profilin-bound actin monomers, pushing the membrane forward (6). Elongation of actin filaments is terminated by capping protein (7), which binds to the growing ends of polymering actin filaments. Older ADP-bound filaments (8) are severed by actin-depolymerizing factor (ADF)/cofilin (9). Depolymerized actin undergoes dissociation of ADP and binds ATP (10), upon which ATP–actin re-binds profilin, renewing the pool of ATP-bound actin monomers available for assembly (11). (Adapted from Pollard TD (2003) *Nature.* 442.).

branching actin network. Multiple sites for actin extension permit a relatively uniform growth pattern during protrusion to be achieved. We will see in the next section that although it is generally accepted that actin branches serve to push the membrane forward by polymerization, the exact physical mechanism by which this may occur is still relatively unclear.

Actin polymerization at the leading edge: involvement of Brownian motion?

A major question about the capacity for actin polymerization to give rise to protrusion at the leading edge is how monomers can be added to the filament ends when they are physically blocked by the membrane? Although it is possible that a polymerizing actin filament may generate sufficient force to deform the cell membrane locally, this does not address how physical space is created between the filament end and the membrane such that there is room for monomers to be added to the growing end of the filament.

Several mechanisms have been proposed to explain how actin monomer addition can occur at the leading edge of a protruding membrane. One such mechanism is often referred to as the Brownian ratchet (Figure 10.17). Here, entropic forces cause

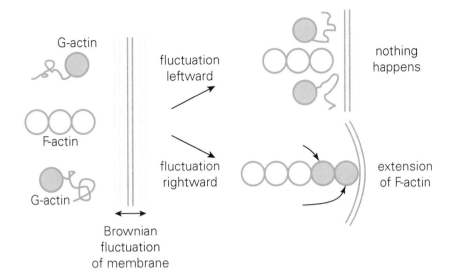

Figure 10.17 The Brownian ratchet depends on thermal fluctuations. As the membrane randomly fluctuates, if the it shifts closer to the tip of the actin filament, nothing happens because the actin monomers are blocked by the membrane. As the membrane shifts away and opens a gap, the actin monomers can extend the actin filament. This has the effect of generating a local protrusion.

the local cell membrane to undergo small thermal fluctuations. If a pool of free actin monomers is available at the leading edge, when the membrane shifts forward, space is temporarily generated that allows polymer extension. This polymerization prevents the membrane from shifting back. As this ratcheting process is repeated, the resulting actin polymerization leads to the development of a protrusion.

An alternative hypothesis that has been advanced is related to the Brownian ratchet in that the actin filaments themselves rather than the membrane are undergoing thermal fluctuations. The filaments randomly fluctuate in end-to-end length. When the end-to-end length decreases slightly (through bending), space is temporarily generated that allows for polymer extension (Figure 10.18). When the filament straightens, this generates a protrusive force that extends the membrane.

Cell motion can be directed by external cues

We have considered the process of cell migration as if it were an aimless process. However, it is clear that, cell migration can be directed to specific locations. Chemical compounds that direct cell migration are called *chemoattractants*. If there is a chemical gradient in a chemoattractant, cells may sense these gradients and direct their migration toward the source. One of the most prominent examples of directed cell migration toward a chemoattractant is the targeting of pathogens by neutrophils. The pathogens release compounds that the neutrophil recognizes as foreign. Upon activation, the neutrophils crawl about, targeting local pathogens. Upon contacting the pathogens, the neutrophils engulf them, removing the source of chemoattractant.

Nota Bene

Somebody stole my actin! *Listeria monocytogenes* is an intracellular bacterium responsible for the disease listeriosis, the most common cause of death associated with food-borne pathogens. As an intracellular bacterium, it resides within the cytoplasm of another cell, remaining relatively hidden from the immune system. To maximize its capacity to move within the cell, the bacteria has developed a mechanism whereby it "hijacks" the actin polymerization machinery in the cell and rides on a wave of polymerizing actin that it induces. This polymerizing actin can often be viewed as a "comet-tail" structure behind the bacterium. The polymerizing actin propels the bacterium to the membrane and generates a protrusion. If the infected cell is adjacent to another cell, the bacterium can then enter the second cell by endocytosis or a similar mechanism. This method of propagation has the advantage that the bacterium can travel from cell to cell without ever leaving the intracellular environment.

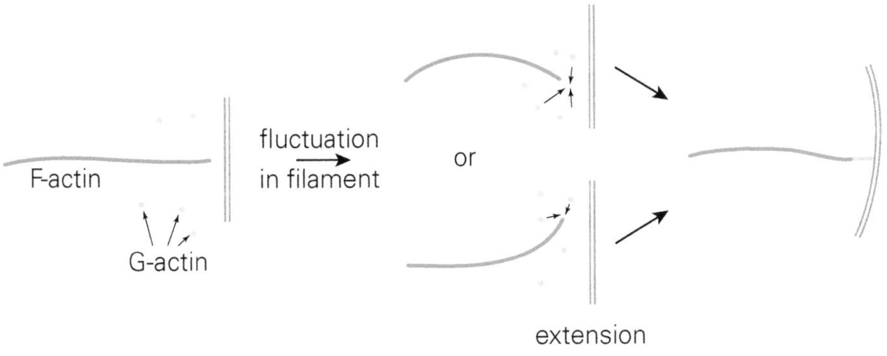

F-actin

G-actin

fluctuation
in filament

or

extension

Figure 10.18 Membrane protrusion formation from polymer thermal fluctuations. In this mechanism, the membrane is considered mostly static, whereas the actin filament undergoes thermal fluctuations, resulting in flexing of the filament. This flexing creates more space at the tip of the filament, allowing for polymerization to extend the filament. Then, when the filament straightens out again, it generates a protrusive force to generate a protrusion.

Figure 10.19 Plot showing amoeba chemotaxis "accuracy" versus chemoattractant (in this case, cyclic AMP [cAMP]) concentration. The accuracy is measured between 0 and 1, with 0 representing random motion and 1 representing fully directed motion. Cells demonstrate the ability to move fairly unerringly toward the source of the chemoattractant given the right concentration. Note the bi-phasic response; at higher concentrations, it is possible that enough chemoattractant exists to engage all cell receptors, thereby "blinding" the cell to existing chemical gradients. (From, Fisher PR, Merld R, & Gerisch G, (1989) *J. Cell Biol.*, Rockefeller University Press.)

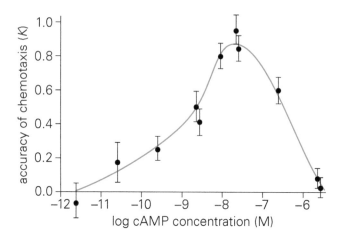

The physical and molecular mechanisms driving chemoattraction have yet to be fully established, though several proposed mechanisms have been put forth. It has been proposed that cells may sense local chemical gradients across the cell surface through the differential activation of receptors from one end of the cell to the other. When one considers that cells are typically only tens of micrometers across, this implies that very small differences in activation from one end of the cell to other must be distinguished. This is a substantial challenge and what the direction-sensing mechanisms might be is still unclear.

Another possible mechanism is that the cell randomly travels over a given distance in many different directions, allowing it to sample the local chemoattractant concentration over a fairly large area. It is not yet known whether cells correct themselves after moving in the direction of decreasing concentration or if they sense whether an increase might be improved upon in a different direction. It is important to note that such a mechanism requires some degree of random migration, suggesting that at times they will be incorrectly oriented. In contrast, most *in vitro* experiments suggest that in the right conditions, cells move fairly unerringly toward the source of the chemoattractant (Figure 10.19).

Cell migration can be characterized by speed and persistence time

In investigating the process of directional cell migration, one question is how do we characterize the degree to which a cell tends to travel in the same direction as it migrates through space? It is often useful to characterize cell trajectories using similar approaches to characterizing a diffusing particle (Section 5.6) or a fluctuating polymer (Section 7.4). Cellular trajectory can be quantified in terms of persistence time (P) and speed (S) (Figure 10.20). In the context of cell migration, speed refers to how fast the cell is moving, whereas persistence time refers to the time a cell spends moving in a given direction. That is, a cell might exhibit large summed displacements with little net translation, which typically correlates with low persistence times.

To better understand the influence of persistence time on cell migration, consider a cell migrating in one dimension with speed S and a change in direction per unit time of λ. The directional persistence time of the cell, defined as the time per direction change, is $P = 1/\lambda$. It can be shown using a random walk model that the differential equation governing the time dependence of the mean square distance of the cell $\langle d^2 \rangle$ is

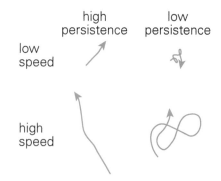

Figure 10.20 Schematic demonstrating persistence versus speed. The traces show different cell trajectories for a given time period. Speed is a measure of how fast a cell is moving. Persistence is related to how likely it is to continue to move in the same direction.

$$\frac{\partial^2 \langle d^2 \rangle}{\partial t^2} + \frac{2}{P} \frac{\partial \langle d^2 \rangle}{\partial t} = 2S. \tag{10.33}$$

The particular solution of Equation 10.33 is

$$\langle d^2 \rangle = S^2 Pt + C_1 + C_2 e^{-2t/P}, \tag{10.34}$$

where the constants $C_1 = -C_2 = -2S^2P^2$ are found by applying the initial conditions, $\langle d^2 \rangle = 0$ at $t = 0$, and $d\langle d^2 \rangle/dt = 0$ at $t = 0$ (the cell initially has no bias). In two dimensions,

$$\langle d^2 \rangle = S^2 \left(Pt - P^2 \left(1 - e^{-t/P} \right) \right). \tag{10.35}$$

Note that in the limit for $t \gg P$

$$\langle d^2 \rangle = 2S^2 Pt = 2Dt, \tag{10.36}$$

where $D = S^2 P$ has the units of m^2/s and can be considered to be an effective "diffusion coefficient" of the cell.

To determine the values of S and P experimentally, one can acquire multiple cell paths for different cells with various time intervals, and compute an average mean square distance for each time interval. With these data established, the speed and persistence can be calculated using standard least squares fitting algorithms. In this approach, it is assumed that every cell is migrating under similar mathematical patterns. Some examples of speed and persistence relationships for various cell types are presented in Figure 10.21.

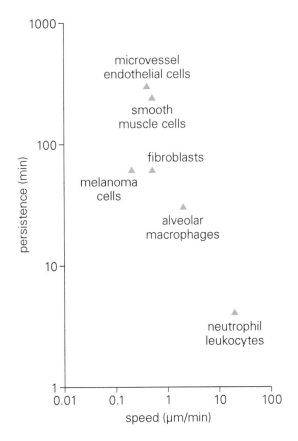

Figure 10.21 Speed–persistence relationships for different cell types. Note that there is a roughly inverse relationship between speed and persistence. It also noteworthy that neutrophils have high speed for low persistence times, suggesting a need to move quickly to reach an infection site but also to change directions quickly to target pathogens. (From Lauffenburger D & Linderman JJ (1993). With permission from Oxford University Press, New York.)

Directional bias during cell migration can be obtained from cell trajectories

Persistence time can be used to directly compare the migration behavior of different cell types. One limitation of persistence time is that it does not take into account whether there is bias in the direction in which a cell travels, under the influence of a chemoattractant. How would we quantify such a tendency?

A relatively simple method is to determine the position of a cell at set time intervals, fit straight-line segments between the points, and determine the change in angle from one segment to another. The average change in angle can then be computed over all the segments. This quantity can be used to distinguish left- or rightward bias during cell migration. Also, it can be used as a fairly simple means to obtain cell speed and a rough indication of persistence during migration.

Example 10.1: Migration paths for two different cells

Consider the trajectories of two cells shown in **Figure 10.22**. Both start at (1, 0), but one cell follows the x-axis (cell 1) while the other does not (cell 2). Determine the average speed of each cell, and show quantitatively that cell 2 has a leftward bias in its migration.

First, we consider the cell speeds. They are not constant throughout their trajectories. For each successive time interval of δt, cell 1 moves 1 unit, then 0.8 units and then 1.1 units. The average speed for cell 1 is then $2.9/3\delta t = 0.97/\delta t$. The average speed for cell 2 can be determined similarly, and is $0.99/\delta t$. The speed of cell 2 is marginally higher than that of cell 1.

Now we seek to demonstrate that cell 2 has a leftward bias. We do this by calculating the change in angle the cell makes at each successive time interval. In doing so, we define the angle relative to a line parallel to the line segment associated with the previous time iteration. Leftward changes in orientation result in negative angles, and rightward changes result in positive angles. For cell 2, we find

angles of $-102°$ between $0 < t < 2\delta t$ and $-99.2°$ between $1 < t < 3\delta t$. The average angle is $-100.6°$. The large magnitude and negative sign indicate a prominent leftward bias in its migration.

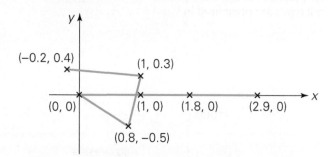

Figure 10.22 Migration paths for two different cells. The values of x in the plot refer to cell positions obtained at time intervals separated by a constant, δt. The (x, y) coordinates of each point is given in parentheses.

10.3 CONTRACTION

In the final section of this chapter, we discuss the generation of intracellular contractile forces. When we talk about cellular contraction, one of the first examples that comes to mind is muscle cells, which possess specialized structures for the generation of contractile forces. However, non-muscle cells also possess the capacity to generate contractile force, such as during locomotion, or in the generation of traction forces. In this section, we give overviews of the molecular structures within muscle and non-muscle cells that allow for contractile force generation, as well as experimental measurements of force generation in both cases. We also present molecular-based mathematical models of contraction that allow us to gain mechanistic insight into a range of observed phenomena.

Muscle cells are specialized cells for contractile force generation

One of the most important examples of cellular force generation is when cells act as an ensemble and generate force within muscle. In this section, we will consider how muscles generate force motivated first by cardiac heart muscle and then skeletal muscle. Both cardiac and skeletal muscles are examples of *striated* muscles. All other muscles, including blood vessels, and the reproductive gastrointestinal, and respiratory tracts, are *smooth* muscles. The basic contractile unit of a striated muscle fiber is called the *sarcomere* (Figure 10.23). It consists of several key structures, including the anchoring z-discs (which are non-contractile structural proteins), a bundle of myosin filaments forming a "thick" filament, and actin filaments that are anchored to the z-discs. In striated muscles the z-discs and M-line (between the z-discs) of the sarcomeres align, giving the muscles a banded appearance. In smooth muscles, this alignment does not occur and, generally, smooth muscles are able to undergo much larger amounts of elastic stretch.

Studying cardiac function gave early insight into muscle function

The foundations of understanding of muscle force generation were laid from basic insights into the pumping action of the heart. These occurred in the early twentieth century, long before we understood the molecular mechanism of actin–myosin force generation or even the physiology of skeletal muscle. The physiologist Ernest Starling, in 1914, anticipated the relationship between force and sarcomere length when he stated, "the mechanical energy set free in the passage from the resting to the active state is a function of the length of the fiber." Specifically, using data collected from intact hearts, he constructed diagrams of the cardiac cycle. He realized that during the diastolic filling phase of the cycle, cardiac muscle fibers are passively stretched, and that during the systolic ejection phase the fibers are actively contracting. Therefore, volume can be interpreted as fiber length, and pressure as fiber tension. The Frank–Starling law (or perhaps more properly the Frank–Starling mechanism) is the observation that an increase in the filling volume of the heart (reflected in a higher-end diastolic volume) causes more stretching of the heart muscle fibers and a more forceful

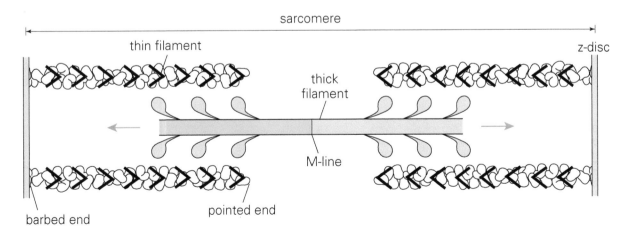

Figure 10.23 A schematic of a sarcomere, the basic contractile unit of a muscle fiber. The z-discs (which are commonly visible in pictures of muscles) anchor the actin filaments. The myosin bundle (thick filament) with the myosin heads sticking out (in opposite directions on either end) is caged inside the actin (thin) filaments. (Adapted from, Bray D (2000) Garland Science, New York.)

ejection and a higher systolic pressure. This led him to the realization that the tension a sarcomere is able to generate must increase with its initial stretch or length (Figure 10.24).

The skeletal muscle system generates skeletal forces for ambulation and mobility

Because of the relative ease of isolating skeletal muscle, much of what we understand about muscle contraction and actin–myosin interaction is derived from the study of skeletal muscle. Using isolated muscles it is possible to quantify the force a muscle generates as a function of excitation, relative length, and contraction velocity. *Isotonic* contractions occur when muscle tension is held constant and length is changed (if the muscle is acting to support or raise a fixed weight). Isotonic contractions might be *concentric* (the muscle is shortened) or *eccentric* (the muscle is lengthened). *Isometric* contractions occur when the muscle is held to a fixed length. Activation can take the form of a single electrical stimulation that produces a transient force known as a *twitch*, or continual stimulation can produce a maximal force in what is termed a *tetanic contraction*.

The two most important mechanical parameters affecting muscle force generation are length and velocity (Figure 10.25). It is perhaps not surprising that the force generated by a muscle depends on length such that if the muscle is stretched or shortened too much, the ability to generate force is compromised. In terms of velocity, the maximum force is generated when the muscle is not allowed to contract, and it is reduced as velocity is increased.

The Hill equation describes the relationship between muscle force and velocity

The classic hyperbolic relationship between muscle force and velocity is the Hill equation,

$$(F + a)(v + b) = k, \tag{10.37}$$

where F is force, v is velocity, and a, b, and k are constants. This equation captures critical aspects of the force–velocity relationship. As force increases, velocity decreases. There is a maximum force at zero velocity and, vice versa, a maximum shortening velocity at zero force. Between these two extremes the relationship is hyperbolic and has been shown to provide a remarkably accurate fit to experimental data.

Interestingly, Hill did not arrive at his famous equation by making measurements of force and shortening velocity directly. He actually made measurements of the heat liberated by muscle as it shortened. These measurements were based on his

Figure 10.24 A schematic of the cardiac cycle. The Frank–Starling law states that as the heart's stroke volume increases owing to more diastolic filling, the muscle contracts with more force, leading to a higher systolic pressure. This is depicted by the blue line with a higher EDV, resulting in a higher systolic pressure, P_{systole}.

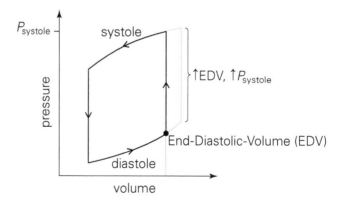

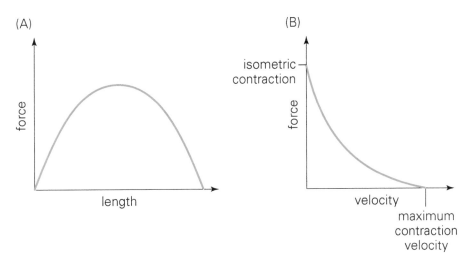

Figure 10.25 Length and velocity control the ability of muscle to generate force. (A) Plot of force versus length. A biphasic dependence of force on length is observed. (B) Plot of force versus velocity. Force generation increases with decreasing velocity. Maximum force generation occurs during *isometric* contraction, or when the muscle is held to a fixed length. Maximum contraction velocity occurs with zero force.

observation that the heat released is proportional to shortening distance x_{sh} through a proportionality constant he denoted as "a". Next, he considered the total energy liberated, E_l, by the muscle during a contraction. This energy contains two parts, the heat released (shortening distance times the proportionality constant) and the mechanical work done (force times the shortening distance), or

$$E_l = Fx_{sh} + ax_{sh}. \tag{10.38}$$

Experimentally he found that the rate at which this energy is liberated decreases linearly with force,

$$(F + a)\frac{dx_{sh}}{dt} = (F + a)v = -bF + c. \tag{10.39}$$

This finding can be easily rearranged to yield,

$$(F + a)(v + b) = c + ab = k. \tag{10.40}$$

Nota Bene

Hill's measurements. The English physiologist Archibald Hill collected meticulous data describing the relationships between muscle force, lengths, and velocity. He made these measurements by quantifying the heat liberated by muscle contraction to a precision of 10^{-3} °C. He also advanced the equation that bears his name to quantify the relationship between muscle force and contraction velocity. His 1938 paper is a classic in terms of the scientific contribution, but also the elegance of the writing is a product of a bygone era.

Non-muscle cells can generate contractile forces within stress fibers

We now shift our focus from contractile force generation in muscle to non-muscle cells. Although the basic molecular machinery responsible is similar, the molecular composition and organization of the structures can be quite different. *Stress fibers* are long thick bundles of actin that were first identified in cells cultured on adhesive surfaces. They span the interior of the cell, interconnecting focal adhesions. Stress fibers are involved in cell motility and can be observed aligned in the direction of movement in slowly migrating cells, but, surprisingly, are transversely aligned in the fast-moving epidermal keratocyte. They are known to have important physiological functions, such as participating in wound closure by fibroblasts and maintaining blood vessel integrity by endothelial cells. Stress fibers can be induced to form with the application of either biochemical or biomechanical stimulation.

The capacity for stress fibers to exert contractile forces resulting in a pre-stress and pre-strain can be demonstrated experimentally. Consider an experiment in which cells are cultured on a flexible substrate that has been pre-stretched in one direction (Figure 10.26), and stress fibers that are oriented in the same direction. If one then releases the substrate stretch by a small amount, such that the resulting contractile strain in the stress fibers is less than its pre-strain, the fiber will retain a

Figure 10.26 If cells are cultured on an elastic membrane that is allowed to uniaxially contract, the stress fibers will exhibit buckling if the contraction is large enough. The length of the fiber under tension is denoted as R_s, length of a straight, but unloaded fiber is L_0, and once it has buckled, it has an end-to-end length of R_b and a contour length of L_c. (From Costa KD, Hucker WJ & Yin FC (2002). *Cell Motil. Cytoskeleton.* 52.)

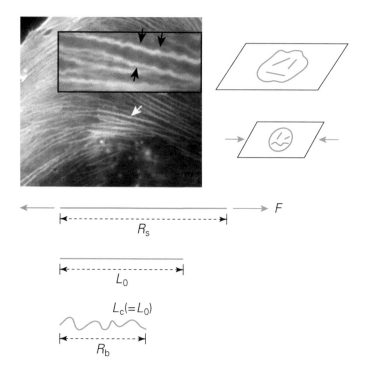

"straight" morphology. However, if the release strain is large such that it exceeds the stress fiber pre-strain, the fiber will exhibit a "wiggly" or "buckled" morphology. An alternative approach to demonstrate pre-stress in stress fibers is to sever a stress fiber using a high-powered laser, or disrupt a focal adhesion anchoring a stress fiber. In both cases, the free end of the stress fiber can be observed to contract, again demonstrating that the stress fiber is under a pre-strain.

Stress fiber pre-strain can be measured from buckling behavior

The buckling of stress fibers can be used to estimate the level of pre-strain. Recall the experiment in Figure 10.26, consider a stress fiber oriented in the same direction of substrate pre-stretch. This stress fiber, is assumed straight with an end-to-end length R_s. We slowly contract the substrate, decreasing the pre-stress in the fiber. Immediately before the stress fiber is found to take on a "buckled" morphology, we measure its end-to-end length and term this L_0, the contour length of the stress fiber in the absence of a pre-stress. As we contract the stress fiber even further, the stress fiber assumes a "buckled" morphology with some end-to-end length R_b that is less than R_0. Using these quantities, the pre-stretch in the fiber can be found as

$$\lambda_f = \frac{R_s}{R_0} \tag{10.41}$$

and the pre-strain as $1 - \lambda_f$. It is noteworthy that λ_f can be calculated in an alternative manner that does not require the determination of R_s. Consider the case where we know R_b and the stretch ratio of the substrate, λ_s. λ_s is related to R_s and R_b as

$$\lambda_s = \frac{R_s}{R_b}. \tag{10.42}$$

Next we define the *tortuosity* as

$$T = \frac{R_b}{R_0}, \tag{10.43}$$

and then λ_f can equivalently be determined as

$$\lambda_f = T\lambda_s. \tag{10.44}$$

Using this approach, the stress fiber pre-strain of cultured human endothelial cells has been estimated to be as high as 25%, although it is highly heterogeneous.

Myosin cross-bridges generate sliding forces within actin bundles

In the previous section, we discussed various structures in muscle and non-muscle cells that allow for the generation of contractile forces. In these cases, although the structures were distinct in molecular makeup and organization, the same basic molecular mechanism was responsible for driving contraction: myosin movement along actin. Actin and myosin together form a cellular force generation system that is critical to an amazingly wide variety of cells and tissues. Given that both muscle cells and non-muscle cells possess the capacity to generate contractile forces using actin and myosin, the question arises, how does this occur? Although it is well established that myosin functions as a molecular motor in converting chemical energy into mechanical work by moving along actin, many details have yet to be elucidated. However, the *cross-bridge* model in which collections of myosin molecules link adjacent actin polymers together provides a remarkably accurate model of actin–myosin interactions both within and outside sarcomeres.

In the cross-bridge model, myosin molecules go though several distinct stages or configurations (Figure 10.27). The process is cyclical in nature and at completion the myosin has returned to its initial configuration. In the first stage, myosin is tightly bound to actin, and its ADP/ATP (adenosine diphosphate/adenosine triphosphate) binding domain is not occupied. This is known as the *rigor* state, because no sliding or motion can occur. To leave the rigor state, myosin must bind a molecule of ATP. When this occurs, myosin releases from actin, and ATP is

Nota Bene

Myosin variants. There is no single myosin protein, but rather a huge *superfamily*. There are over 40 different genes and about 100 different protein products that are grouped into 18 major variants known as classes. This immense diversity is generally the result of various splice variants. Myosin II is the type found in muscle; the tails of this myosin allows it to form polymer filaments. Other myosins remain as monomers and are known as *unconventional* myosins because myosin was first identified in muscles. Most of the variability is found in the tail region, associated with binding to different cargos or assembling into fibers, whereas the head regions tend to be much more conserved. All myosins move toward the + or barbed end of the actin polymer, except myosin VI, which moves backward!

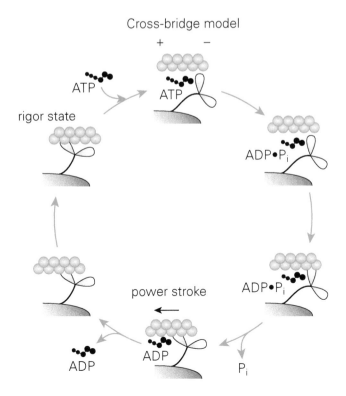

Cross-bridge model

Figure 10.27 The myosin configurations of the cross-bridge model. In this case the actin molecule is sliding from right to left.

hydrolyzed into ADP. This releases chemical energy, allowing myosin to change conformation into what is known as a *cocked* configuration. The hallmark of this configuration is that the myosin head has moved toward the (–) or barbed end of the actin polymer. In the third stage, myosin rebinds to the actin, but because it is in the cocked configuration, it rebinds at a site more toward the (–) end. In the fourth stage, rebinding of actin induces the myosin to release its phosphate. In this process, it also undergoes the *power stroke*. It returns to the uncocked configuration, and in the process drags the actin molecule toward the center of the myosin cross-bridge. The power stroke is where mechanical work is produced. In the fifth and final stage, the spent molecule of ADP is released. This returns the myosin to the rigor state.

Myosin molecules work together to produce sliding

An interesting aspect of actin–myosin contraction is that the myosin heads are not able to bind just anywhere along the polymer. In fact, each actin subunit has only one binding site for the myosin heads. As we learned in Section 7.1, actin subunits polymerize in a helical structure. The *pitch* of this helix has been measured to be $\Delta = 36$ nm; in other words, the polymer completes one full turn of the helix for every 36 nm along its length. Interestingly, the distance that the myosin molecule stretches when it is cocked in preparation for a power stroke, δ, is only around 5 nm. Because binding sites can only occur once in each full twist, it would be impossible for a myosin molecule to reach far enough forward to grab another binding site each time through the cross-bridge cycle. A simple explanation for this apparent paradox is that many myosin molecules must work together to produce sliding. At any given point in time, most are not engaged with actin at all, whereas a few are bound and tugging on the actin molecule, much like the legs of a millipede.

The implication of this cooperative behavior is that any given myosin molecule, spends most of its time bound to ATP and, as a result, unbound from actin. For example, let the average time per cycle that the myosin is not bound to actin be t_{off}, and the time that myosin is bound to actin be t_{on}. We might estimate the fraction of time myosin is engaged (t_{on}/t_{total}), a quantity we will call the duty ratio, r_{duty}, by assuming that is it proportional to the ratio of δ and Δ,

$$r_{duty} = \frac{\delta}{\Delta} = \frac{5\,\text{nm}}{36\,\text{nm}} = 0.14. \qquad (10.45)$$

However, even this seems to require much more frequent binding to actin than is actually observed. In fact, myosin seems to be quite a "lazy" molecule. The rationale for this statement comes from considering the force generation capacity of entire actin–myosin bundles. The total tension should be approximately the number of cross-bridges multiplied by the force generated by each cross-bridge. The number of cross-bridges can be estimated, and the time-average force per cross-bridge is a function of t_{on} and t_{off},

$$\langle F \rangle = \frac{t_{on}\langle F_{on}\rangle + t_{off}\langle F_{off}\rangle}{t_{on} + t_{off}}. \qquad (10.46)$$

However, the force generated when myosin is not engaged to actin must be zero ($<F_{off}> = 0$). Therefore,

$$\langle F \rangle = \frac{t_{on}}{t_{on} + t_{off}}\langle F_{on}\rangle = r_{duty}\langle F_{on}\rangle \qquad (10.47)$$

or

$$r_{\text{duty}} = \frac{\delta}{\Delta} = \frac{t_{\text{on}}}{t_{\text{on}} + t_{\text{off}}} = \frac{\langle F \rangle}{\langle F_{\text{on}} \rangle}. \qquad (10.48)$$

The power-stroke model is a mechanical model of actomyosin interactions

Given what we know about the mechanisms of actin–myosin contraction, we now analyze the mechanical aspects of this process. To do this, we use the *power-stroke* model, which quantitatively describes the molecular mechanics. This model treats the myosin heads as simple linear springs (Figure 10.28) with an effective spring constant of k_{m}. The model considers three possible positions of the myosin spring. The relaxed position is the zero-force position. The spring moves forward by a distance δ_+ when it is cocked and, consequently, where it binds to actin. Finally, the model does not assume that the actin is released at the zero position. Rather, it assumes that the spring may be compressed by some amount δ_- past the zero position before it is released. This phenomenon is sometimes referred to as *drag* (Figure 10.29).

We now compute the average force being generated by the spring. Recall that $\langle F_{\text{on}} \rangle$ is the average force over one cross-bridge cycle when myosin is bound to actin. This force occurs as the spring travels from the δ_+ position, when it first binds, through the zero position, to the δ_- position, where it unbinds. The average force is

$$\langle F_{\text{on}} \rangle = \frac{k_{\text{m}} (\delta_+ - \delta_-)}{2}. \qquad (10.49)$$

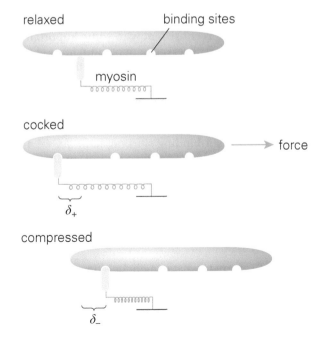

relaxed

binding sites

myosin

cocked

force

δ_+

compressed

δ_-

Nota Bene

The famous Huxleys. Sir Andrew Huxley is credited with developing the power-stroke model, the paradigm that established the equations upon which our quantitative understanding of actin–myosin force generation is based. However, this contribution is overshadowed by his work on the Hodgkin–Huxley model of nerve conduction, for which he was awarded the Nobel Prize in Physiology or Medicine in 1963. This established the *action potential*, and suggested the existence of ion channels. Remarkably, his half-brother Aldous Huxley is also very well known. He wrote about the dehumanizing aspects of scientific progress and was the author of *Brave New World*.

Figure 10.28 Schematic diagram of the three power-stroke model configurations. In the relaxed position there is no force and the myosin is not engaged with actin. Myosin is modeled as a simple linear spring. The spring moves forward by a distance δ_+ when it is cocked and imparts a force of $k_{\text{m}}\delta_+$ on the actin filament. In the "compressed" position, the spring compresses by δ_- past the zero position before releasing actin.

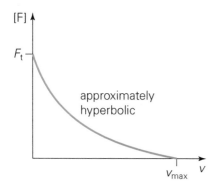

Figure 10.29 The force–velocity relationship predicted by the power-stroke model. Although not strictly hyperbolic, for physiological values of the physical parameters, the power-stroke model is similar to the prediction of Hill's equation.

Combining this with our expression for duty ratio and making the observation that the total stroke distance $\delta = \delta_+ + \delta_-$, we obtain

$$\langle F \rangle = r_{\text{duty}} \langle F_{\text{on}} \rangle$$

$$= \frac{\delta}{\Delta} \frac{k_{\text{m}}}{2} (\delta_+ - \delta_-)$$

$$= \frac{k_{\text{m}}}{2\Delta} (\delta_+ + \delta_-)(\delta_+ - \delta_-) \qquad (10.50)$$

$$= \underbrace{\frac{k_{\text{m}}}{2\Delta} \delta_+^2}_{F_+} - \underbrace{\frac{k_{\text{m}}}{2\Delta} \delta_-^2}_{F_-},$$

where the first term is the positive elastic force, F_+, and the second term is the negative drag force, F_-.

Let us consider the form of Equation 10.50 in a bit more detail and see if we can relate it to sliding velocity. The first term depends on the distance that the myosin heads move forward when they are cocked. As we described, most of the myosin head groups are already bound with ATP and therefore cocked at any point in time. Therefore, there is no reason to expect that this would depend on velocity. On the other hand, the second term depends on the distance the heads overshoot the zero force position before releasing, in other words the distance the myosin springs get compressed during each cycle. It would be reasonable for this quantity to depend on sliding velocity. In fact, this compression distance should increase linearly with velocity. If we introduce the release time as a new parameter t_{r}, we obtain,

$$\langle F \rangle = F_+ - \frac{k_{\text{m}}}{2\Delta} (v t_{\text{r}})^2. \qquad (10.51)$$

Notice that the first term is the maximum positive force the myosin is capable of generating. It is a constant, depending only on structural parameters. Also, note that the maximum force occurs when the velocity is zero, consistent with the concept of a stall force, at which point the second term (the drag force) is also zero. The power-stroke model is consistent with some important features of the force–velocity behavior we have observed, a maximum force at zero velocity and a maximum contraction velocity that occurs when the force is zero. However, the drag predicted with this model increases with the square of the sliding velocity. Upon inspection of Equation 10.51, we can see that this prediction is not consistent with the predictions of the Hill model, which does not have a square dependence on velocity. To resolve this issue, let us again consider the parameter Δ, the distance separating the myosin-binding sites on actin. As we alluded to previously, the minimum possible value of Δ is the physical spacing of the sites, which we will now term d_{s}. From the duty ratio argument, we expect Δ to exceed d_{s} by some amount, but the myosin heads and the actin-binding sites must align at some point. Therefore,

$$\Delta = n d_s, \qquad (10.52)$$

where n is an integer. To estimate what n might be, consider that each binding site has an effective "sweet spot" of width w_{bs}. If, in order to bind, the myosin head needs to be within this width, the time available for binding is inversely related to velocity such that the time available to form a bond with the binding site is

$$t_{\text{bs}} = \frac{w_{\text{bs}}}{v}. \qquad (10.53)$$

Further, we assume that while the myosin is in the sweet spot, it will bind actin at a constant rate of k_+ such that

$$\frac{d[M]}{dt} = -k_+[M].$$ (10.54)

Note that Equation 10.54 is entirely analogous to Equation 10.31, which described the time evolution of adhesive bonds in the Bell model. As before, the solution to Equation 10.54 is an exponential of the form

$$[M](t) = [M](0)e^{-k_+t}.$$ (10.55)

From the perspective of the individual myosin molecule, the probability of transitioning from an unbound to a bound state in the time available for binding is

$$p(t_{bs}) = 1 - e^{-k_+t_{bs}}$$

$$= 1 - e^{-k_+\frac{w_{bs}}{v}}.$$ (10.56)

The probability that a myosin head will bind in the time available is an exponential. To find the total probability over the period, we must integrate from 0 to t_{bs}.

$$p(t_{bs}) = \int_0^{t_{bs}} k_+e^{-k_+t}dt = \Big.^{t_{bs}}_0 \left(-e^{-k_+t}\right) = 1 - e^{-k_+t_{bs}}.$$ (10.57)

You can see from this expression that if the velocity is very high, this probability gets quite small, and the myosin heads would pass by many potential binding spots without binding. In this limiting case $\Delta \gg d$ and n is large. In other words, relatively few of the binding sites are occupied. Because each binding site is separated from the next by a distance δ, and the distance between an occupied site is Δ, then the probability of a given head being bound is also given by

$$p(t_{bs}) = \frac{1}{n} = \frac{d_s}{\Delta}.$$ (10.58)

Setting Equation 10.56 equal to Equation 10.58 we obtain

$$\Delta = \frac{d_s}{1 - e^{-k_+\frac{w_{bs}}{v}}}$$ (10.59)

and

$$\langle F \rangle = \frac{k_m}{2\Delta}\left(\delta_+^2 - \delta_-^2\right)$$

$$= \frac{k_m\delta_+^2}{2\Delta}\left(1 - \frac{\delta_-^2}{\delta_+^2}\right)$$ (10.60)

$$= \frac{k_m\delta_+^2}{2d_s}\left(1 - e^{-k_{on}\frac{w_{bs}}{v}}\right)\left(1 - \frac{v^2t_-^2}{\delta_+^2}\right).$$

Equation 10.59 gives behavior that resembles the hyperbolic relationship observed experimentally by Hill. To see this, consider some realistic values for the parameters. Molecular measurements of myosin predict a value of k_m of roughly 5 pN/nm and $t_- = 0.6$ ms. As discussed, molecular structures suggest a binding-site spacing of $d_s = 36$ nm and myosin stroke distance of $\delta_+ = 5$ nm. The term k_+ would change depending on ATP concentration, but from statistical mechanics a theoretical maximum limit is the rate at which ATP will encounter its myosin-binding site ($\sim$21 s^{-1}) if there is an abundance of ATP. w_{bs} is also difficult to measure, but again a theoretical maximum would be the binding site spacing, d_s. As can be seen in Figure 10.29, for these parameters, the force velocity curve is not precisely hyperbolic, but within a physiological range it is quite close.

Key Concepts

- Cell adhesion involves multiple proteins interacting to form structures such as focal adhesions in cell–matrix adhesion and similar adhesive plaques with cell–cell adhesion.

- Adhesiveness is difficult to model precisely without knowing the number and energy of receptors. However, lumped parameter models based on energy of adhesion and surface tension can be quite informative.

- During inflamation, neutrophils attach to vessel walls in stages characterized by endothelial cell activation, selectin-mediated rolling, neutrophil activation, and integrin-mediated firm adhesion.

- The Bell model describes the effect of force on bond kinetics with an exponential increase in off-rate with force.

- Catch bonds exhibit increased adhesion with load. Catch-slip bonds exhibit decreased adhesion once a certain threshold force is reached.

- Cell migration can be observed *in vivo* but is more easily quantified *in vitro*. The net migration of a cell is the result of the speed and persistence. They can be quantified individually from cellular trajectories.

- Cell migration involves a series of coordinated processes, including protrusion, adhesion, displacement/contraction, and release/retraction. Brownian ratchet processes occur at the interface of polymerizing actin and membrane, and depend on thermal fluctuations forming a gap for the insertion of a new monomer.

- Actin–myosin interactions generate cellular force in an ATP-dependent fashion. They are responsible for force generation in smooth and striated muscles as well as non-muscle cells in stress fibers and during cellular motility.

- The Hill equation describes the inverse hyperbolic relationship between force and velocity.

- Cross-bridge models describe the cyclic series of states that myosin goes through in converting ATP to force. The duty ratio of myosin is the fraction of time that it is engaged with actin, which is surprisingly low.

- The power-stroke model describes the actin–myosin interaction mechanics at a molecular level by treating myosin as a simple spring that is shortened during cocking. It predicts force–velocity relationships in terms of cocking distance, drag, and the size and frequency of actin-binding sites.

Problems

1. During cell migration, a cell does work that is typically dissipated into viscous losses within and outside the cell. *In vitro*, the outside of the cell is typically much less viscous than the cytoplasm, so the losses outside the cell can be neglected. Treat the cell as a thin pancake of height h and radius R, and treat the cytoplasm as an incompressible Newtonian fluid of viscosity μ and density ρ. The cell attaches to the substrate with an adhesion force per unit area of F_a (you may assume this is uniform across the cell's basal surface). If the velocity of the cell v is constant, derive an expression for v in terms of the other parameters and determine whether the cell would move faster or slower if it were to thin by a small amount (h decreases).

2. Consider a cell attached to a surface, as shown below. Other than the flattened contact region, the cell is spherical with radius R. You have deployed a shear stress device to determine the adhesion strength of the cell to the surface, by gradually increasing the shear force applied to the cell. The adhesion force per unit area is denoted F_a, exerting its force roughly uniformly over the adhesion area, which is approximately R_a in diameter. The shear stress device consists of a chamber with an upper platform moving with velocity v, which is increased very slowly so that the flow pattern can be considered fully developed Couette flow at any time. The chamber height is h. At some critical velocity v_c the cell detaches.

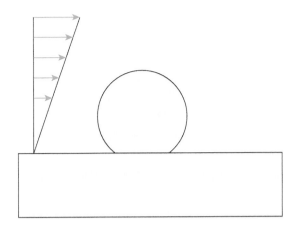

The fluid has viscosity μ and density ρ, and the flow can be considered laminar. Use a scaling relationship based on (a) dimensional analysis and (b) a (scaling) moment balance to derive an expression for v_c based on the provided parameters. (c) Show that the expression obtained from (b) can be derived using the functional form from (a).

3. In discussing actin–myosin contraction, we briefly described what occurs during rigor mortis. Examine this process more closely by considering the following:

   (a) At which stage of the contraction cycle does rigor mortis occur?
   (b) Why does this make the body stiff? If you just lie there and do not move around, are you considerably less stiff than a dead body? But are your muscles not in the same position?
   (c) After a brief time, rigor mortis dissipates. This occurs fast enough that experts can sometimes tell when death occurred based on the body stiffness. Why does rigor mortis dissipate?

4. (a) From Equation 10.14, $\left(EI\dfrac{d^4y}{dx^4} + F_a = T\left(\dfrac{d^2y}{dx^2}\right)\right)$,

   determine a critical peel length (along the membrane) at which the bending component of the membrane roughly balances the stretching component of the membrane.
   (b) Assume that stretching is much more dominant than bending. Derive an expression for the membrane tension as the cell is being peeled.

5. (a) Sketch out sample trajectories for a cell with the four combinations of speed and persistence (high and low of each). Alternatively, you may use Figure 10.20 as a reference. Discuss the differences among the different combinations.
   (b) Assume a cell travels along a sine wave $(y = \sin(2\pi t))$ at constant velocity. If you sample every half second (0.5 s), what is the speed and

persistence of the cell? What if you sample every tenth of a second (0.1 s)?

6. Consider a spherical cell. If you split the sphere with a flat plane, you get two components as shown below (the component that is labeled B has been inverted for clarity).

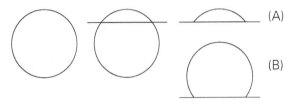

Show that the ratio of adhesion energies for cell component A and cell component B can be expressed as $\tan^2\dfrac{\theta}{2}$, where θ is the contact angle of cell component A to the surface. You may treat the cell components as droplets without receptor interactions.

7. From geometric arguments and the form of Equation 6.5, derive Equation 10.19.

8. Consider the cell migration process, and focus on the Brownian ratchet model as applied to the membrane or the actin filaments (or both). What would you expect to happen to cell migration as the temperature is increased or decreased? Why is a fever advantageous for neutrophils?

9. Derive the probability of a single myosin head binding (Equation 10.55) from the sweet spot argument.

10. In terms of the Hill equation parameters, a, b, and k, what is the maximum shortening velocity?

11. Typical values for the Hill parameters for frog muscle are $a = 37.49$ mN/mm^2, $b = 0.317$ mm/s, and $k = 47.14$ mN/mm^2 s. Plot the predicted force–velocity curve and determine the isometric contraction force and maximum contraction velocity.

12. For the power-stroke model (Equation 10.59) prepare a force–velocity plot for typical values. What is the maximum velocity? Calculate work $\langle F\rangle\Delta$, working distance δ, and duty ratio $r_{duty} = (\delta/\Delta)$ as a function of sliding speed, v.

13. Consider stress fiber actin-myosin activity. What would happen if the fiber was cut with a laser? How would δ_+ and δ_- change. What would happen to drag?

Annotated References

Ananthakrishnan R & Ehrlicher A (2007) Forces behind cell movement. *Int. J. Biol. Sci.* 3(5), 303–317. *This journal review article presents a nice overview of cell adhesion and migration as they work together to move cells.*

Bell GI (1978) Models for the specific adhesion of cells to cells. *Science* 200, 618–627. *The original presentation of Bell's model of the effect of force on adhesion bonds.*

Boal D (2002) Mechanics of the Cell. Cambridge University Press. *Boal's text covers many aspects of cell mechanics mathematically, including some of the adhesion derivations presented in this chapter.*

Costa KD, Hucker WJ & Yin FC (2002) Buckling of actin stress fibers: a new wrinkle in the cytoskeletal tapestry. *Cell Motil. Cytoskel.* 52, 266–274. *This article is the source for the method of determining stress fiber pre-stress by culturing on a flexible substrate.*

Deguchi S, Ohashi T & Sato M (2005) Evaluation of tension in actin bundle of endothelial cells based on preexisting strain and tensile properties measurements. *Mol. Cell. Biomech.* 2, 125–33. *Direct measurement of stress fiber pre-stress using cantilevers.*

Dembo M, Torney DC, Saxman K & Hammer D (1988) The reaction-limited kinetics of membrane-to-surface adhesion and detachment. *Proc. R. Soc. Lond. B.* 234, 55–83. *Some of the first clear evidence suggesting that catch bonds might exist.*

Dillard DA & Pocius AV (2002) The Mechanics of Adhesion. Elsevier Science. *Dillard's text provides some mathematical background in adhesion. It complements but also overlaps some of Kendall's material. Dillard's material includes the discussion of the JKR described in Section 10.1.*

Evans EA (1985) Detailed mechanics of membrane-membrane adhesion and separation I&II. *Biophys. J.* 48, 175–192. *An early report of micropipette cell–cell adhesion experiments and peeling analysis.*

Evans EA & Calderwood DA (2007) Forces and bond dynamics in cell adhesion. *Science* 316, 1148–1153. *A review of the role of forces in cellular adhesion.*

Finger EB, Puri KD, Alon R et al. (1996) Adhesion through L-selectin requires a threshold hydrodynamic shear. *Nature* 379, 266–268. *A quantitative characterization of cell rolling via selectin binding.*

Hill AV (1938) The heat of shortening and dynamics constants of muscles. *Proc. R. Soc. Lond. B* 126, 136–195. *This elegant paper describes the original Hill experiments describing measurements of changes in temperature of $1 \times 10^{-3}\,°C$. It is the seminal presentation of the Hill model.*

Holmes JW (2006) Teaching from classic papers: Hill's model of muscle contraction. *Adv. Physiol. Edu.* 10, 67–72. *A description of a simulation-based teaching module that takes students through Hill's original reasoning process, eventually allowing them for formulate his model.*

Howard J (2001) Mechanics of Motor Proteins and the Cytoskeleton. Sinauer Associates. *A great introduction to cytoskeletal polymerization and mechanics. This book provides a solid foundation of the mechanics of the cytoskeleton and motor protein force generation. Much of the development for the cross-bridge model, as well as some problems, are adapted from this text.*

Huxley AF (1957) Muscle structure and theories of contration. *Prog. Biophys. Biophys. Chem.* 7, 255–318. *Original presentation of the power-stroke model of force generation in terms of molecular parameters.*

Huxley AF & Niedergerke R (1954) Structural changes in muscle during contraction: interference microscopy of living muscle fibres. *Nature* 173, 971–973. *Early description of muscle striations and how they change with contraction.*

Johnson KL (1985) Contact Mechanics. Cambridge University Press. *A good reference for Hertz contact and other contact problems in mechanics.*

Kamm RD and Lang M. Molecular, Cellular, and Tissue Biomechanics (course material from course number 20.310, Massachusetts Institute of Technology, Cambridge, MA available online at http://ocw.mit.edu). *This course covered basic principles of biomechanics, including scaling analysis and engineering analysis, as applied to biological molecules, cells and tissues. Much of the class material served as inspiration for many topics throughout this chapter. Some of the lecture materials can be found online in OpenCourseWare.*

Kendall K (2001) Molecular Adhesion and Its Applications. Kluwer Academic/Plenum Publishers. *This textbook presents many fundamentals of adhesion with a section on cell adhesion. Some of the basic laws of adhesion are presented in this book.*

Lauffenburger D & Linderman JJ (1993) Receptors: Models for Binding, Trafficking and Signaling. Oxford University Press. *This text covers both biology and mathematics of cell interactions, migration, adhesion and signaling. The speed–persistence relationship and membrane peeling presented in this chapter is based, in part, on the material in this book.*

Maheshwari G & Lauffenburger DA (1998) Deconstructing (and reconstructing) cell migration. *Micro. Res. Tech.* 43, 358–368 *This journal article reviews many aspects of cell migration, including chemical gradients, path tracing, and the various steps involved in cell motion.*

Marshall BT, Long M, Piper JW et al. (2003) Direct observation of catch bonds involving cell-adhesion molecules. *Nature* 423, 190–193. *More recent direct observations of the existence of catch bonds.*

Pollard TD, Blanchoin L & Dyche Mullins R (2000) Molecular mechanisms controlling actin filament dynamics in nonmuscle cells. *Annu. Rev. Biophys. Biomol. Struct.* 29, 545–576 *Pollard's review article discusses various aspects of actin remodeling and dynamics in the context of the leading edge of a cell undergoing migration.*

Pollard TD (2002) The cytoskeleton, cellular motility and the reductionist agenda. *Nature* 422, 741–745. *This insight article (similar to a focused review) discusses what's known about the role of the cytoskeleton in describing cell mobility.*

Ramachandran V, Williams M, Yago T, et al. (2004) Dynamic alterations of membrane tethers stabilize leukocyte rolling on P-selectin. *Proc. Natl. Acad. Sci. USA* 101, 13519–13524. *This research article describes the behavior of leukocytes as they contact a surface coated with P-selectin. The cell velocities and tether formations are discussed.*

Tanner K, Boudreau A, Bissell MJ & Kumar S (2010) Dissecting regional variations in stress fiber mechanics in living cells with laser nanosurgery. *Biophys. J.* 99, 2775–2783. *This article describes the use of high-power lasers to disrupt stress fibers.*

Tees DF & Goetz DJ (2003) Leukocyte adhesion: an exquisite balance of hydrodynamic and molecular forces. *News Physiol. Sci.* 18, 186–190. *Review of balance and interplay between the forces of flow and adhesion in leukocytes.*

CHAPTER 11

Cellular Mechanotransduction

I n this chapter, we focus on the process of mechanotransduction, in other words the process by which a cell transduces a mechanical stimulus into a specific cellular response. In biology, cellular signal *transduction* is typically defined as the process by which a cell senses and responds to a chemical stimulus. For example, binding of a ligand to a cell receptor may lead to a conformational change in that receptor, triggering an intracellular signaling cascade that ultimately alters cellular function. Mechanotransduction can be viewed as a completely analogous process to chemical signal transduction in that application of mechanical forces may induce a conformational change in mechanosensing molecules that activate an intracellular signaling pathway, which leads to altered function. As discussed in Section 1.1, an incredibly diverse range of tissues and organs depend on the cellular sensation of extrinsic mechanical signals for proper development or function. Disruptions in this process have been implicated in contributing to the onset or progression of several critically important diseases.

We will discuss cellular mechanotransduction mechanisms in a variety of cell types. This chapter is divided into four parts, organized to reflect the sequentially occurring stages of cellular mechanotransduction. First, the cell must be exposed to mechanical loading. In other words, force must be transmitted from the environment to the level of the cell. Second, the force must be detected by mechanosensitive molecules. That is, these loads must be transmitted to a molecule or a complex of molecules that can activate (that is, undergo conformational change). Third, a conformational change in these molecules must be sufficient to activate intracellular signaling pathways (Figure 11.1). Finally, a cascade of intracellular signaling events occur that ultimately result in altered cellular function. In the first section, we provide a survey of various mechanical stimuli that regulate cell function. Next, we discuss various structures that have been proposed to transmit the loads arising from these stimuli to putative mechanically sensitive molecules. In the third section, we discuss ways in which force-induced conformational changes in proteins may allow the conversion of mechanical forces into biochemical events. Finally, we conclude the chapter by discussing mechanisms by which intracellular signaling events ultimately alter cellular function. As we will see, this is an active area of investigation, with many molecular aspects of the process not fully defined.

11.1 MECHANICAL SIGNALS

We begin our discussion of cellular mechanotransduction with a survey of mechanical signals that have been demonstrated to alter cellular function. We first introduced the notion of fluid flow as a regulatory signal in cells in Section 6.3. The body is filled with fluid: air, blood, cerebrospinal fluid, etc. When fluid-filled tissues and organs are subjected to mechanical loads that generate pressure gradients, fluid flow results, subjecting the cells within to several physical signals. These signals include altered transport of nutrients and/or signaling

Figure 11.1 Schematic of a bone cell subjected to fluid flow, demonstrating that mechanotransduction occurs in four fundamental steps. (A) The first step involves transfer of load from the environment to the cell. (B) Second, the load is transmitted to a "flow-sensing" molecule, resulting in a conformational change. (C) Third, an intracellular signaling cascade is initiated, by altered levels of a second messenger molecule. (D) Finally, the procession of the intracellular signaling events leads to altered cellular function, specifically new bone formation.

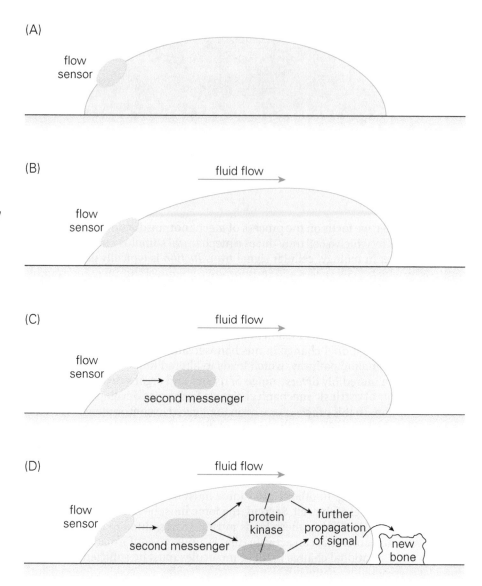

molecules, streaming potentials, and fluid shear stress. As we will see, cells in a variety of tissues sense and respond to fluid flow, even in tissues in which transport of fluid is not a primary function.

Vascular endothelium experiences blood-flow-mediated shear stress

The vascular system is one of the earliest systems in which fluid flow was implicated as an important signal for regulating cell behavior. This is due in large part to the fact that *endothelial* cells (cells lining the vessels) exhibit dramatic, observable changes in morphology in response to fluid shear stress. In particular, when exposed to steady flow, they typically adopt a morphology that is aligned in the direction of flow (**Figure 11.2**), although this can be dependent on the location of the cells within the vasculature. Endothelial cells isolated from heart valves tend to realign perpendicular to the direction of shear stress (**Figure 11.3**). *In vitro* studies investigating the responses of endothelial cells to flow have revealed the capacity for fluid shear to not only regulate all morphology, but also to regulate the levels of a variety of other regulatory molecules such as vasodilators/vasoconstrictors and growth factors. Recirculant, or "disturbed," flows have been found to result in monolayer disruptions and alterations in cell division that have been linked to sites of atherosclerotic plaque formation in humans (Figure 11.4). This suggests that the onset of atherosclerosis may be linked to flows that are unsteady. This hypothesis has ample support from *in vitro* and *in vivo* experiments,. Disturbed flows tend to occur near vascular bifurcations, and these bifurcations are where plaques tend to form. Fluctuating, fluid shear stress may contribute to the development and progression of atherosclerosis.

Lumen-lining epithelial cells are subjected to fluid flow

Similar to endothelial cells lining blood vessels, lumen-lining epithelial cells such as those lining kidney tubules are subjected to fluid shear stress. In kidneys, the ability of epithelial cells to sense urine flow is fundamental to proper kidney function. A link between renal flow sensing and polycystic kidney disease (PKD), which is characterized by the formation of renal cysts, has been established and will be discussed in Section 11.2. Flow sensing has similarly been implicated in maintaining liver health. Cholangiocytes, epithelial cells that line the lumen of the liver bile duct, are exposed to passive movement of bile. These cells have been found to sense and respond to fluid flow by a mechanism that appears to be similar to that of the kidney.

Fluid flow occurs in musculoskeletal tissues

Fluid flow also regulates cells in tissues that do not transport fluid as part of their primary function. Bone tissue incorporates a fluid-filled network of voids (*lacunae*) and channels (*canaliculi*). Within the lacunae reside bone cells called *osteocytes* that extend long, slender *processes* through the canaliculi to neighboring osteocytes. Mechanical loads associated with habitual loading (ambulation—crawling, walking or running) generate pressure gradients

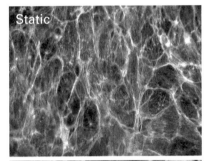
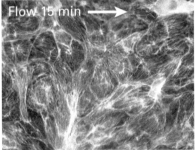
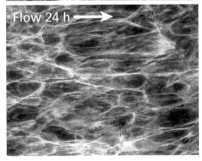

Figure 11.2 Pulmonary endothelial cells adapt their actin cytoskeleton to flow. The cells in these figures have been stained with fluorescently conjugated phalloidin, which binds to actin. Under flow, the cells first quickly reinforce their actin cytoskeleton, as seen in the middle panel. More gradually, they reorganize their cytoskeleton and overall morphology in the direction of flow. (From, Birokov KG et al. (2002) *Am. J. Respir. Cell Mol. Biol.* 464.)

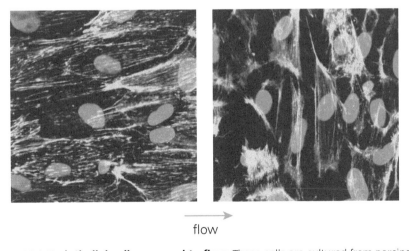

flow

Figure 11.3 Endothelial cells exposed to flow. These cells are cultured from porcine aortas (left) and aortic valves (right) exposed to 20 dynes/cm^2 of shear flow for 48 hours. They exhibit vastly different realignment properties demonstrating that cell source is a critical factor in the mechanobiologic response. See www.garlandscience.com for a color version of this figure. (from Butcher JT et al. (2004) *Arterioscler. Thromb. Vasc. Biol.* 24.)

Figure 11.4 Regions near bifurcations are more prone to atherosclerotic formation (shaded black). Several groups hypothesized that because of recirculant or other disturbed flow patterns near these regions, the endothelial cells alter their behavior, leading to atherosclerotic formations. (From, Thubrikar MJ et al. (1995) *Ann. Thorac. Surg.* With permission from Elsevier)

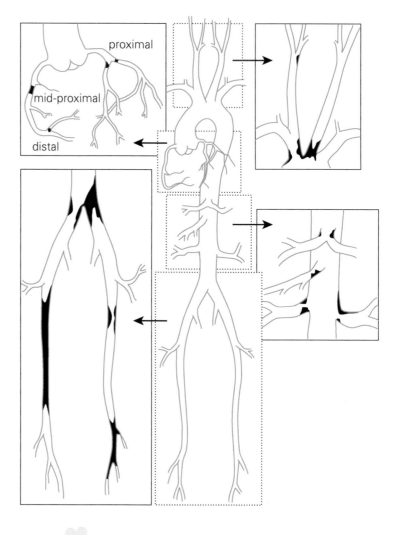

Figure 11.5 Mechanical loading of bone induces interstitial fluid flow within the lacunar–canalicular system in which osteocytes reside. When bone is loaded fluid flows from regions of compression to regions of tension.

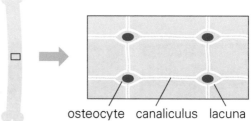

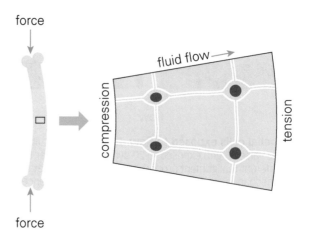

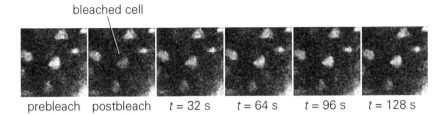

bleached cell

prebleach postbleach *t* = 32 s *t* = 64 s *t* = 96 s *t* = 128 s

Figure 11.6 Fluorescence recovery after photobleaching in a single lacuna within a mouse femur. By monitoring increases in the recovery rate in the presence of mechanical loading, one may determine whether enhanced transport of unbleached dye due to convective flow is occurring.

within the lacunar–canalicular system that drives flow from areas of compression to areas of tension (Figure 11.5), exposing osteocytes to fluid shear stress. An important distinction of osteocyte flow sensing compared with luminal flow sensing is that although luminal flow (in blood vessels or within kidney tubules) can be readily observed, experimental measurement of bone flow is very difficult. This is due to the technical challenges associated with the microscopic fluid spaces of the lacunar–canalicular system (~0.1–1 μm). However, the recent use of fluorescence recovery after photobleaching (FRAP) has provided some insight (Figure 11.6). In this approach, a fluorescent tracer is injected into the animal and allowed to equilibrate within the lacunar–canalicular system. A bone surface is exposed, and a laser scanning confocal microscope is used to photobleach a single lacuna. Infilling of unbleached tracer molecules from surrounding lacunae and canaliculi results in gradual recovery of fluorescence in the photobleached lacuna. In the presence of flow, infilling is enhanced by *convective* transport (the motion of tracer carried by the fluid). We can determine whether flow is occurring by comparing transport rates in the presence or absence of loading. Further, the convective velocity can be estimated by mathematical modeling. Several experiments in cultured bone cells have demonstrated that they sense and respond to levels of fluid flow predicted to occur in habitual activities of daily living. Fluid flow within the musculoskeletal system is not limited to bone. Like bone, articular cartilage is a mechanosensitive tissue that is well known to adapt its properties to the loading it experiences. Within cartilage, chondrocytes are embedded in an extracellular matrix consisting primarily of proteoglycans, collagens, and water. With mechanical loading, a variety of physical signals are generated including hydrostatic pressure, matrix strain, and fluid flow. Several experiments in cultured chondrocytes as well as in perfused cartilage and tissue-engineered constructs indicate that fluid flow is a potent regulator of cartilage metabolism.

Fluid flow during embryonic development regulates the establishment of left–right asymmetry

We have so far discussed the role of fluid flow in regulating tissues and organs in the adult. However, the regulatory role of fluid flow is also important during development. In the mammalian embryo, the generation and subsequent sensing of flow is critical for the establishment of left–right asymmetry. During gastrulation, a triangular-shaped indentation called a *node* appears on the surface of the embryo. The cells lining the surface of the node are ciliated, and some of these cilia are motile (Figure 11.7). They move in a vortical pattern, generating a leftward flow of extra-embryonic fluid (termed *nodal flow*). As will be discussed in more depth later in this chapter, this nodal flow is sensed by cells on the left margin of the node, resulting in an asymmetric intracellular Ca^{2+} signal. Superimposing a rightward flow (by loading the embryos in a flow chamber) can reverse the left–right asymmetry such that the left and right sides of the body are exchanged.

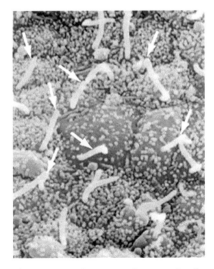

Figure 11.7 Electron micrograph of primary cilia (highlighted by white arrows) within the embryonic node. These cells sense flow that is critical to establishing the right/left directions in the embryo (From, Nonaka S et al. (1998) *Cell.* 95.)

Advanced Material: Fluid shear stress or chemotransport?

Fluid flow exposes cells to several different physical signals, and it is often desirable to distinguish which of these signals are driving flow-induced cellular responses. Flow exposes cells to convective chemotransport, in other words the carriage of nutrients and/or signaling molecules by the fluid motion. It also exposes cells to fluid shear stress. Because these signals are coupled, one question is how do we determine whether cells are being stimulated by chemotransport or fluid shear?

One way to isolate the effects chemotransport in parallel plate flow chambers is to alter shear stress parametrically while keeping flow rate constant, and vice versa. Recall from Equation 10.2 that, in these chambers, shear stress is related to flow rate as

$$\tau = \frac{6\mu Q}{bh^2} \qquad (11.1)$$

where Q is the flow rate, μ is the fluid viscosity, b is the chamber width, and h is the chamber height. Equation 11.1 implies that one can subject cells to the same shear stress but different flow rates by adjusting the viscosity of the flow media (this can be achieved by adding neutral dextran to the media). The same shear stress would result from a doubling of the fluid viscosity and halving of the flow rate, or a halving of the viscosity and a doubling of the flow rate. If cell responsiveness increases with flow rate but not shear stress the cells are responding to chemotransport. Alternatively, if the responsiveness of cells increases with shear stress but not flow rate, it suggests that the cells are responding to fluid shear.

Strain and matrix deformation function as regulatory signals

We now turn our attention from fluid flow as a regulatory signal to substrate or matrix strain. As we learned in Section 3.2, whenever a solid material is loaded (or exposed to stress), it deforms and undergoes strain. As the matrix deforms, the cells embedded in these tissues experience loads at the sites of adhesion to the matrix; as a result, they are subjected to strain (Figure 11.8). We present some examples in which substrate or matrix strain function as critical mediators of biological processes.

Smooth muscle cells and cardiac myocytes are subjected to strain in the cardiovascular system

We previously learned that, within blood vessels, endothelial cells sense and respond to hydrodynamic forces arising from blood flow. Strain is another regulatory signal within the cardiovascular system. Pulsatile changes in blood pressure induce oscillatory stretch within the walls of blood vessels, whose major cellular component is vascular smooth muscle cells. Under normal circumstances, these cells do not come into contact with blood directly. However, they are sensitive to stretch. Mechanical stretch alters several functions in vascular smooth muscle cells, such as cell alignment, migration, proliferation, and apoptosis. Stretch also regulates secretion and production of several paracrine and endocrine factors.

Figure 11.8 Mechanical coupling of cell to matrix. Because the cell is well connected within the extracellular matrix, deformation of a matrix will transmit loads to the cells residing within. This will result in cellular strain.

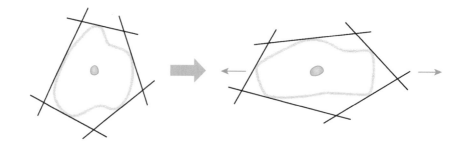

Cardiac myocytes are also subjected to stretch-induced stimulation. Under hypertension, this stimulation is increased, and this overstimulation is believed to lead to cardiac hypertrophy. Elucidation of relevant regulatory pathways could have therapeutic implications. We know that pressure overload leads to altered mechanical stimulation of cardiac myocytes and hypertrophy, which is unhealthy in the long term. Current therapies focus on preventing pressure overload, but this is not always possible. A better understanding of how this pathway becomes mechanically activated may reveal new targets for pharmacological inhibition of hypertension-induced hypertrophy.

Cellular strain in the musculoskeletal system is dependent on tissue stiffness

Given the function of the musculoskeletal system to support and generate forces during movement, cells within these tissues are also subjected to mechanical strain. Articular cartilage is subjected to cyclic compression during ambulation, which deforms the matrix and the cells residing within. Because such loading is important physiologically there are extensive efforts being made to characterize these deformations both experimentally and numerically. The local mechanical environment of chondrocytes is determined in large part by the interaction of the chondrocyte with its pericellular matrix, which possesses a high level of type VI collagen and proteoglycans (the combination of the chondrocyte and its pericellular matrix is sometimes referred to as a chondron). When cartilage is compressed, chondrocytes have been demonstrated to undergo changes both in shape and volume. These deformations are nonuniform, owing in large part to mismatches in stiffness between the cell and pericellular matrix.

The lung and bladder are hollow elastic organs that are regulated by stretch

Sensation of deformation within elastic, hollow organs such as the lung and bladder is critical for their proper function. Within the lung, breathing generates cyclic matrix stretch. *In vitro*, stretch has been shown to regulate lung cell growth, remodeling of the cytoskeleton, activation of signaling molecules, and phospholipid secretion. Within the bladder, during the storage phase, stretch stimulates afferent neurons. Bladder epithelium is also sensitive to stretch, as release of adenosine triphosphate (ATP), acetylcholine, and nitric oxide occurs under mechanical stimulation. Elucidation of these mechanosensitive pathways may provide potential pharmacological targets for future treatment of lung or bladder dysfunction.

Cells can respond to hydrostatic pressure

In addition to fluid shear and substrate strain, cells can sense and respond to hydrostatic pressure. The cells within bone, articular cartilage, intervertebral discs, and the cardiovascular system are exposed to cyclically varying hydrostatic pressure. Compared with fluid shear and substrate strain, the body of studies investigating the mechanisms by which cells sense pressure is much smaller. This is perhaps due, in part, to the fact that cells are fluid-filled structures, and their compressibility under pressure is typically small. Red blood cells have been estimated to undergo a change in cellular volume of only about 0.1% when subjected to a relatively high 10 MPa load (note that typical peak systolic blood pressures of 120 mmHg corresponds to 0.02 MPa, or about three orders of magnitude less). It may also be due to the insensitivity in some cell types when exposed to physiological pressures. Within the bone marrow cavity, pressures are typically on the order of 10 mmHg but may be on the order of 100 mmHg during impact loading. However, several *in vitro* studies have demonstrated that bone cells are

Nota Bene

Smooth muscle cells respond to fluid shear. Shear stress has been shown to inhibit the migratory and proliferative behavior of smooth muscle cells. *In vivo*, shear stresses experienced by the endothelial cells are transmitted to the smooth muscle cells directly underneath them. Because smooth muscle and endothelial cells are in contact and likely receive and transmit mechanical and biochemical stimuli to and from each other, co-culture mechanotransductive studies are a promising approach. Similarly, an active area of research is examining the response of endothelial cells to stretch. It turns out that stretch and shear stress can be competing factors, with endothelial cells sometimes orienting in an "undecided" fashion if they experience contrary input from different mechanical stimuli.

unresponsive to such pressures. It is interesting to note that, despite the apparent insensitivity of cells to physiological pressures, almost all cells undergo a similar alteration in cytoskeletal structure and morphology under high hydrostatic pressure (>100 MPa). When subjected to these pressures, the cytoskeleton rapidly disassembles and the cells become rounded. In many instances, apoptotic pathways become activated, and cell death occurs.

Advanced Material: Parasitic flows

Because cells must be immersed in culture media during substrate stretch experiments, some degree of fluid movement (sometimes referred to as parasitic flow) is unavoidable. Careful attention must be paid to minimize parasitic flow, as it can severely confound interpretation of experimental results. A classic example can be found in the field of bone mechanotransduction. Bone is a relatively stiff material, so the strains it is subjected to are relatively small: on the order of hundreds to thousands of microstrains (abbreviated as µε). Currently, the consensus of the scientific community is that these small strains are generally insufficient to stimulate bone cells *in vitro*, but this view was not always accepted. For many years, strains of one thousand µε were believed to be sufficient to stimulate bone cells because *in vitro* cell-straining systems were based on bending (Figure 11.9A). Because bending resulted in a large displacement of the substrate, these systems generated a significant amount of parasitic fluid flow. In particular, the faster the displacement through the media, the higher the levels of fluid flow to which the cells were subjected.

To isolate the effects of substrate strain versus fluid flow, investigators exploited the dependence of strain on substrate thickness during four-point bending. In particular, the beam-deflection equation for four-point bending is

$$\varepsilon = \frac{td}{\alpha}(L - 1.33\alpha), \qquad (11.2)$$

where ε is strain on the substrate surface, t is the substrate thickness, d is the displacement, L is the length between inner supports, and α is the length between the inner and outer supports. By seeding cells on slides of different thicknesses, investigators were able to apply different strains while keeping displacement rate (and parasitic flow) constant, and vice versa. They found that the cells were insensitive to changes in strain, but were highly sensitive to the displacement rate. This shows that cells were responding to fluid flow rather than substrate strain. Because of these and other similar studies, alternative approaches to cell stretching that minimize parasitic flows have been developed (Figure 11.9B).

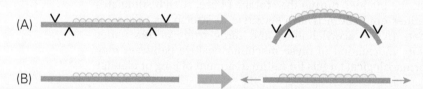

Figure 11.9 Modalities for inducing substrate strain *in vitro*. (A) Cells are seeded on a substrate subjected to four-point bending. The cells undergo a large degree of movement within the culture medium, exposing them to parasitic flow. (B) Cells are subjected to substrate strain in a manner that reduces parasitic flow.

11.2 MECHANOSENSING ORGANELLES AND STRUCTURES

In the last section, we reviewed different types of mechanical stimuli that act as regulatory signals in cells in a variety of tissues and organs. We now shift the focus to cellular structures and organelles that have been implicated in sensation of mechanical forces. Much of what we have learned in the next few sections has been a direct result of the *in vitro* reductionist approaches discussed in the previous section. As we will see, cells may possess very specialized structures to facilitate load transmission. We begin our survey with one of the most elegantly designed structures for this purpose, which is found in hair cells in the inner ear.

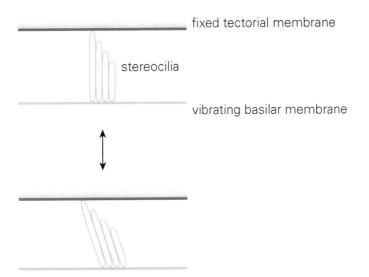

fixed tectorial membrane

stereocilia

vibrating basilar membrane

Figure 11.10 Inner ear hair cells. The hair cells of the inner ear are responsible for hearing. They transduce vibrations of the cochlear basilar membrane into nerve impulses.

It is also worth noting that, by virtue of the topic, this chapter is more biological and less quantitative than the others in this text.

Stereocilia are the mechanosensors of the ear

The inner ear is one of the best-characterized mechanosensitive structures in the mammalian body. Its function is to convert physical vibrations into the electro-chemical nerve impulses that mediate hearing. The eardrum receives sound waves and transmits the vibration to three bones (*hammer, anvil,* and *stirrup*), which in turn transmit the vibrations into the *cochlea* (Figure 11.10). Inside the cochlea, there is a basilar membrane, which has varying frequency sensitivities. For a given frequency, a specific part of the membrane resonates and undergoes more motion than other parts of the membrane (frequency range 20 Hz to 20 kHz). Within the cochlea reside mechanosensory cells (hair cells) that transduce this membrane motion into nerve impulses. Hair cells have evolved a remarkable structure to sense this movement. Bundles of stereocilia (hair bundles) project from the apical surface and are linked together by coupling structures. The longest stereocilia are in direct contact with the membrane. When the membrane reso-nates as a result of sound, the hair bundle deflects, leading to an influx of cations into the stereocilia. Mechanosensing in these cells can be incredibly fast. Frog cochlear hair cells have been shown to produce a nerve impulse in 40 μs. In con-trast, visually evoked potentials in the retina occur on the order of tens of millisec-onds, or several orders of magnitude slower.

Example 11.1: Deflection of hair-cell bundles

Consider the hair-cell bundles depicted in Figure 11.10. Recall from Figure 1.1 the arrangement of distal tip links that mechanically open membrane ion channels. Given that the stiffness of hair-cell bundles has been measured to be about 500 μN/m, estimate the tip force required to open a channel.

The amount of channel displacement required for its activation, in other words to allow the transition from a closed to an open configuration, has not been measured directly. However, a reasonable estimate would be the diameter of a typical molecule that can pass through the channel pore. Five nanometers is a reasonable guess and likely accurate within an order of magnitude. Because the stiffness of hair-cell bundles has been measured to be about 500 μN/m, a reasonable estimate for the tip force would be 2.5 pN. For comparison, the binding force of a single integrin molecule is on the order of tens of piconewtons.

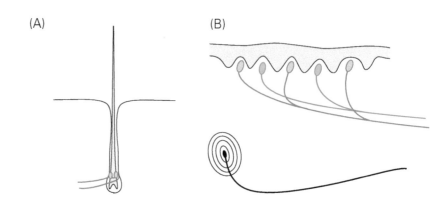

Figure 11.11 The mechanosensing of touch. Mechanosensory structures and cells involved in touch sensation in hairy (A) and non-hairy (B) skin all involve nerves that terminate in specialized effector cells. These include slowly adapting pressure-sensing Merkel cells (grey), rapidly adapting vibration-sensing Pacinian corpuscles (black), and Meissner's corpuscles (blue) that are nerve endings sensitive to light touch.

Specialized structures are used in touch sensation

Touch, the sensory experience resulting from force transmission to the skin, is one of the five major senses by which humans are able to experience their environment. Of these five senses, the cellular and molecular mechanisms underlying touch are the least well understood. Touch sensation is initiated by mechanical deformation of nerve endings within specialized structures. In hairy skin, oval-shaped Merkel cells envelope hair follicles, allowing them to be activated by hair movement (Figure 11.11). Alternatively, nerve endings may be free, or encapsulated within structures called corpuscles. Meissner's corpuscles are a type of nerve ending involved in the sensation of light touch. They consist of flattened support cells arranged in horizontal lamellae and encapsulated by a connective-tissue capsule. Within the corpuscle, a single nerve fiber meanders between the lamellae. Other similar structures include Pacinian corpuscles, which are involved in sensing vibration, and Ruffini corpuscles, which are slow-response mechanoreceptors activated by skin stretch.

Primary cilia are nearly ubiquitous, but functionally mysterious

A projecting structure that is well suited for mechanosensing in non-sensory cells is the primary cilium. It is a rod-like structure that projects from the surface of the plasma membrane into the extracellular environment. Primary cilia are found in almost all cells, and despite having been initially described near the turn of the century, their function has only recently begun to be elucidated. Unlike motile cilia, primary cilia are solitary and nonmotile. However, like motile cilia, primary cilia are microtubule-based structures with *axoneme* cores. The axonemes of primary cilia are composed of nine microtubule doublets, but lack the central pair of doublets that motile cilia have. They are described as having a 9 + 0 (as opposed to 9 + 2 for motile cilia) microtubule architecture. By extending into the extracellular space and possessing mechanical characteristics that allow flow-induced bending, they are uniquely situated to serve a mechanosensory role. Several investigators have demonstrated that cilia passively bend under fluid flow and recoil after cessation of flow (Figure 11.12). Ciliary bending by micropipette suction is sufficient to increase intracellular Ca^{2+} concentration in kidney epithelial cells. Cilia have also been implicated as sensors of fluid flow in kidney, liver, bone, blood vessels, and the embryonic node.

Because flow-induced deflections would result in the generation of membrane tension that could serve to open stretch-activation ion channels, efforts have been made to identify channels that localize to the cilium. One such channel is the polycystin-1/2 complex. Polycystin-1 is an integral membrane protein encoded by the gene *PKD1*, and polycystin-2 is a cation channel encoded by the gene *PKD2* that heterodimerizes with polycystin-1. Mutations in either *PKD1* and *PKD2* result in autosomal dominant polycystic kidney disease (PKD). Polycystin-1 and polycystin-2 co-distribute in renal primary cilia, and loss of polycystin-1 and polycystin-2 function disrupts flow sensing in kidney cells, suggesting that aberrant

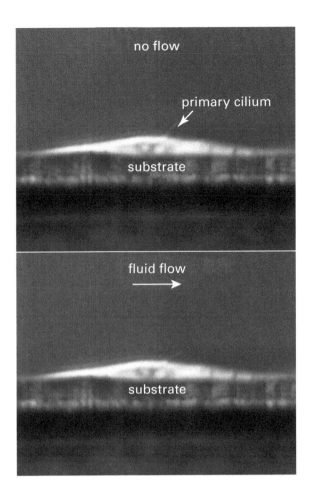

Figure 11.12 Primary cilia bending with flow. This side view of a bone cell and its primary cilium in the absence (top) and presence (bottom) of fluid flow. Deflection of the cilium under flow can be observed. (From Gefen A (2011) Cellular and Biomolecular Mechanics and Mechanobiology. Copyright permission Springer Science, New York.).

mechanosensing due to dysfunction in this complex may contribute to PKD pathogenesis.

Cellular adhesions can sense as well as transmit force

Adhesion sites, or plaques, are thought to serve as mechanosensory structures because they would likely experience high stress under both substrate strain and fluid shear. Most connective tissues are composed primarily of structural polymers such as collagen or fibronectin. Cells embedded within these tissues are anchored within this fiber network at discrete sites of adhesion (sometimes referred to as focal contacts or focal adhesions). If the fiber network were deformed, the cells would also deform, with the loads transferred from the network to the cell through these focal contacts (see Figure 11.8). A variety of linker proteins mechanically couple focal adhesions to the cell cytoskeleton, so, during fluid shear, loads would be transmitted from the apical surface of the cell, through the cytoskeleton to points of cell–substrate attachment (see Figure 11.1).

Several proteins within adhesion complexes that are directly involved in this load-bearing path also possess signaling functions. Such proteins would be ideally suited to undergo force-induced conformational changes that could serve to transduce mechanical force into an intracellular biochemical signal. Such proteins include focal adhesion kinase, vinculin, talin, tensin, paxillin, and others.

Cell–cell adhesions have also been suggested as sites of mechanotransduction, because they incorporate signaling molecules and would likely be sites of high stress. Cell–cell junctions are particularly important in endothelial cells, where they form a tight barrier that prevents vessel leakage. In these cells, PECAM-1 localizes to the cell–cell junction, and is rapidly activated upon exposure to shear stress. There is also evidence that cadherins, the primary component of adherens junctions, and vascular endothelial growth factor receptor 2 (VEGFR-2), the receptor for VEGF, forms a complex with PECAM-1 that is activated by fluid shear.

The cytoskeleton can sense mechanical loads

As described in Section 7.1, the cytoskeleton is composed of three components: actin filaments, microtubules, and intermediate filaments. Of these components, the actin cytoskeleton has received much attention as a mechanosensing structure. There are several ways in which the actin cytoskeleton may contribute to mechanosensing. First, it may play an indirect role, transmitting loads to sites of mechanotransduction. In cells exposed to fluid flow, the cytoskeleton could serve to transmit shear forces to mechanotransduction sites, such as focal adhesions, where force-induced conformational changes could occur. Alternatively, the actin cytoskeleton has been proposed to serve a mechanosensory function by transmitting loads to the nucleus and directly manipulating DNA, exposing previously hidden transcription sites to initiate the synthesis of transcription factors or other proteins. Structural nuclear matrix proteins including *nesprin* and *lamins* have been implicated as being required for sensing of some mechanical loads.

In terms of direct mechanosensing, the cytoskeleton may itself serve as a structure in which mechanical forces are transduced into a biochemical signal. Loading may cause elements of the network to slide or deform, liberating trapped signaling molecules or exposing (or hiding) sites of enzymatic activity. Indeed several signaling molecules as well as other structural molecules have been found to associate with the actin cytoskeleton. Structural rearrangements of the network under load could lead to activation or inactivation of intracellular signaling by bringing two molecules closer together or farther apart (Figure 11.13). Microtubules have been shown to buckle when loaded with external compressive loads, leading to loss of the ATP cap and entry into the depolymerization catastrophe. Many of the cytoskeletal transduction mechanisms are more theoretical than accepted. It is a compelling research direction because rarely are so many proteins with both signaling and structural roles in such close proximity.

Example 11.2: Microtubule buckling

One proposed model of microtubule-based mechanotransduction is that externally applied forces cause buckling that acts as an intracellular signaling trigger. What magnitude of pressure could be detected by such a mechanism?

To estimate an upper bound on sensitivity, consider a cell responding to an external pressure. Ignoring the contributions of the cytoplasm, imagine that this pressure is supported by a microtubule network radiating from the perinuclear region to the cell periphery like the spokes of a bicycle wheel. Recall that the buckling load is given by Equation 3.38 as

$$F_b = \frac{\pi^2 EI}{L^2}. \tag{11.3}$$

Experimental measurements of flexural rigidity for microtubules are highly variable and span orders of magnitude ranging from 1 to 200 pN/μm². Using a typical value (Table 3.3) for a 10 μm cell, the buckling load for each molecule is

$$F_b = \frac{(3.142)^2 \ 360 \times 10^{-25} \, \text{Nm}^2}{(10 \times 10^{-6} \, \text{m})^2} \approx 3.5 \text{pN}. \tag{11.4}$$

If we have 1000 microtubules in a typical cell, the resulting pressure would be

$$P_b = \frac{14 \text{nN}}{4\pi(5 \times 10^{-6} \, \text{m})^2} \approx 45 \, \text{Pa}. \tag{11.5}$$

To lend some scale, typical blood pressures are on the order of tens of millimeters of mercury or kilopascals. Microtubule bucking has the potential to be an extremely sensitive mechanism. Of course, in reality this is the theoretical lower bound on the detectable pressure. Fluid pressurization of the cytoplasm would be expected to carry a significant amount of external load.

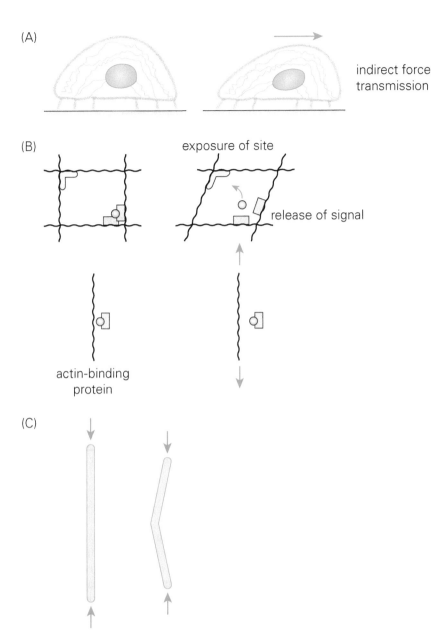

(A)

indirect force transmission

(B)

exposure of site

release of signal

actin-binding protein

(C)

Figure 11.13 Potential roles of the cytoskeletal network in mechanotransduction. These include (A) indirect force transmission, (B) exposure (or hiding) of signaling domains, release (or capture) of signaling molecules with load, and (C) even simple buckling-induced depolymerization. Mechanotransduction involves converting a physical signal to a biochemical one, so the cytoskeleton is a natural candidate because the network brings structural and signaling roles together.

Mechanosensing can involve the glycoproteins covering the cell

The glycocalyx, mentioned briefly in Section 9.5 as a source of bending stiffness, is a layer of membrane-bound macromolecules that surrounds several cell types. Its composition and thickness depend on tissue type, and, even within similar tissues, on anatomical location. It consists of a meshwork of proteoglycans, glycosaminoglycan side chains, and associated proteins from the surrounding fluid that bind this meshwork. Physically, it is ideally suited to serve a mechanosensory role, as it acts as the interface between the extracellular fluid and the cell membrane/cytoskeleton, particularly in cells that respond to fluid flow. In addition, because the proteoglycans, with their numerous side chains, project into the luminal region, they have the potential to deflect under shear flow. Through their associations with the apical membrane/cytoskeleton, they may indirectly contribute to mechanotransduction by transmitting forces to the cell. Conversely, the glycocalyx could serve to decrease cell sensitivity to fluid flow by shielding the

Figure 11.14 Longitudinal (A) and transverse (B) view of pericellular matrix surrounding an osteocyte process in bone. The cell process and the mineralized bone tissue are both dark. The light-colored space between the two is occupied by a glycoprotein gel as well as discrete linker proteins. (Courtesy of Lidan You, University of Toronto.)

(A)

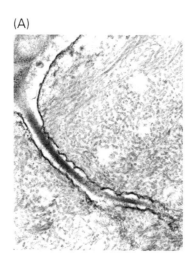

(B)

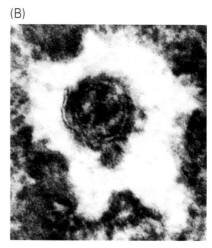

membrane from shear stresses. So far, studies investigating the role of the glycocalyx in flow sensing have focused primarily on endothelial cells.

Though the role of the glycocalyx in flow sensing has been studied primarily in the context of endothelial mechanotransduction, membrane-associated proteins and glycoproteins have also been proposed as facilitators of mechanotransduction in bone. Within the lacunar–canicular system, osteocyte processes are surrounded by a pericellular matrix that is believed to be similar in composition to the endothelial glycocalyx (Figure 11.14). In the presence of fluid flow, drag forces are generated on the pericellular matrix that are transmitted to the cell surface. Theoretical calculations predict that these forces are much higher than those generated by fluid shear alone.

The cell membrane is ideally suited to sense mechanical loads

Although evidence for cytoskeletal mechanosensing is relatively scarce, the role of the membrane in mechanosensing is much better developed. The membrane is an ideal site, as it contains specific structures that confer sensitivity to the environment while maintaining the integrity of the intracellular components. Recall that the bilayer represents a hydrophobic layer of a specific thickness (about 7 nm). Any protein that contains a hydrophobic region, known as membrane-spanning regions, of the same thickness can become embedded within the membrane. Proteins such as receptors that have binding or activation domains on both the extracellular and intracellular sides of the cell can facilitate signaling through the cell membrane by conformational changes that occur when a ligand binds or dissociates from one side. Analogously, changes in the physical and chemical properties of the lipid bilayer after exposure to mechanical force can play a critical role in supporting mechanotransduction by potentially inducing conformational changes in membrane proteins.

Perhaps one of the best-studied mechanisms of mechanosensing, membrane tension, is well known to contribute to increased probability of ion channel opening. *Stretch-activated channels* have been studied in detail in membrane *patches* or fragments of bilayers anchored to the end of a micropipette. Pipette pressure can be changed to regulate the areal strain in the patch, while opening and closing of the channels is assayed by measuring electrical conductance.

Other than tension, alterations in other membrane physical properties may be involved in mechanosensing. Shear-induced alterations in the membrane may initiate mechanotransduction through enhancing/inhibiting interactions between signaling molecules. Several studies have demonstrated that membrane fluidity can be altered by fluid flow. Membrane fluidity refers to the changes in

Nota Bene

Membrane-spanning regions.
Membrane-spanning regions are commonly α-helices and can often be identified by looking for a series of hydrophobic amino acids of the correct length from a genetic sequence. This can be a critical step in estimating a protein's structure and function from sequence information alone. In the case of multiple membrane-spanning domains, the protein will fold back on itself, with each membrane transition forming intracellular or extracellular loops. If a membrane-spanning protein has an odd number of membrane-spanning domains, the carboxy (C)- and amino (N)-terminal ends of the protein will be on opposite sides of the membrane. Likewise, if the protein has an even number of membrane-spanning domains, the C and N termini will be on the same side of the membrane, either intracellularly or extracellularly.

membrane viscosity that may arise from local ionic transients coupled with aggregation of proteins. Alterations in membrane fluidity may induce conformational changes in transmembrane proteins, or enable a change in the rate of interaction of membrane proteins. In particular, because proteins with membrane-spanning domains float within the cell membrane, they can be thought of as being confined within a two-dimensional space. This confinement can greatly alter their chemical kinetics. Proteins are much more likely to encounter and interact with each other if they are constrained to the two-dimensional space of the membrane than if they must find each other in the full three-dimensional cytoplasm. Mechanically induced changes in membrane fluidity may serve to alter normal interactions within this space, or allow aggregation of new groups of proteins that breach the typical compartmentalization.

Finally, membrane-bound proteins may also act as direct mechanical receptors independent of changes in membrane physical properties. Fluid shear has been shown to be sufficient to activate heterotrimeric G-proteins reconstituted into liposomes, even in the absence of any other potential mechanotransducing molecules. As described briefly in Section 2.2, G-protein-coupled receptors are membrane-spanning receptors that are linked to G-proteins. When activated, the receptor undergoes a conformational change that allows the guanosine triphosphate (GTP) to exchange for the guanosine triphosphate (GDP) on the G-protein, resulting in activation of the G-protein.

Lipid rafts affect the behavior of proteins within the membrane

As discussed previously, protein interaction kinetics within the membrane may be enhanced by their confinement within a two-dimensional space. These interactions may be enhanced even further by the inherent heterogeneity in fluid properties within the membrane. There are separate phases or domains within the lipid bilayer. One such microdomain, 20–200 nm in size, occurs because of the presence of cholesterol, and other lipids known as sphingolipids. These microdomains were termed *lipid rafts* in the 1970s. The hydrophobic tails of cholesterol have a slightly different structure than the rest of the membrane phospholipids. Specifically, their tail domains are a bit longer. This causes cholesterol to self-aggregate into small regions with lower fluidity than the bulk of the phospholipids. Proteins suspended in these domains tend to remain within them and have difficulty passing laterally in and out of the raft. The reaction rates of proteins confined to lipid rafts can be higher as a result of having a locally higher reactant concentration. These types of lipid raft are referred to as planar to distinguish them from a second type of lipid raft that occurs because of the action of *caveolin*. Caveolin is a small protein with cytoplasmic C and N termini. The termini are linked, causing an increase in curvature (Figure 11.15). In caveolin-rich areas of

Nota Bene

Small GTPases regulate cytoskeletal behavior. G-proteins are a particular type of GTPase. Another type of GTPase linked to many mechanobiological functions includes the small GTPases, consisting of Rho, Rac, Rap, and others. These GTPases are activated when bound to a GTP. Knocking out, or mutating, various GTPases leads to alterations in cell morphology, actin organization, migratory capability, spreading, and adhesion. Though these molecules have been implicated in the control of several pathways related to motility and adhesion, the role of these small GTPases in mechanotransductory pathways is not well understood.

(A)

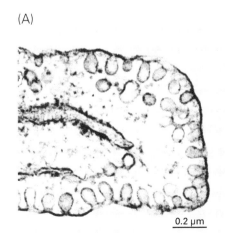

(B)

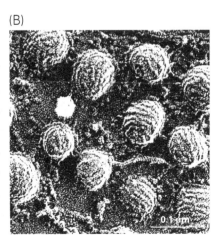

0.2 µm

0.1 µm

Figure 11.15 Electron micrograph of a thin section (A) and freeze-fracture (B) of the cell membrane. The protein caveolin causes the membrane to bend, forming infolded regions. These microdomains form subregions within the membrane with altered chemical kinetics. (From, Anderson RGW (1998) *Annu. Rev. Biochem.* 67.)

the membrane, tiny invaginations of the membrane known as *caveolae* (Latin for "small cave") occur. Both forms of lipid raft have been demonstrated to mediate mechanotransduction in a variety of cell types, although the specific mechanism has not been established. However, similar to the mechanisms described above in the membrane, it has been proposed that mechanical force could alter the physicochemical properties of the lipid rafts, leading to the direct activation of individual signaling molecules residing within them, or facilitating interactions between molecules.

11.3 INITIATION OF INTRACELLULAR SIGNALING

In the first section we provided a survey of the types of mechanical stimulus that cells are exposed to, and in the second section we discussed structures in cells that transmit loads from the environment to mechanically sensitive molecules. This section focuses on the final event of mechanotransduction, load-induced protein conformational changes and the generation of an intracellular signaling cascade. In particular, we discuss two potential mechanisms by which mechanical force can be transduced into a biochemical signal at the molecular level, opening of mechanosensitive ion channels, and exposure of cryptic binding sites.

Ion channels can be mechanosensitive

Many membrane-bound proteins are involved with cellular exchange of molecules between the cytoplasm and the extracellular environment. Ion channels are complexes of proteins that form small passages in the membrane for ion flow. Channels can be selective based on species, size, charge, and chemical interactions. They are passive transport mechanisms in the sense that they only allow ion passage in the direction of their chemical gradient (from high to low concentration). Examples of ions that move through channels are hydrogen, calcium, sodium, chloride, and potassium. Channels can be *gated* and change configurations from an open to a closed state and back. Many biochemical factors can affect the gating characteristics of a channel. As we have touched upon in this chapter, mechanical signals can alter channel behavior, allowing the transduction of forces into a biochemical event. In particular, mechanosensitive ion channels are membrane-bound, pore-forming proteins that open in response to mechanical forces. Mechano-neurosensing in both hearing and touch response is almost exclusively due to mechanically gated ion channels. They are thought to mediate mechanosensation in several nonsensory cells as well.

The identification of mechanosensitive ion channels has generally relied on one of two experimental approaches. In the first approach, membrane tension is applied to cells or pieces of isolated membrane containing candidate channels. Such an approach led to the first demonstration of mechanical gating of one of the most well-characterized mechanosensitive ion channels, the bacterial mechanosensitive channel of large conductance (MscL). It has been shown that when there is a large pressure imbalance between the outside and inside of a bacterium carrying this channel, the MscL opens and allows the bacterium to jettison some of its innards and relieve the pressure. In the second approach, randomly generated genetic mutations are screened for alterations in specific touch responses (such as gentle touch or pressure sensing). The specific mutated genes responsible for the changed response are identified through approaches such as linkage mapping. Examples of mechanosensing mutants found with screens include touch-insensitive mutants in the worm *Caenorhabditis elegans*, defects in bristle mechanosensation in *Drosophila melanogaster* mutants, and lateral line mechanosensation mutants in zebrafish. Three classes of mechanosensitive ion channel have been identified to mediate mechano-neurosensing: epithelial sodium channel (ENaC); transient receptor potential (TRP) channel; and the two-pore-domain potassium channel protein.

Example 11.3: Is membrane tension sufficient to open a channel?

To see if bilayer tension is a feasible mechanism for regulating ion channel opening, estimate how much work is performed on a channel opening under membrane surface tension.

The channel *alamethicin* is known to increase its effective two-dimensional area when it transitions from closed to open. If the membrane is exerting a surface tension, n, on the channel when this happens, there will be work done. Specifically,

$$W = n\Delta A, \tag{11.6}$$

where ΔA is the change in area. Next, we can estimate the change in area. *Thioredoxin* is a molecule known to pass through alamethicin channels and has a radius of

approximately 3.5 nm. A lower bound on the change in area would be

$$\Delta A = \pi r^2 = 38.5 \text{ nm}^2. \tag{11.7}$$

In Section 1.5 we estimated the surface tension in a neutrophil to be about 35 pN/μm. The associated work is

$$W = 38.5 \text{ nm}^2 \times (10^{-9} \text{m/nm})^2 \, 35 \text{ pN/μm}$$
$$(10^{-12} \text{ N/pN}) \, (\text{μm}/10^{-6}\text{m}) = 1.35 \times 10^{-21} \text{ J} \tag{11.8}$$

or about one zeptojoule. Although this is a tiny amount of energy, it is on the same scale as the channel activation energy, which has been estimated at $14k_B T$ or 50 zJ. Thus, particularly at higher membrane tensions, this may be an important mechanotransduction mechanism.

Conceptually, the mechanism whereby membrane tension could lead to channel opening is attractive, owing to its simplicity. Because the channel is embedded within the bilayer, it exists in a two-dimensional environment. In-plane membrane tension could directly pull on the proteins that make up the channel and cause it to open. Of course, on this scale entropic effects need to be considered. Channel kinetics are often described in terms of the probability of the channel being in one of two configurations, open or closed, with a transition or activation energy required to move between the two. Open and closed transitions can be observed directly in experiments examining single channels, or the aggregate effects of many channels can be superimposed to modulate membrane conductance.

Hydrophobic mismatches allow the mechanical gating of membrane channels

Another level of complexity in bilayer–protein interactions involved in mechanical gating of membrane channels is *hydrophobic mismatch*. This phenomenon involves the mechanical deformation of the membrane after protein insertion. When the thickness of the hydrophobic region of the membrane protein is different than the thickness of the bilayer, it will cause localized distortion, including squashing, stretching, and/or tilting of the lipid chains (Figure 11.16). This effectively results in a mechanical coupling of the lipids to the protein. In this situation, in-plane stretching of the bilayer will lead to changes in membrane thickness, potentially altering the conformation of the membrane protein. In support of this hypothesis, some channels are known to get shorter when they are in the open configuration. They would be a better fit within a membrane that had been thinned due to stretching. In fact, experimental evidence has shown that when certain channels are placed in a thinner bilayer, they are more likely to be in the open configuration.

Nota Bene

The MscL Channel. Although much insight into mechanosensitive ion channels has been gained through studying the MscL channel, it is important to note that the channel is found only in prokaryotes. Because the cell wall of bacteria is reinforced with a structural component known as *peptidoglycan*, there are important differences between this channel and those found in eukaryotic bilayers.

Figure 11.16 Hydrophobic mismatch. When a hydrophobic protein or a protein domain that has the same thickness as a membrane is inserted, it is energetically favored. However, if it is longer or shorter it will distort the adjacent bilayer, increasing the free energy.

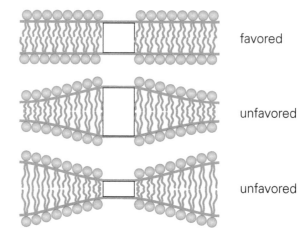

favored

unfavored

unfavored

Example 11.4: Can membrane thinning contribute to channel opening?

Estimate the increase in free energy incurred by membrane thinning and how much hydrophobic area this would expose in a membrane-bound protein.

Consider the energetics of the hydrophobic mismatch ion-channel mechanism. When a hydrophobic protein is transferred from an organic solvent to an aqueous environment, the free energy increases by an amount proportional to the exposed area, roughly 17 mJ/m². So the energy available to open a channel would be roughly the free energy increase due to the hydrophobic surface exposed. Again considering an upper bound, the maximum membrane thinning will occur at the *lytic limit* or when the membrane is stretched just to the point of rupture. This corresponds to an areal strain of roughly 3%. However, for the volume to be preserved, the membrane will need to thin by a corresponding 3% or 0.15 nm for a 5 nm thick membrane. Next we need to determine how much hydrophobic area this would expose on the protein. For the MscL channel, it is roughly cylindrical in shape, with a 5 nm diameter. So, 2.4 nm² of area is exposed to water, corresponding to an increase in free energy of 40 zJ. This is quite comparable to the channel activation energy.

Advanced Material: Incompressibility of bilayer leads to thinning with stretch.

Hydrostatic pressures ($\sigma_{xx} = \sigma_{yy} = \sigma_{zz}$) have been applied as high as 100 atmospheres (10^7 N/m²) with no significant effect on lipid density. Compressibility in response to hydrostatic pressures is quantified by the bulk modulus E_B, which is the ratio of the hydrostatic stress and the dilatational strain.

$$E_B = \frac{\sigma_h}{s_d} \tag{11.9}$$

where

$$\sigma_h = \frac{\sigma_{xx} + \sigma_{yy} + \sigma_{zz}}{3} \quad \text{and} \quad \varepsilon_d = \frac{\varepsilon_{xx} + \varepsilon_{yy} + \varepsilon_{zz}}{3}. \tag{11.10}$$

Results from these studies have produced estimates of bilayer bulk modulus on the order of 10^{10} N/m². Essentially, for physiological pressures the bilayer is incompressible. Therefore any in-plane stretching resulting in an area increase is accompanied by a corresponding decrease in thickness.

Mechanical forces can expose cryptic binding sites

In addition to force-induced activation of membrane ion channels, another molecular mechanism that has the potential to transduce mechanical loads is force-induced exposure of cryptic (that is, hidden) binding sites. Consider a protein coupled to or within a load-bearing structure, such as a focal adhesion or the cytoskeleton. When the cell is loaded, mechanical forces are transmitted to the protein and may cause conformational changes (such as partial unfolding) that

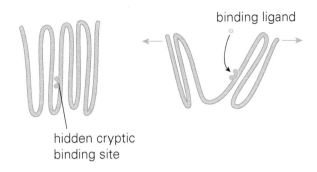

binding ligand

hidden cryptic
binding site

**Figure 11.17 Cryptic binding sites
can act as mechanotransducers.**
Force-induced unraveling of a protein
can expose cryptic binding sites. This
allows them to become enzymatically
active or to bind their ligands.

expose new binding sites at which chemical reactions can occur (Figure 11.17).
Certain structural motifs exist that may facilitate force-induced conformational
changes in a predictable and force-dependent manner. Some proteins that
mechanically link integrins to the cytoskeleton (and would likely be exposed to
large forces during fluid shear and/or cell stretching) have been found to possess
repeating sequences of structural motifs or *modules* (such as α-actinin and talin).
The mechanical stability of these modules would dictate the sequence in which
the protein unravels under force, with the more unstable modules unfolding
before the stable ones. These intermodule differences in mechanical stability
could result in sequential exposure of cryptic binding sites with increasing load
magnitudes. A molecule that undergoes such sequential conformational changes
under increasing loads could allow the cell to sense force magnitude.

Bell's equation describes protein unfolding kinetics

Protein folding and unfolding occurs on very small length scales, so entropic
effects need to be considered. In fact, folding and unfolding are more stochastic
than deterministic. In Section 10.1 we described how Bell's model can be used to
describe adhesive bond rupture and breakage. Perhaps, not surprisingly, Bell's
equation is also effective in describing protein unfolding. Specifically, the kinetic
rate of unfolding is given by

$$k = k_0 e^{\frac{F\Delta x}{k_B T}}, \tag{11.11}$$

where k_0 is the unfolding rate of an unloaded protein, F is the force, and Δx can be
related to the extension of the protein with unfolding but is more correctly termed
the effective *energy barrier width*.

Molecular conformation changes can be detected fluorescently

The capacity for force to induce conformational changes through the mechanisms
described above has primarily been investigated using one of two approaches. In
the first approach, *molecular dynamics* can be used to directly simulate the molecular rearrangements that occur during an unfolding trajectories. The second
approach is to use genetically encoded fluorescent sensors to investigate these
phenomena experimentally. In this approach, a biosensor is constructed by fus-
ing two different fluorescent proteins to a candidate mechanosensing protein.
The fluorescent proteins are chosen such that the emission spectrum of one of the
molecules (*donor*) overlaps the excitation spectrum of the other molecule (*acceptor*). If the proteins are sufficiently close to one another and the donor molecule is
excited, a physical phenomenon called *Förster resonance energy transfer* (FRET)
occurs. Specifically the donor and acceptor are nonradiatively coupled, leading to
excitation of the acceptor. For any donor–acceptor pair, the FRET efficiency is
dependent on the distance between the two as well as their relative orientation to

Figure 11.18 FRET biosensors can change fluorescence depending on proximity or orientation. For construction of a FRET biosensor, fluorescent proteins must be chosen such that the emission spectrum of one of the molecules (in this case cyan fluorescent protein (CFP)) overlaps the excitation spectrum of the other molecule (in this case yellow fluorescent protein (YFP)). FRET efficiency changes, depending on the relative distance and orientation between the two molecules.

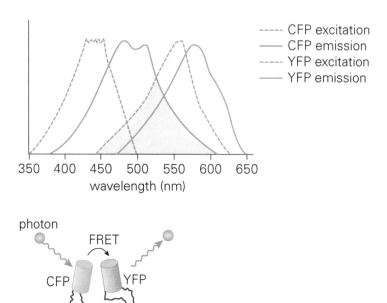

Enzymes strain their substrates. It is well known that enzymes exert strain on their substrates upon binding, thereby catalyzing the underlying reaction. Straining the enzymatic substrate could therefore potentially regulate enzyme activity. In this way, mechanical force has been hypothesized to modulate enzymatic activity independently of exposure of cryptic binding sites. However, this mechanism has been relatively unexplored.

one another (typically, FRET occurs at distances of 10 nm or less). By monitoring changes in FRET efficiency (the ratio of acceptor–donor emission intensity) in a cell under mechanical load, one can probe whether mechanical forces can be transformed into a conformational change in the candidate protein of interest (Figure 11.18).

11.4 ALTERATION OF CELLULAR FUNCTION

After the initial molecular mechanotransduction event, an intracellular biochemical signaling cascade must be initiated if these conformational changes are ultimately to lead to alterations in cellular function. In Section 2.2, we provided a survey of intracellular signaling mechanisms and pathways. Many of these have been shown to be activated by mechanical stimulation. In general, the investigation of specific signaling pathways is driven, in large part, by physiological relevance for a given cell type. In vascular endothelial mechanotransduction, we might be interested in a pathway that ultimately leads to regulation of vascular tone. In bone cells, we might be interested in a pathway that has been shown to lead to increased mineralization. Many detailed reviews of mechanically regulated intracellular signaling pathways are available (many of which are specific for certain cell or tissue types); we will not try to recapitulate them here. Rather, we focus on responses to mechanical stimuli that appear common across the many cell types. We note that, the material presented is somewhat general, but these are very active areas of research.

Intracellular calcium increases in response to mechanical stress

One of the most commonly investigated intracellular signaling systems in mechanotransduction is intracellular calcium signaling. Calcium signaling is important during mechanotransduction both in excitable cells (in other words, cells that exhibit action-potential propagation such as nerves and cardiac myocytes) as well as non-excitable cells (epithelial cells, bone cells, chondrocytes, etc.). One reason for its widespread investigation in mechanotransduction is that it is a ubiquitous second messenger molecule, capable of eliciting a wide range of effects that can affect many different downstream signaling molecule pathways.

Intracellular Ca^{2+} signaling can be visualized in real time through the use of fluorescent Ca^{2+} indicators (calcium *ionophores* or *chelators*). *In vitro*, such dyes can be loaded into cells relatively easily, and, when combined with fluorescent imaging, allow the visualization of the initiation and propagation of intracellular Ca^{2+} waves in response to mechanical stimuli. Mechanical activation of intracellular calcium signaling, in general, occurs very quickly (generally on the order of seconds after the initial mechanical stimulus), and the calcium waves can spread throughout the cell rapidly (on the order of milliseconds). Mechanically induced intracellular Ca^{2+} elevation is thought to be mediated by the rapid opening of mechanically sensitive calcium channels in the cell membrane, or release of calcium from intracellular Ca^{2+} stores. Once elevated, intracellular calcium then goes on to activate other, downstream pathways.

Nitric oxide, inositol triphosphate, and cyclic AMP, like Ca^{2+}, are second messenger molecules implicated in mechanosensation

Although intracellular calcium signaling is the most well-studied of the second messenger systems, several other second messenger molecules have been shown to exhibit rapid changes soon after exposure to mechanical stimulation. Recall from Section 2.2 that second messengers can be broadly classified into one of three categories: hydrophobic molecules associated with the cell membrane, dissolved gases, and molecules that do not freely cross lipid membranes. Like Ca^{2+}, *cyclic adenosine monophosphate* (cAMP) is a second messenger molecule that falls into the latter category. cAMP is synthesized from ATP by a process that is catalyzed by *adenylyl cyclase*. Activation of adenylyl cyclase is generally associated with activation of G-protein-coupled receptors. Levels of cAMP can also be regulated by cyclic nucleotide phosphodiesterases, which degrade cAMP. cAMP has been shown to be rapidly regulated after exposure to mechanical loading in a variety of cell types. Adenylyl cyclase exists in several different isoforms whose expression can be highly tissue-specific. The activity of many of these isoforms can be modulated by Ca^{2+}, allowing cross-talk between the Ca^{2+} and cAMP pathways.

Recall from Section 2.2 that inositol triphosphate (IP_3) is a second messenger molecule associated with the cell membrane. It is synthesized by hydrolysis of the molecule phosphatidylinositol 4,5-bisphosphate (PIP_2) by phospholipase C, a phospholipid that is localized in the plasma membrane. Upon cleavage of PIP_2 to IP_3 by phospholipase C, IP_3 diffuses to the endoplasmic reticulum and, after binding to IP_3 receptors, initiates intracellular calcium release. Several studies have implicated IP_3 in mediating mechanically stimulated intracellular Ca^{2+} release.

The dissolved gas nitric oxide can be a potent second messenger because of its ability to quickly diffuse though the cytoplasm and across lipid membranes. Shear-induced nitric oxide production in endothelial and bone cells has been widely observed. Production of nitric oxide is catalyzed by nitric oxide synthase (NOS). In mammals, the endothelial isoform of NOS, eNOS (also known as NOS-3) is a primary regulator of vascular tone. In particular, eNOS-derived nitric oxide has been shown to be a potent vasodilator, relaxing smooth muscle within blood vessels.

It is noteworthy to point out that second messenger molecules like those described above have often been used as primary outcome measures for assessing the mechanosensory role of a particular protein. In general, implicating a molecule as having a mechanosensory function by inhibitory strategies can be more straightforward when assaying a cellular response that occurs fairly rapidly after stimulation. This is because after long exposures to loading, numerous molecular

Nota Bene

Intracellular calcium and cell mechanics. A variety of processes associated with the generation or resistance to mechanical force depend on intracellular calcium. Muscle contraction depends on calcium interacting with the troponin/tropomyosin system for actin–myosin contraction to occur. Further, calcium is required for certain types of cell adhesion, such as the calcium-dependent adhesion molecule cadherin.

Nota Bene

Ionophores and chelators. An ionophore allows the transfer of ions through a hydrophobic barrier such as the bilayer, in which they would normally be insoluble. Ionophores typically encase the ion in a polar interior while exposing a hydrophobic exterior to the outside. A chelator typically forms multiple stable bonds with metal ions, inactivating them from their normal function or effect. It is derived from the Greek word for "lobster claw", *Chelè*.

interactions often occur, and thus it becomes harder to interpret whether effects on mechanosensation caused by inhibition of a particular molecule are because the molecule in question has a direct mechanosensory function, or whether it is merely involved somewhere "downstream" in the signaling pathway.

Mitogen-activated protein kinase activity is altered after exposure to mechanical stimulation

Downstream of second messengers, protein–protein signaling cascades involving phosphorylation and dephosphorylation can be critical (see Section 2.2). A particularly prominent signaling mechanism often activated within minutes after exposure to mechanical stimulation is the group of the mitogen-activated protein kinases (MAP kinases). MAP kinase phosphorylation in response to most (including mechanical) stimuli can be highly dynamic, occurring soon after stimulation and remaining in the phosphorylated state for many minutes. One well-studied MAP kinase is extracellular-signal-regulated kinase 1 and 2 (ERK1/2), which regulates cell growth and differentiation. Another is c-jun N-terminal kinase (JNK), which is a so-called *stress-activated* protein kinase. Note that "stress-activated", in this context, refers to systemic stresses and not necessarily mechanical ones (such as, heat shock or chemical shock). Because JNK is implicated in apoptosis and inflammation, it is particularly of interest in the study of chronic conditions such as atherosclerosis, which is likened to a slow-acting inflammation (similar to how metal rusting is a slow combustion process).

One interesting aspect of MAP kinases is that they are capable of directly activating transcription factors, the DNA-binding proteins that control gene transcription. ERK1/2 is known to activate the transcription factor Elk1. JNK has been shown to activate several transcription factors, including c-Jun, Elk1, SMAD4, ATF2, and NFAT1. By possessing the capacity to be quickly activated and to directly regulate transcription factors, MAP kinases are ideally suited to participate in pathways involved in early gene expression. Indeed, several components of the MAP kinase pathway have been implicated in the expression of *primary response* genes, in other words genes whose expression is altered soon after stimulation and that do not require *de novo* protein synthesis.

Mechanically stimulated cells exhibit prostaglandin release

Another well-characterized response of cells to mechanical stimulation is prostaglandin release. Prostaglandins are lipid compounds that are enzymatically derived from fatty acids and mediate several cellular functions. There are several types of prostaglandin, of which prostaglandin E2 (PGE2) is arguably the most well studied in cellular mechanotransduction. PGE2 is synthesized from arachidonic acid, which is derived from membrane phospholipids. This arachidonic acid is converted to prostaglandin G2 and subsequently prostaglandin H2 by the enzyme cyclooxygenase (COX). A final isomerization step converts prostaglandin H2 into the biologically active PGE2. COX exists in *constitutive* (COX-1) and *inducible* (COX-2) isoforms and is considered the rate-limiting enzyme in the PGE synthesis process. Several studies have demonstrated that mechanical stimulation activates *COX2* gene expression, elevates COX-2 protein levels, and induces PGE2 release into the extracellular environment. Upon its release, PGE2 may initiate signaling cascades in an autocrine fashion or in other cells by binding PGE2 receptors on the cell surface, such as the receptor for PGE2 (EP2).

Mechanical forces can induce morphological changes in cells

The cascade of early biochemical responses discussed above ultimately results in functional alterations downstream. One that can be readily observed is altered cell morphology. It has been widely demonstrated that when cells are subjected to substrate stretch or fluid shear (generally over an extended period, on the order of hours or sometimes days), cells may actively remodel their cytoskeletons and change their gross shape. As described in Section 11.1, endothelial cells subjected to fluid flow have been observed to align in the direction of flow. These cells have also been demonstrated to align perpendicular to the direction of uniaxial stretch. Other cells, such as smooth muscle cells and fibroblasts may exhibit alignment parallel or perpendicular to major stretch directions, depending on the type of cell used, the plating conditions, and the exact nature of the stretch (such as percentage strain and applied frequency).

In some cases, cells may exhibit no gross changes in morphology but may undergo extensive cytoskeletal remodeling. Mechanically induced formation of actin stress fibers has been observed in many cells. The stress fibers can exhibit a preferential alignment relative to the primary direction of stretch or flow. It has been proposed that these alignments may serve to minimize intracellular stresses (Figure 11.19).

Mechanical stimulation can induce extracellular matrix remodeling

In Section 1.1, we learned about several examples in which local matrix remodeling occurs in response to mechanical stimuli. When bone cells do not experience proper mechanical stimulation, bone formation ceases and bone resorption is initiated. Another example is onset of osteoarthritis after changes in mechanical signals experienced by chondrocytes. Given these examples, it is unsurprising that mechanoregulation of matrix generation and matrix degradation pathways are of particular importance. A prominent degradation mechanism is through regulation of matrix metalloproteinases, proteins that cleave and degrade matrix molecules. Generally, mechanical stresses are thought to provoke a "protective" response (down-regulation of matrix metalloproteinases) in cells where the immediate extracellular environment is structurally reinforced. This serves to reduce mechanical loads on the cells within. However, this protective response may not necessarily be physiologically beneficial. For a heart subjected to high

20x magnification

(A)

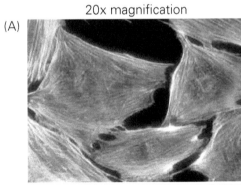

control (no flow)

(B)

steady flow (1.2 Pa)

Figure 11.19 Fluorescent staining of the actin cytoskeleton in flowed bone cells. Cells were either kept in static control conditions (A) or steady flow (B) to show that flow exposure induces stress fiber formation. Notice that this stress fiber formation is induced in the absence of any preferred fiber orientation. (From, Malone AM et al. (2007) *Am. J. Physiol. Cell. Physiol.*)

blood pressure, fibrosis may be induced. In this case, the heart can become much stiffer, resulting in increased workload for the heart muscles.

Cell viability and apoptosis are altered by difference processes

Cell viability is another functional outcome that can be altered by mechanical stimuli. Cell death can occur by two distinct processes, necrosis or apoptosis. Necrosis is cell death due to cell damage, and is a process that can be considered unexpected or accidental. In contrast, apoptosis is cell death due to programmed events. Unlike necrosis, apoptosis is a normal preprogrammed event, and does not induce an inflammatory response. It has been demonstrated in a variety of cell types that mechanical stimulation can alter cell apoptosis. It has been shown that endothelial cells may exhibit decreased apoptosis with fluid shear, although this may depend on the flow profile. Shear resulting from disturbed flows or oscillatory shear stresses has been found to be less effective than steady shear or pulsatile shear in reducing cell apoptosis.

Key Concepts

- The sensing of mechanical signals by cells, or cellular mechanotransduction, is critical to many aspects of physiology and understanding disease. Four distinct phases are involved: conversion of tissue or organ-level loads into cell-level physical signals; force detection by mechanosensitive molecules undergoing a conformational change; activation of intracellular signaling systems; and altered cell metabolism.

- Many cellular-level physical signals have been shown to be potent regulators of cell metabolism, including fluid flow, stretch, and pressure.

- Specialized excitable cells mediate our senses of touch and hearing. They are some of the best-understood mechanosensing cells and often have highly sensitive cellular structures capable of exquisite sensitivity.

- In nonexcitable cells, structures such as the glycocalyx, cell membrane, cytoskeleton, focal adhesions, the nucleus, and primary cilia have all been implicated as potential sites of mechanosensing.

- Mechanosensitive ion channels are a major class of mechanosensitive molecules. They are known to respond to membrane tension as well as to thinning due to hydrophobic mismatch.

- Other proteins can change their enzymatic potential by buckling, unfolding, and exposure of cryptic binding sites. FRET is a powerful way to detect such changes fluorescently.

- Second messenger signaling including IP_3, CAMP, and Ca^{2+} signaling have been shown to be activated by mechanical signals. This leads to activation of protein signaling cascades such as MAPK signaling, and intercellular signaling such as PGE2. Ultimately, these cascades lead to altered gene expression, modification of the extracellular matrix, and changes in cell viability.

Problems

1. Estimate the force required to remove a membrane-bound protein from the bilayer if it has a cylindrical hydrophobic region that is 2 nm in diameter and 5 nm in length. You can do this by assuming that the mechanical work done by the pulling force is equal to the change in free energy caused by exposing the hydrophobic domain to water. You will also need to assume that 5 nm of displacement is sufficient to remove the protein.

2. Consider the bending hair bundle from Example 11.1. How much energy was put into deflecting the tip?

   Would that be sufficient to overcome the activation energy of a channel?

3. Figure 11.11 depicts a primary cilium undergoing bending. From the figure, estimate the magnitude of deflection and strain in the membrane surrounding the cilium at its base. You may assume that the cilium diameter is 200 nm. Finally, use the result from Problem 8.4 to estimate what force applied to the tip would produce this magnitude of bending.

4. One of the roles of the cytoskeleton is to connect the nucleus to the rest of the cell. This force transmission is thought to be involved in cellular mechanotransduction. If a cell is exposed to an externally imposed deformation, describe how you would expect the nucleus to deform in response, if (a) the cytoskeleton is removed, (b) the cell cytoplasm acts like a continuum solid without a cytoskeleton per se, or (c) the cytoskeleton acts like a tensegrity structure, with normal cytoplasm. Specifically, how would the magnitude of the nuclear deformation compare with that of the cell overall?

5. Ion channels are sometimes gated by the passage of ions themselves. Specifically, they can lose conductance as ions flow through them. One putative mechanism is that the ions moving through the channel supply the energy to close the channel. How much energy is associated with a monovalent ion moving through the channel along its electrical gradient? How does this compare with the channel activation energy? Assume a typical resting potential of -70 mV.

6. In Example 11.3, the change in free energy for a channel moving from a closed to an open configuration was estimated to be $14k_{B}T$. Using the Boltzmann equation, predict the fraction of the time that the channel would be open owing only to thermal fluctuations. How would this change if you account for the work done by membrane tension as the channel opens? Assume that the cell is at its rupture or lytic strain (3%) and a typical areal expansion modulus, K_{A}, of 0.5 N/m.

7. For a bilayer with an areal expansion modulus $K_{A} = 1.0$ N/m, what would the stiffness in the transverse direction be? In other words, for a force applied through the thickness, what is the slope of the force–deflection curve?

Annotated References

Anderson RG (1998) The caveolae membrane system. *Annu. Rev. Biochem.* 67, 199–225. *Gives a comprehensive review of the cell biology of caveolae.*

Birukov KG, Birukova AA, Dudek SM et al. (2002) Shear stress-mediated cytoskeletal remodeling and cortactin translocation in pulmonary endothelial cells. *Am. J. Respir. Cell Mol. Biol.* 26, 453–464. *This journal research article describes the response of pulmonary endothelial cells to applied shear stress. The study also describes the effects of certain GTPases and early response mechanoresponsive pathways to shear.*

Brown TD, Bottlang M, Pedersen DR & Banes AJ (1998) Loading paradigms—intentional and unintentional—for cell culture mechanostimulus. *Am. J. Med. Sci.* 316, 162–168. *A numerical analysis of parasitic flows in bending-based systems for studying the strain response in vitro.*

Chalfie M (2009) Neurosensory mechanotransduction. *Nat. Rev. Mol. Cell Biol.* 10, 44–52. *Comprehensive review of molecular and cellular mechanisms of mechanosensing in sensory cells.*

Chancellor TJ, Lee J, Thodeti CK & Lele T (2010) Actomyosin tension exerted on the nucleus through Nesprin-1 connections influences endothelial cell adhesion, migration, and cylic strain-induced reorientation. *Biophys. J.* 99, 115–123. *Provides recent evidence of the role of nucleus loading and nuclear matrix proteins in mechanotransduction.*

Corey DP & Hudspeth AJ (1979) Response latency of vertebrate hair cells. *Biophys. J.* 26, 499–506. *Early data on the molecular mechanism of hair cell mechanosensing, particularly the incredibly fast response time.*

DeBakey ME, Lawrie GM & Glaeser DH (1985) Patterns of atherosclerosis and their surgical significance. *Ann. Surg.* 201, 115–131. *This journal article presents an analysis of common atherosclerotic lesion development sites as noted in clinical examinations. The article further discusses the clinical aspects of atherosclerosis, including classification, progression and recurrence of disease.*

Ernstrom GG & Chalfie M (2002) Genetics of sensory mechanotransduction. *Annu. Rev. Genet.* 36, 411–53. *Mechanotransduction examined from a genetics perspective, particularly in sensory cells.*

Eyckmans J, Boudou T, Yu X & Chen CS (2011) A hitchiker's guide to mechanobiology. *Dev. Cell* 21, 35–47. *A summary of mechanotransduction mechanisms, primarily focused on non-sensory cells.*

Gefen A (2011) *Cellular and Biomolecular Mechanics and Mechanobiology. An edited text describing recent advancements in cell and molecular mechanics and mechanobiology.*

Hamill OP & Martinac B (2001) Molecular basis of mechanotransduction in living cells. *Physiol. Rev.* 81, 685–740. *A review of molecular mechanotransdcution mechanisms with a focus on the membrane and channels; the primary source for the material on hydrophobic mismatch and channel-membrane coupling.*

Jacobs CR, Temiyasathit S & Castillo AB (2010) Osteocyte mechanobiology and pericellular mechanics. *Annu. Rev. Biomed. Eng.* 12, 369–400. *Provides a comprehensive review of mechanosensory structures and mechanisms identified in bone cells.*

Knothe Tate ML, Steck R, Forwood MR & Niederer P (2000) In vivo demonstration of load-induced fluid flow in the rat tibia and its potential implications for processes associated with functional adaptation. *J. Exp. Biol.* 203, 737–745. *Describes quantification of loading-induced flow in bone.*

Kooppel DE, Axelrod D, Schlessinger J et al. (1976) Dynamics of fluorescence marker concentration as a probe of mobility. *Biophys. J.* 16, 1315–1329. *An early description of fluorescence recovery after photobleaching used to measure lateral diffusion in the cell membrane.*

Kung C (2005) A possible unifying principle for mechanosensation. *Nature* 436, 647–654. *Provides a concise summary of the roles of mechanosensitive ion channels in touch sensation and hearing.*

Malone AM, Batra NN, Shivaram G et al. (2007) The role of actin cytoskeleton in oscillatory fluid flow-induced signaling in MC3T3-E1 osteoblast. *Am. J. Physiol. Cell Physiol.* 292, C1830–C1836. *Demonstrated differential cytoskeletal remodeling in cells exposed to static and dynamic fluid flow.*

Nauli SM, Alenghat FJ, Luo Y et al. (2003) Polycystins 1 and 2 mediate mechanosensation in the primary cilium of kidney cells. *Nat. Genet.* 33, 129–137. *A discussion of the polycystins and their putative role in primary-cilium-based mechanosensing.*

Nishiyama M, Shimoda Y, Hasumi M et al. (2010) Microtubule depolymerization at high pressure. *Ann. N. Y. Acad. Sci.* 1189, 86–90. *A demonstration that high hydrostatic pressure can induce microtubule depolymerization in vitro.*

Nonaka S, Tanaka Y, Okada Y et al. (1998) Randomization of left-right asymmetry due to loss of nodal cilia generating leftward flow of extraembryonic fluid in mice lacking KIF3B motor protein. *Cell* 95, p. 829–837. *Seminal study demonstrating that flow generated by nodal cilia is critical for left–right determination.*

Olsen B (2005) Nearly all cells in vertebrates and many cells in invertebrates contain primary cilia. *Matrix Biol.* 24, 449–450. *An editorial on the ubiquitous nature of primary cilia and their potential physiological function.*

Owan I, Burr DB, Turner CH et al. (1997) Mechanotransduction in bone: osteoblasts are more responsive to fluid forces than mechanical strain. *Am. J. Physiol.* 273 (3 Pt 1), C810–815. *Provides evidence that osteoblasts subjected to four-point bending were responsive to fluid flow rather than substrate strain.*

Qin YX, Lin W & Rubin C (2002) The pathway of bone fluid flow as defined by in vivo intramedullary pressure and streaming potential measurements. *Ann. Biomed. Eng.* 30, 693–702. *Some of the earliest evidence that fluid flow is a critical cell-level physical signal in bone mechanobiology.*

Reilly GC, Haut TR, Yellowley CE et al. (2003) Fluid flow induced PGE2 release by bone cells is reduced by glycocalyx degradation whereas calcium signals are not. *Biorheology* 40, 591–603. *Supplies some of the only evidence that the cellular glycocalix is critical for mechanosensing in bone cells.*

Simons K & van Meer G (1988) Lipid sorting in epithelial cells. *Biochemistry* 27, 6197–6202. *An early description on microdomains in lipid membranes.*

Tabouillot T, Muddana HS & Butler PJ (2011) Endothelial cell membrane sensitivity to shear stress is lipid domain dependent. *Cell. Mol. Bioeng.* 4, 169–181. *Direct evidence that microdomains in the bilayer, including rafts and caveolae, are modified with fluid shear stress and are potentially involved in mechanosensing.*

Vogel V & Sheetz M (2006) Local force and geometry sensing regulate cell functions. *Nat. Rev. Mol. Cell. Biol.* 7, 265–275. *Provides a concise review of potential mechanisms by which force-induced conformational changes may lead to exposure of cryptic binding sites.*

Wang Y, McNamara LM, Schaffler MB & Weinbaum S (2007) A model for the role of integrins in flow induced mechanotransduction in osteocytes. *Proc. Natl Acad. Sci. USA* 104, 15941–15946. *Describes a proposed mechanism for mechanosensing of flow by integrins.*

Abbreviations

Chapter 1

IRDS	infant respiratory distress syndrome
RBC	red blood cell

Chapter 2

AFM	atomic force microscopy
cAMP	cyclic adenosine monophosphate
cGMP	cyclic guanosine monophosphate
DAG	diacylglycerol
IP$_3$	inositol triphosphate
NO	nitric oxide
CO	carbon monoxide
ER	endoplasmic reticulum
GDP	guanosine diphosphate
GEF	guanine nucleotide exchange factor
GPCR	G-protein-coupled receptors
GTP	guanosine triphosphate
DAPI	4',6-diamidino-2-phenylindole
GFP	Green fluorescent protein
STM	scanning tunneling microscopy
SDS	sodium dodecyl phosphate
PAGE	polyacrylamide gel electrophoresis
PCR	polymerase chain reaction
siRNA	small inhibitory RNA
SSR	site-specific recombinase
Cre	cyclic recombinase

Chapter 4

SI	Système International d'Unités

Chapter 6

AFM	atomic force microscopy
TFM	traction force microscopy
PDMS	polydimethylsiloxane

Chapter 7

ADP	adenosine diphosphate
ATP	adenosine triphosphate
FJC	freely jointed chain
GDP	guanosine diphosphate

GTP	guanosine triphosphate
MF	microfilament
MT	microtubule
WLC	wormlike chain

Chapter 9

AM	acetoxymethyl
CMC	critical micelle concentration
FRAP	fluorescence recovery after photobleaching

Chapter 10

RGD	arginine–glycine–aspartate
VCAM	vascular cellular adhesion molecule
ICAM	intercellular adhesion molecule
NCAM	nerve cellular adhesion molecule
Arp	actin-related protein

Chapter 11

VEGFR-2	vascular endothelial growth factor receptor 2
PIP$_2$	phosphatidylinositol 4,5-bisphosphate
PKD	polycystic kidney disease
ATP	adenosine triphosphate
GTP	guanosine triphosphate
GDP	guanosine diphosphate
MscL	mechanosensitive channel of large conductance
ENaC	epithelial sodium channel
TRP	transient receptor potential
FRET	Förster resonance energy transfer
cAMP	cyclic adenosine monophosphate
IP$_3$	inositol triphosphate
NOS	nitric oxide synthase
eNOS	endothelial isoform of nitric oxide synthase
MAP	mitogen-activated protein
ERK1/2	extracellular-signal-regulated kinase 1 and 2
JNK	c-jun N-terminal kinase
PGE2	prostaglandin E2
COX	cyclooxygenase
EP2	prostaglandin receptor for PGE2
CFP	cyan fluorescent protein
YFP	yellow fluorescent protein

List of variables and units

M	moment
m	mass
n	surface tension
P	pressure
$\mathbf{P'}$	transformed vector
$\mathbf{P}$	original vector
$\boldsymbol{Q}$	rotation matrix
R	radius
$\boldsymbol{RU}$ or $\boldsymbol{VR}$	rotation and stretch components of polar decomposition
$\boldsymbol{S_x}$	resultant force
S_{Xx}	components of $\boldsymbol{S_x}$
S_{Xy}	components of $\boldsymbol{S_x}$
S_{Xz}	components of $\boldsymbol{S_x}$
u, v, w	displacements
$\mathbf{v}$	eigenvector
$\mathbf{v}_1, \mathbf{v}_2$, and $\mathbf{v}_3$	principal directions
w	beam displacement
x	spring deformation
$\mathbf{x}$	deformed vector
$\mathbf{X}$	undeformed vector
x, y, z	spatial coordinates

Chapter 4

α	constant scaling exponent
β	constant viscous factor
δ	phase lag
γ	shear strain
γ^*	complex shear strain
μ	viscous coefficient
μ_{eff}	effective viscosity
ω	frequency
ρ	density
$\boldsymbol{\sigma}$	stress vector
τ^*	complex shear stress
τ	shear stress
ξ	structural damping coefficient
E	elastic modulus or storage modulus
E^*	complex modulus
g	acceleration due to gravity
G^*	complex shear modulus
h	height
i	imaginary unit, $\sqrt{-1}$
k	spring constant
k_{B}	Boltzmann's constant

L	length
m	mass
n	surface tension
P	pressure
Re	Reynolds number
t	time
u	fluid velocity
x, y, z	spatial coordinates

Chapter 5

γ	shear strain
$\boldsymbol{\sigma}$	stress vector
τ	shear stress
A	cross-sectional area
β	$1/k_{\text{B}}T$
b	distance traveled in each step, Kuhn length
D	diffusion coefficient
ϵ	energetic cost per hairpin
E	Young's modulus
k	spring constant
k_{B}	Boltzmann's constant
L	contour length
m	microstate
n	surface tension
n, n_+	number of flips, number of flips that come out heads
N	number of particles
N_h	number of hairpin sites
$p(m)$	probability of microstate m
q	heat
$Q_s(m_s)$	energy of microstate m_s
R	end-to-end length
S	entropy
t	time
T	temperature
V	volume
Y	helmholtz free energy
z	single partition function
Z	partition function

Chapter 6

α	angle
γ	shear strain
δ	phase lag
ε	strain

$\varepsilon_{xy}, \varepsilon_{yx}, \varepsilon_{xz}$	components of strain		k_{on}, k_{off}	rate of reaction
η	viscous friction coefficient		K	dissociation constant
λ	stretch ratio		ℓ_p	persistence length
μ	dynamic viscosity		$\mathscr{L}$	Langevin function
ν	Poisson's ratio		L	length
ρ	density		n	surface tension
τ	time constant, η_2/k		P	probability or pressure
τ	shear stress		P_{loop}	probability of looping
υ	kinematic viscosity		Q_{loop}	energy of looping
$\boldsymbol{\sigma}$	stress vector		R	end-to-end distance or radius
ω	rotational speed		$\mathbf{r}_i$	segment vector
c	speed of light		$\mathbf{R}$	end-to-end vector
$\boldsymbol{C}$	$\boldsymbol{F}^T\boldsymbol{F}$, right Cauchy-Green deformation tensor		s	arc length
$\boldsymbol{F}$	deformation gradient		S	entropy
$\boldsymbol{G}(r)$	Green's function		t	time
h	height		υ	elongation/shrinking
I	intensity of the trapping light		x, y, z	direction
$\mathbf{I}$	identity tensor		z	single partition function
k	spring constant		Z	partition function
k_B	Boltzmann's constant			
L	length		**Chapter 8**	
l	wavelength of light		γ	shear strain
n	surface tension		δ	transverse displacement
$n_b\, n_m$	indices of refraction		λ	stretch ratio
R	radius		ρ_n	number of polymers per unit volume
Re	Reynolds number		ρ_{vol}	volume fraction
t	time		$\boldsymbol{\sigma}$	stress vector
V	velocity		τ	shear stress
w	displacement		Ψ	free energy
$\mathbf{x}$	deformed vector		w_a	areal strain energy density
$\mathbf{X}$	undeformed vector		A	area
			d	depth
Chapter 7			E	Young's modulus
γ	shear strain		F	force
θ	angle		G	shear modulus
$\boldsymbol{\sigma}$	stress vector		I	second moment of inertia
τ	shear stress		k_B	Boltzmann's constant
Ψ	free energy		k_{sp}	spring constant
$\Omega(\mathbf{R})$	density of microstates		K_s	membrane shear modulus
b	Kuhn length		ℓ_p	persistence length
$\cos\Delta\theta(s)$	orientation correlation function		L	length
E	Young's modulus		P	pressure
I	mass or first moment of inertia		$\boldsymbol{r}$	vector
k	spring constant		$\boldsymbol{R}$	vector
k_B	Boltzmann's constant		t	time

V	volume
w	beam displacement
x, y, z	spatial coordinates

Chapter 9

α	scaling factor
γ	shear strain
γ_{int}	interfacial energy
ε	strain
θ	angle
λ	length
ρ	number of molecules per unit volume
$\boldsymbol{\sigma}$	stress vector
τ	shear stress
$\varphi(R)$	excitation intensity
Ψ	free energy
$\hbar$	Planck's constant
A	area
c	concentration
D	diffusion constant
E	Young's modulus
EI	flexural rigidity
G	shear modulus
J	flux
k	spring constant
k_B	Boltzmann's constant
K_A	areal expansion modulus
K_B	bending stiffness
K_S	shear stiffness
l	hydrocarbon chain length
l_c	average bond length between carbon atoms
ℓ_p	persistence length
L	length/dimension
m	mass of molecule
M, m	moments
n	force
n	surface tension
n_c	number of carbon atoms in chain
n_c	number of molecules
N	number
P	pressure
P_0	laser power
q	heat
R	radius

S	entropy
t	thickness
T	temperature
T	time
u^{tot}, v^{tot}	total deformation
w	transverse displacement
W	energy
W	density of states
x, y, z	spatial coordinates

Chapter 10

γ	shear strain
Δt	time interval
Δ	pitch of polymerized actin helical structure
θ	angle that a membrane makes with a surface
λ_b	stretch ratio of substrate
μ	fluid viscosity
n	membrane tension
ρ	density
$\boldsymbol{\sigma}$	stress vector
τ	shear stress
a	constant
A	area
d_s	binding-site spacing
D	diffusion constant
E	Young's modulus or energy
F	force
F_-	negative drag force
F_+	positive elastic force
h	height
J	adhesion energy density
k	spring constant
k_-	unbinding rate constant
k_+	binding rate constant
k_B	Boltzmann's constant
k_d	equilibrium constant
k_{on}	rate of reaction
K_B	bending modulus
n	surface tension
n_b	area density of bonds
p	probability
R	radius or end-to-end length
S	speed
t	time

t_{off}	average time per cycle that myosin is not bound to actin	$\boldsymbol{\sigma}$	stress vector
t_{on}	time per cycle that myosin is not bound to actin	τ	shear stress
		ΔA	change in area
t_r	release time	Δx	effective energy barrier width
T	temperature or tortuosity	A	area
v or V	velocity	d	displacement
w	displacement	E_B	bulk modulus
w_{bs}	width of binding site 'sweet spot'	F	force
W	density of states	F_b	buckling load
W	work	k	spring constant
W_{adh}	adhesion energy	k_B	Boltzmann's constant
W_{def}	strain energy associated with deformation	K_A	areal expansion modulus
x	maximum length of spring	n	surface tension
x, y, z	spatial coordinates	P	pressure
x_{sh}	shortening distance	Q	flow rate
		R	radius
		T	temperature

Chapter 11

γ	shear strain	W	density of states
ε	strain	W	work
		x, y, z	spatial coordinates

Index

T - #0935 - 101024 - C350 - 279/216/16 - PB - 9780815344254 - Gloss Lamination